International Workshop on Statistical-Mechanical Informatics 2009

(IW-SMI 2009)

Journal of Physics: Conference Series Volume 197

Kyoto, Japan
13-16 September 2009

Editors:

Masato Inoue
Shin Ishii
Yoshiyuki Kabashima
Masato Okada

ISBN: 978-1-61738-280-2

International Workshop on Statistical-Mechanical Informatics 2009

(IW-SMI 2009)

Journal of Physics: Conference Series Volume 197

Kyoto, Japan
13-16 September 2009

ISBN: 978-1-61738-280-2

IOP Institute of Physics

Printed from e-media with permission by:

Curran Associates, Inc.
57 Morehouse Lane
Red Hook, NY 12571

Some format issues inherent in the e-media version may also appear in this print version.

Printed by Curran Associates, Inc. (2010)

For permission requests, please contact the Institute of Physics
at the address below.

Institute of Physics
Dirac House, Temple Back
Bristol BS1 6BE UK

Phone: 44 1 17 929 7481
Fax: 44 1 17 920 0979

techtracking@iop.org

Additional copies of this publication are available from:

Curran Associates, Inc.
57 Morehouse Lane
Red Hook, NY 12571 USA
Phone: 845-758-0400
Fax: 845-758-2634
Email: curran@proceedings.com
Web: www.proceedings.com

TABLE OF CONTENTS

Statistical mechanical informatics (SMI) is an approach that applies statistical-mechanics techniques developed for many-body problems in natural science to information sciences. SMI has been extensively applied to various research fields of information sciences, including in particular information and communication theory, probabilistic inference, and combinatorial optimization problems, and has brought about novel advances in these fields. From a formal point of view, the most significant advantage of the SMI approach would be that it facilitates our look at hierarchical organization of a complex system, on the basis of which reduction of many `microscopic' degrees of freedom is performed to obtain a `macroscopic' description of the system with a small number of parameters, which is in turn utilized to analyze the system microscopically. Hierarchy, reduction, and macroscopic description should also be essential in order to understand biological systems. These observations imply that the conceptual framework of SMI should be extensible to various research fields in bioinformatics, and new theoretical frameworks for bioinformatics will be established from the system science viewpoint. To advance research activities in these directions, however, interactions between SMI and bioinformatics are needed.

The International Workshop on Statistical-Mechanical Informatics has been organized in order to explore such new directions for SMI and bioinformatics through interactions between various research disciplines, and to incubate joint research initiatives. This workshop, held at Mielparque-Kyoto, Kyoto, Japan, 13–16 September 2009, and sponsored by the Grant-in-Aid for Scientific Research on Priority Areas `Deepening and Expansion of Statistical Mechanical Informatics (DEX-SMI)' (Head investigator: Yoshiyuki Kabashima, Tokyo Institute of Technology) (Project http://dex-smi.sp.dis.titech.ac.jp/DEX-SMI), was intended to provide leading researchers in the fields of DNA microarray analysis, brain science, medical engineering and mathematical biology, to exchange advanced findings on and to detail achievements in bioinformatics and statistical mechanics.

We would like to thank the contributors of the workshop as well as all the participants, who have enjoyed the workshop as well as their stay in Kyoto. This successful workshop will stimulate further development of the interdisciplinary research field of bioinformatics and SMI.

Editors
Masato Inoue
Shin Ishii
Yoshiyuki Kabashima
Masato Okada

The IW-SMI 2009 Organizing Committee
Masato Okada, General Chair (University of Tokyo)
Yoshiyuki Kabashima, Vice-General Chair (Tokyo Institute of Technology)
Shin Ishii, Program Chair (Kyoto University)
Masato Inoue, Publications Chair (Waseda University)

Kazuyuki Tanaka (Tohoku University)
Toshiyuki Tanaka (Kyoto University)

All papers published in this volume of *Journal of Physics: Conference Series* have been peer reviewed through processes administered by the proceedings Editors. Reviews were conducted by expert referees to the professional and scientific standards expected of a proceedings journal published by IOP Publishing.

International Workshop on Statistical-Mechanical Informatics 2009 (IW-SMI 2009) IOP Publishing
Journal of Physics: Conference Series **197** (2009) 012001 doi:10.1088/1742-6596/197/1/012001

Sparse linear models: Variational approximate inference and Bayesian experimental design

Matthias W Seeger

Saarland University and Max Planck Institute for Informatics, Campus E1.4, 66123 Saarbrücken, Germany

Abstract. A wide range of problems such as signal reconstruction, denoising, source separation, feature selection, and graphical model search are addressed today by posterior maximization for linear models with sparsity-favouring prior distributions. The Bayesian posterior contains useful information far beyond its mode, which can be used to drive methods for sampling optimization (active learning), feature relevance ranking, or hyperparameter estimation, if only this representation of uncertainty can be approximated in a tractable manner. In this paper, we review recent results for variational sparse inference, and show that they share underlying computational primitives. We discuss how sampling optimization can be implemented as sequential Bayesian experimental design. While there has been tremendous recent activity to develop sparse estimation, little attendance has been given to sparse approximate inference. In this paper, we argue that many problems in practice, such as compressive sensing for real-world image reconstruction, are served much better by proper uncertainty approximations than by ever more aggressive sparse estimation algorithms. Moreover, since some variational inference methods have been given strong convex optimization characterizations recently, theoretical analysis may become possible, promising new insights into nonlinear experimental design.

1. Introduction

Most real-world signal classes of interest are highly structured. While precise high level statistical modelling of such structure is intractable at present, it imprints robust traces onto low level representations, which are in the feasible range of today's statistical methodology. For example, an image measured by a digital camera or a magnetic resonance scanner exhibits coherent edges, textures, and smooth areas, revealed by applying derivative or wavelet filters to the image bitmap: the filter responses show remarkably robust statistical behaviour far from Gaussianity (which would be obtained by filtering random noise) [1]. This low level statistical understanding, known as *sparsity* of image statistics, is used widely today in order to compress image bitmaps and video frame sequences. Similarly robust deviations from high-entropy Gaussianity occur for speech or music audio waveforms, leading to efficient compression schemes which can be decoded in real time. The statistical mechanisms underlying such low level applications are sparse linear models. The predominance of this technology for direct applications, such as coding and decoding (estimation), highlights the relevance of formulating and developing Bayesian frameworks, with which higher-order problems become accessible. In this paper, we will outline such a program, reviewing recent progress, as well as pointing out directions for research.

A running example in this paper will be sampling design optimization for magnetic resonance imaging (MRI) [2, 3], which has recently been addressed in [4]. In MRI, Fourier coefficients of selected image slices are acquired in a sequential fashion: if images of desired quality can be reconstructed from fewer acquisitions, we achieve a reduction in scan time, arguably today's most relevant limitation of MRI technology. Iterative nonlinear *compressive sensing* methods can be used to robustly undercut the classical Nyquist-Shannon sampling limit [5, 6, 7, 8]. The statistical idea behind these techniques is to import knowledge about "sparse" low level image statistics by way of a prior distribution, with which redundancies of dense sampling applied to images are exposed. Within this line of work, two major questions have to be faced: how to reconstruct given a fixed sampling design, and how to choose the design in the first place? The first question is addressed by *sparse estimation*. Given fixed design and data, a single most sensible image is produced by way of point estimation, minimizing a criterion composed of a data fit and a sparsity prior (or regularization) term. A wide array of convex and non-convex sparse estimation algorithms have been proposed. The second question, sampling design optimization, is different and has seen little attention so far, while there is evidence that the choice of design can be more relevant for successful real-world image reconstruction than the estimation algorithm or sparsity penalization used [9].

The nonlinear design optimization problem for realistic images has not been optimally solved, nor has a fully satisfying characterization been provided. Recent compressive sensing theory [5, 6] applies to unstructured, highly exactly sparse signals, but entirely fails to support design choices for natural or real medical images [9, 10, 4]. Short of optimal answers and conclusive theory, it is sensible to address the problem in an adaptive, data-driven manner, using concepts from Bayesian machine learning [4]. In this article, design optimization is phrased as search over realizable sampling patterns, optimizing a statistical criterion based on prior knowledge and real-world training data, with the aim of tractably detecting a pattern that generalizes well to future reconstruction problems. How could such a criterion look like? Which information from the data should it be based on? Which tools are available today in order to tractably evaluate it along the search? In this paper, we argue that these questions *can* successfully and tractably be addressed by adopting an approximate Bayesian viewpoint, extending the presently prevailing sparse estimation approaches by representations of *uncertainty*. In order to improve a given sampling design, a natural idea is to quantify the present shortcomings on real training data in a way that allows scoring additional candidate measurements as to how much novel information they are able to provide. This is not an image reconstruction problem per se, and it is hard to see how sparse point estimation techniques could be used to address it. Rather, it is an instance of nonlinear *Bayesian experimental design*, if the uncertainty representation in this picture is provided for by the Bayesian posterior, a distribution over all possible image reconstructions. Novel variational inference approximations can be used in order to implement this idea in MRI practice. Sparse *estimation*, reporting the mode of the Bayesian posterior, is not well suited to address higher-order design optimization. The latter is successfully driven by sparse *inference*, quantifying shape and covariance of the posterior beyond its mode.

At present, approximate Bayesian inference is hardly ever applied to low level imaging problems[1], for which much more efficient and well-characterized point estimation technology is preferred. Variational inference technology for continuous-variable image models is developed mainly in machine learning or Bayesian statistics. Since issues such as scalability, numerical robustness, or reductions to standard optimization primitives are not paid much attention there, previous methods run orders of magnitude slower than reconstruction techniques on the same model, so that sampling optimization for full high-resolution images cannot sensibly be addressed with them. A second goal of this article is to raise awareness of these issues. We show that at least some variational Bayesian approximations can be solved by scalable, well-characterized algorithms, which reduce to underlying standard image computations. As reviewed

[1] Bayesian *sparse estimation* techniques *are* applied to such problems, for example sparse Bayesian learning [11]. The crucial differences to Bayesian *sparse inference* will be highlighted below.

here, the major variational inference relaxations known today share similarities, and scalable reformulations should be attempted for all of them. Approximate Bayesian inference applied to low level real-world problems may come with new concepts and help to overcome obstacles in novel ways. Yet being based on the same model, prior assumptions and basic primitives, it should not have to come with entirely new computational methodology. If convex optimization and numerical mathematics technology is highly successful for imaging, in practice as well as in theoretical understanding, higher order Bayesian methods would ideally make use of these advances directly, instead of going different uncharted ways altogether.

The structure of the paper is as follows. The sparse linear model is introduced in Section 2, along with Bayesian nonlinear experimental design, motivated by MRI sampling design optimization. Current variational approximations to sparse Bayesian inference are reviewed in Section 3, along with convexity characterizations and scalable algorithms. This paper draws on material from previous publications, notably [12, 9, 4, 13], and on the report [14].

2. Sparse linear model. Experimental design

Let us formulate the setup for the MRI sampling optimization problem. Denote the desired MR image by $\boldsymbol{u} \in \mathbb{R}^n$, where n is the number of pixels. Under ideal conditions, the raw data $\boldsymbol{y} \in \mathbb{R}^m$ from an NMR scanner consists of Fourier coefficients of $\boldsymbol{u}$ at specific k-space spatial frequencies, so that $\boldsymbol{y} = \boldsymbol{X}\boldsymbol{u} + \boldsymbol{\varepsilon}, \quad \boldsymbol{\varepsilon} \sim N(\boldsymbol{0}, \sigma^2\boldsymbol{I})$. Rows of $\boldsymbol{X}$ are Fourier filters (k-space is synonimous with 2D Fourier space)[2]. Elementary units of the design (blocks of rows of $\boldsymbol{X}$) are called *phase encodes*. For example, in Cartesian MRI, a phase encode is a complete column in k-space. A Nyquist-dense design contains encodes to cover k-space completely, while for an undersampled design $\boldsymbol{X}$, certain encodes are not acquired ($m < n$). For undersampled reconstruction, a prior $P(\boldsymbol{u})$ is chosen which represents low level statistics of (MR) images, distinctly super-Gaussian "sparse" distributions. The posterior has the form

$$P(\boldsymbol{u}|\boldsymbol{y}) \propto N(\boldsymbol{y}|\boldsymbol{X}\boldsymbol{u}, \sigma^2\boldsymbol{I}) \prod_{j=1}^{q} e^{-\tau_j|s_j/\sigma|}, \quad \boldsymbol{s} = \boldsymbol{B}\boldsymbol{u}. \tag{1}$$

The likelihood $P(\boldsymbol{y}|\boldsymbol{u}) = N(\boldsymbol{y}|\boldsymbol{X}\boldsymbol{u}, \sigma^2\boldsymbol{I})$ is Gaussian, while the prior $P(\boldsymbol{u})$ is the normalized product of q Laplacians on linear projections s_j of $\boldsymbol{u}$, among them the image gradient and wavelet coefficients (see [9] for details), so that typically $q > n$. The Laplace distribution encourages sparsity of $\boldsymbol{s}$ [12]. This posterior is log-concave: it has a single mode, and all level sets are convex. We refer to this setup as *sparse linear model* (SLM). Note that we omit (parts of) $\boldsymbol{X}$ from distribution notation. These are control variables (covariates), and distributions are conditioned on them implicitly[3].

This Bayesian setup can be used for *sparse estimation* or *sparse inference* respectively. Sparse estimation concerns the reconstruction of a single image, which is often done by maximizing the posterior, or *MAP estimation*: $\hat{\boldsymbol{u}} = \mathrm{argmin}_{\boldsymbol{u}} -\log P(\boldsymbol{y}|\boldsymbol{u}) - \log P(\boldsymbol{u})$. For our setup here, this is a convex quadratic program, a special case of which ($\boldsymbol{B} = \boldsymbol{I}$) is known as the Lasso [15]. *Sparse inference* goes beyond finding the posterior mode, trying to approximate posterior moments such as mean $\mathrm{E}_P[\boldsymbol{u}|\boldsymbol{y}]$ or covariance $\mathrm{Cov}_P[\boldsymbol{u}|\boldsymbol{y}]$, or its log-partition function $\log P(\boldsymbol{y})$. Reviewing the numerous good reasons for doing so is not in the scope of this paper (see for example [16]): we will concentrate solely on design optimization. We will see below that Bayesian

[2] In practice, both $\boldsymbol{u}$ and $\boldsymbol{y}$ are complex-valued, which was neglected (for $\boldsymbol{u}$) in [4], but is accounted for properly [14] by doubling the number of real-valued variables (using the $\mathbb{C} \to \mathbb{R}^2$ embedding), and placing sparsity potentials on $|s_j|$ instead of s_j (potentials are even functions). This issue, which does not add implementational complexity, is ignored in the remainder of this paper, while it is crucial for making the MRI application work. More generally, norm potentials can be placed on $\|\boldsymbol{s}_j\|$ [14], using minor modifications.

[3] It is useful to address uncertainties in $\boldsymbol{X}$ itself, but this is not done here.

experimental design in our setting here is mainly driven by the posterior covariance, largely unrelated to the mode $\hat{\boldsymbol{u}}$. It is widely accepted today that the computational simplicity of finding the latter does not transfer to Bayesian inference, which remains intractable even for log-concave posteriors. Approximate sparse inference can be done by Markov chain Monte Carlo [17] or by variational approximations, both originally proposed in statistical physics. While most Bayesian statisticians prefer to use MCMC, little is known about how to obtain rigorous finite-time, limited-resources performance guarantees. In this paper, we concentrate on inference approximations which can be relaxed to image reconstruction technology in place already (such as convex optimization), thus on variational approximations.

In the remainder of this section, we briefly review Bayesian sequential experimental design and motivate how it can be used to address MRI sampling optimization. More general reviews are found in [18, 19, 20]. Recalling the basic idea from Section 1, we have to score extension candidates $\boldsymbol{X}_* \in \mathbb{R}^{d,n}$ (phase encodes) with respect to the decision of appending them to the present design $\boldsymbol{X}$. A generally useful measure is the *information gain*

$$\Delta(\boldsymbol{X}_*) := \mathrm{H}[P(\boldsymbol{u}|\boldsymbol{y})] - \mathrm{E}_{P(\boldsymbol{y}_*|\boldsymbol{y})}\left[\mathrm{H}[P(\boldsymbol{u}|\boldsymbol{y},\boldsymbol{y}_*)]\right], \tag{2}$$

where $P(\boldsymbol{u}|\boldsymbol{y})$ is the Bayesian posterior before, $P(\boldsymbol{u}|\boldsymbol{y},\boldsymbol{y}_*)$ after including $(\boldsymbol{X}_*,\boldsymbol{y}_*)$, $P(\boldsymbol{y}_*|\boldsymbol{y}) = \mathrm{E}_{P(\boldsymbol{u}|\boldsymbol{y})}[P(\boldsymbol{y}_*|\boldsymbol{u})]$ is the predictive posterior, and $\mathrm{H}[P(\boldsymbol{u}|\boldsymbol{y})] := \mathrm{E}_{P(\boldsymbol{u}|\boldsymbol{y})}[-\log P(\boldsymbol{u}|\boldsymbol{y})]$ is the Shannon (differential) entropy [21]. The information gain measures the decrease in uncertainty about $\boldsymbol{u}$, averaged over the "soft" prediction $P(\boldsymbol{y}_*|\boldsymbol{y})$. It is not the only criterion that can be used [20], especially if more specific utilities can be defined for the task at hand. What *is* characteristic of Bayesian experimental design, is that averages over $P(\boldsymbol{u}|\boldsymbol{y})$ and $P(\boldsymbol{y}_*|\boldsymbol{y})$ are taken: a principal and important difference to plugging in "best estimates", the preferred option with purely estimation-based techniques. The highest-scoring encode is appended to $\boldsymbol{X}$, and the process is iterated sequentially. When applying this procedure to MRI sampling optimization [4], many candidates $\boldsymbol{X}_*$ can be scored in each round without performing costly NMR measurements for them: an important advantage over "trial-and-error" approaches.

Given that we have to approximate sparse inference, it is important to understand *which* posterior moments are required in order to evaluate the information gain (2). In this paper, we concentrate on Gaussian posterior approximations $Q(\boldsymbol{u}|\boldsymbol{y}) \approx P(\boldsymbol{u}|\boldsymbol{y})$, fitted by means of a variational optimization problem. For a Gaussian, $\mathrm{H}[Q(\boldsymbol{u}|\boldsymbol{y})] = (1/2)\log|\mathrm{Cov}_Q[\boldsymbol{u}|\boldsymbol{y}]| + C$. Moreover, we approximate $P(\boldsymbol{u}|\boldsymbol{y},\boldsymbol{y}_*)$ by the Gaussian $\propto N(\boldsymbol{y}_*|\boldsymbol{X}_*\boldsymbol{u},\sigma^2\boldsymbol{I})Q(\boldsymbol{u}|\boldsymbol{y})$, so to avoid having to re-run the variational optimization for each candidate $\boldsymbol{X}_*$. With these approximations,

$$\Delta(\boldsymbol{X}_*) \approx -\log|\boldsymbol{A}| + \log\left|\boldsymbol{A} + \boldsymbol{X}_*^T\boldsymbol{X}_*\right| = \log\left|\boldsymbol{I} + \boldsymbol{X}_*\boldsymbol{A}^{-1}\boldsymbol{X}_*^T\right|, \quad \mathrm{Cov}_Q[\boldsymbol{u}|\boldsymbol{y}] = \sigma^2\boldsymbol{A}^{-1}. \tag{3}$$

The notation in terms of $\boldsymbol{A}$ will be convenient below. Therefore, for the methods of interest here, $\Delta(\boldsymbol{X}_*)$ most strongly depends on the posterior covariance $\mathrm{Cov}_P[\boldsymbol{u}|\boldsymbol{y}]$. More precisely, (3) is dominated by the largest eigenvectors and eigenvalues of $\mathrm{Cov}_Q[\boldsymbol{u}|\boldsymbol{y}]$, which fits into the intuitive picture of Section 1: we assess the directions of largest posterior spread of uncertainty, then align novel measurements with these directions of maximum uncertainty in what $\boldsymbol{u}$ should be. While we do not even know how to represent the exact SLM posterior, its leading covariance eigendirections constitute a definite approximation target in practice as well as for theoretical investigations.

3. Variational inference for sparse linear models

In this section, we review variational approximation techniques which have been applied to Bayesian inference for the sparse linear model. We will be interested in properties of the underlying variational optimization problems which allows for reductions to commonly used scientific computing techniques. We rely on [22, 12].

The posterior $P(\boldsymbol{u}|\boldsymbol{y})$ is intractable to integrate over for two reasons coming together. First, $\boldsymbol{u}$ is very high-dimensional, so that black-box techniques like quadrature are impossibly expensive. Second, $P(\boldsymbol{u}|\boldsymbol{y})$ is not Gaussian, due to the presence of non-Gaussian potentials $t_j(s_j) = e^{-\tau_j|s_j/\sigma|}$. Gaussian integrals are analytically tractable linear algebra expressions, which can often be approximated reliably even at large scales, by exploiting structure in the matrices $\boldsymbol{X}$, $\boldsymbol{B}$. Variational inference techniques relax the problem of integrating against $P(\boldsymbol{u}|\boldsymbol{y})$, by replacing the posterior (or relevant moments thereof) by approximations which allow for tractable computation. This goes hand in hand with introducing new variational parameters, and the variational problem is to optimize over these in order to fit the approximation to the posterior. Once this is done, the posterior approximation is used in place of the true posterior, when it comes to Bayesian queries such as design score computations (2). In this paper, we concentrate on *Gaussian* posterior approximations $Q(\boldsymbol{u}|\boldsymbol{y}) \approx P(\boldsymbol{u}|\boldsymbol{y})$, of the form

$$Q(\boldsymbol{u}|\boldsymbol{y}) = N(\boldsymbol{u}|\boldsymbol{u}_*, \sigma^2\boldsymbol{A}^{-1}) \propto P(\boldsymbol{y}|\boldsymbol{u}) \prod_{j=1}^{q} e^{-\sigma^{-2}(s_j^2/(2\gamma_j)-b_j s_j)}, \quad \boldsymbol{A} := \boldsymbol{X}^T\boldsymbol{X} + \boldsymbol{B}^T\boldsymbol{\Gamma}^{-1}\boldsymbol{B}, \quad (4)$$

where $\boldsymbol{\Gamma} := \operatorname{diag}\boldsymbol{\gamma}$. The family $\{Q(\boldsymbol{u}|\boldsymbol{y})\}$ is formally obtained from $P(\boldsymbol{u}|\boldsymbol{y})$ by replacing potentials $t_j(s_j)$ with Gaussian functions, introducing variational parameters $\boldsymbol{\gamma} = (\gamma_j)$, $\boldsymbol{b} = (b_j)$. Most of the methods discussed below do not impose the structure of (4), but rather start with a general idea of how to approximate $P(\boldsymbol{u}|\boldsymbol{y})$. Given that, the form (4) is *implied*.

What is the role of the variational parameters? For sparse linear models, the $\boldsymbol{b}$ parameters seem much less relevant. In some techniques, they are fixed to zero or other constant values. For other models not discussed here, $\boldsymbol{b}$ plays a more significant role (they allow to shift the mean without affecting the covariance). The variance parameters $\boldsymbol{\gamma}$ are instrumental. If $\boldsymbol{\gamma} \succeq \boldsymbol{0}$, it is easy to show that[4] $\mathrm{Var}_Q[s_j|\boldsymbol{y}] \le \sigma^2\gamma_j$ [14], so that the approximate posterior variance along s_j is bounded in terms of γ_j. Sparse linear models embody *selective shrinkage*: most $|s_j|$ are shrunk strongly towards zero, while some $|s_j|$ are hardly penalized at all. This property captures low level statistical properties of real-world signals discussed above[5]. Selective shrinkage is controlled by the γ_j parameters, which will be small for coefficients that are shrunken under the posterior, but large for such that are not strongly penalized. Readers familiar with image modelling may be sceptical about the proposal to approximate sparse image model posteriors by Gaussians. Is it not generally accepted that images have strong non-Gaussian statistics, for example along edges? Is it not a main argument of this paper that non-Gaussian linear models are to be used? This "dilemma" is solved by the presence of variational parameters. While $Q(\boldsymbol{u}|\boldsymbol{y})$ is a Gaussian, it is heavily parameterized (typically, q is larger than n) in a very non-stationary manner. While an "average" member of $\{Q(\boldsymbol{u}|\boldsymbol{y})\}$ may not represent image statistics, the closest fit to $P(\boldsymbol{u}|\boldsymbol{y})$ from this large family does so indeed[6].

How will we approximate the posterior covariance matrix $\mathrm{Cov}_P[\boldsymbol{u}|\boldsymbol{y}]$? In contrast to discrete variable variational relaxations, where only very simple distributions can be represented exactly, the Gaussian family $\{Q(\boldsymbol{u}|\boldsymbol{y})\}$ is rich enough to represent a wide range of *global* covariances, to the extent that for a log-concave (uni-modal) posterior $P(\boldsymbol{u}|\boldsymbol{y})$, the covariance of a close global fit $Q(\boldsymbol{u}|\boldsymbol{y}) \approx P(\boldsymbol{u}|\boldsymbol{y})$ can be a useful proxy for the leading directions of posterior covariance (see also end of Section 3.4).

[4] Strictly speaking, the proof assumes that $\boldsymbol{\gamma} \succ \boldsymbol{0}$. By continuity, it holds also if coefficients of $\boldsymbol{\gamma}$ become zero, as long as $\boldsymbol{A}$ stays positive definite.

[5] These properties are often referred to as *sparsity* (hence sparse linear model), but *super-Gaussianity* (introduced below) would be a better term in the context of natural images, which are *not* piecewise constant (which sparsity would imply), but filter coefficients exhibit a super-Gaussian (power law) decay.

[6] If $\boldsymbol{\gamma}$ for this closest fit is plotted accordingly for large enough m, edges of the underlying image are typically revealed.

Journal of Physics: Conference Series **197** (2009) 012001 doi:10.1088/1742-6596/197/1/012001

3.1. Super-Gaussianity. Scale mixtures. Direct site bounding

How to replace single potentials $t_j(s_j)$ by Gaussian functions in a principled manner? Two general ideas are reviewed in [22]: writing $t_j(s_j)$ as a *scale mixture*, and lower-bounding $t_j(s_j)$ by a Gaussian function. In the former case, $t_j(s_j) = \int_{>0} N(s_j|0, \sigma^2\gamma_j) f_j(\gamma_j)\, d\gamma_j$. In the latter case, $t_j(s_j) = \max_{\gamma_j>0} e^{-(s_j/\sigma)^2/(2\gamma_j)-h_j(\gamma_j)/2}$. Both ideas are restricted to potentials $t_j(s_j)$ for which such representations exist [22]. Potentials for which the latter max-representation exists, are called *super-Gaussian*: they can be lower bounded by Gaussian functions of any width. This definition is equivalent to $t_j(s_j)$ being even, positive, and $s_j^2 \mapsto \log t_j(s_j)$ being convex and decreasing. Namely, if $x_j := (s_j/\sigma)^2$, $\lambda_j := -1/(2\gamma_j)$, then $\log t_j(s_j) = \max_{\lambda_j<0} x_j\lambda_j - h_j(-1/(2\lambda_j))/2$, convex as maximum over affine functions (this is a Legendre/Fenchel duality, see [23]). Many potential functions used in statistics are super-Gaussian. Remarkably, it is shown in [22] that all scale mixture potentials are super-Gaussian: the latter concept is more general. For Laplacian potentials (which are scale mixtures), $h_j(\gamma_j) = \tau_j^2\gamma_j$. Further examples and closure properties can be found in [22, 14]. In the context of SLMs, super-Gaussianity is precisely the property associated with sparsity-favouring potentials: for a Gaussian $t_j(s_j)$, $s_j^2 \mapsto \log t_j(s_j)$ is affine, so that convexity of this function translates into heavier tails and stronger concentration of mass close to zero. The concept can be extended to non-even potentials, whenever $t_j(s_j)e^{\kappa_j s_j}$ is even for some κ_j. For example, Bernoulli (or logistic) potentials used as binary classification likelihoods are super-Gaussian, which was noted and exploited in [24].

An early Bayesian treatment of SLMs was sparse Bayesian learning (SBL) [11], motivated in terms of scale mixture decompositions for Student's t potentials, but derived in a somewhat heuristic manner. The maximum a posteriori arguments used to motivate SBL are problematic, because by changing the parameterization (say, γ_j versus $\pi_j = 1/\gamma_j$), we arrive at different algorithms, which may address different problems (sparse estimation, sparse inference), and whose convergence properties and speed can be very different. SBL, which in the form applied in [11] is a sparse *estimation* technique, has found widespread use, in the wake of which the concepts of different variational relaxation principles, sparsity priors, and update algorithms have been mixed in a confusing manner. Confusions of this kind can be avoided by focussing on approximation *principles* (independent of *algorithms* for solving them) that are invariant to parameterizations, which is what we restrict ourselves to in this paper. Among sparse estimation techniques, SBL (also known as ARD) works remarkably well, often improving on MAP estimation significantly [25].

For super-Gaussian potentials $t_j(s_j)$, *direct site bounding*[7] (DSB) is obtained by plugging site bounds into the log partition function:

$$\log P(\boldsymbol{y}) \geq \max_{\boldsymbol{\gamma}\succ\mathbf{0}} -\phi_{\mathrm{DSB}}(\boldsymbol{\gamma})/2, \quad \phi_{\mathrm{DSB}}(\boldsymbol{\gamma}) := -2\log Z_Q + h(\boldsymbol{\gamma}),$$
$$Z_Q := \int P(\boldsymbol{y}|\boldsymbol{u}) e^{-\frac{1}{2}\sigma^{-2}\boldsymbol{s}^T\boldsymbol{\Gamma}^{-1}\boldsymbol{s}}\, d\boldsymbol{u}, \quad \boldsymbol{\Gamma} := \operatorname{diag}\boldsymbol{\gamma},\ h(\boldsymbol{\gamma}) := \sum_j h_j(\gamma_j). \tag{5}$$

The variational problem is $\min_{\boldsymbol{\gamma}\succ\mathbf{0}} \phi_{\mathrm{DSB}}(\boldsymbol{\gamma})$, motivated by tightening the lower bound to $\log P(\boldsymbol{y})$. Why should this lead to a closer fit of $Q(\boldsymbol{u}|\boldsymbol{y})$ to $P(\boldsymbol{u}|\boldsymbol{y})$? The log partition function $\log P(\boldsymbol{y})$ is the general approximation target of variational techniques. It is the moment generating function of $P(\boldsymbol{u}|\boldsymbol{y})$, while the lower bound plays the same role for $Q(\boldsymbol{u}|\boldsymbol{y})$. A more direct interpretation in terms of relative entropy divergence is given in Section 3.2. However, at present there is no finite-size theory we are aware of that firmly links improvements in lower bound tightness to approximation quality of posterior moments. Note that Z_Q is the normalization constant of $Q(\boldsymbol{u}|\boldsymbol{y})$, whose form does not depend on specifics of the potentials $t_j(s_j)$.

[7] As opposed to SBL, DSB is not a commonly used terminology, but was chosen for convenience here.

As a Gaussian integral, $\phi_{\mathrm{DSB}}(\boldsymbol{\gamma})$ can be evaluated tractably. A simple optimization strategy is coordinate descent, updating each γ_j in turn while keeping all others fixed. For Laplacian potentials, this algorithm is due to [26], see [22, 12] for more general discussions. The update for γ_j is a simple function of the posterior marginal $Q(s_j|\boldsymbol{y})$. As shown in [22], this algorithm can be implemented without knowing $h_j(\gamma_j)$ explicitly. However, in order to update all parameters, posterior means *and* variances are required for all s_j. In cases where the precision matrix $\sigma^{-2}\boldsymbol{A}$ of $Q(s_j|\boldsymbol{y})$ exhibits a sparse graphical model structure, such *single site updating* algorithms can be implemented by message passing. Once a potential has been updated, the information is propagated to the next site to be visited. In large scale situations of interest here, where $Q(s_j|\boldsymbol{y})$ does not have useful graphical model structure, a $n \times n$ linear system has to be solved from scratch in order to access any single marginal $Q(s_j|\boldsymbol{y})$: message passing is not attractive. We will come back to this point in Section 3.4.

3.2. Variational mean field approximations

A variational mean field (VMF)[8] characterization of Bayesian inference is given by

$$\log P(\boldsymbol{y}) = \max_{Q(\boldsymbol{u}|\boldsymbol{y})} \mathrm{E}_Q\left[\log P(\boldsymbol{y},\boldsymbol{u}) - \log Q(\boldsymbol{u}|\boldsymbol{y})\right], \tag{6}$$

where the maximum is taken over all distributions [27]. The maximizer is given by $Q(\boldsymbol{u}|\boldsymbol{y}) = P(\boldsymbol{u}|\boldsymbol{y})$, the true posterior. The slack in the lower bound for any fixed $Q(\boldsymbol{u}|\boldsymbol{y})$ is the relative entropy $\mathrm{D}[Q(\boldsymbol{u}|\boldsymbol{y}) \,\|\, P(\boldsymbol{u}|\boldsymbol{y})]$. VMF relaxations are obtained by restricting the maximum to a subset of distributions, for which the lower bound to $\log P(\boldsymbol{y})$ can be evaluated tractably. In other words, the closest member $Q(\boldsymbol{u}|\boldsymbol{y})$ from this subset in terms of $\mathrm{D}[\cdot \,\|\, P(\boldsymbol{u}|\boldsymbol{y})]$ divergence is pursued[9]. In this section, we will discuss two relaxations, which have been used for SLMs.

First, we may simply restrict ourselves to unconstrained *Gaussian* distributions $Q(\boldsymbol{u}|\boldsymbol{y})$, which allows for tractable lower bound maximization (this involves computing Gaussian expectations over $\log t_j(s_j)$). Interestingly, stationary points of this optimization have the form prescribed by (4) (see Appendix), so that $Q(\boldsymbol{u}|\boldsymbol{y})$ can be restricted to lie in this family without loss of generality. The lower bound becomes

$$\begin{aligned} &\log P(\boldsymbol{y}) \geq \max_{\boldsymbol{\gamma},\boldsymbol{b}} -\phi_{\mathrm{MF}}(\boldsymbol{\gamma},\boldsymbol{b})/2, \quad \phi_{\mathrm{MF}}(\boldsymbol{\gamma},\boldsymbol{b}) := -2\log Z_Q + \sum_j h_j^{\mathrm{MF}}(\gamma_j, b_j, Q(s_j|\boldsymbol{y})), \\ &h_j^{\mathrm{MF}}(\gamma_j, b_j, Q(s_j|\boldsymbol{y})) := -2\mathrm{E}_Q[\log t_j(s_j) + (s_j/\sigma)^2/(2\gamma_j) - b_j s_j/\sigma^2], \end{aligned} \tag{7}$$

where Z_Q is the normalization constant of $Q(\boldsymbol{u}|\boldsymbol{y})$. This criterion has a more complicated structure than (5), because h_j^{MF} depends on $Q(s_j|\boldsymbol{y})$. More can be said if $\boldsymbol{b}$ is fixed to zero and all $t_j(s_j)$ are super-Gaussian: $h_j^{\mathrm{MF}} = -2\mathrm{E}_Q[\log t_j(s_j) + (s_j/\sigma)^2/(2\gamma_j)]$. Since $\log t_j(s_j) + (s_j/\sigma)^2/(2\gamma_j) \geq -h_j(\gamma_j)/2$ for all s_j, $\gamma_j > 0$, we have that $h_j^{\mathrm{MF}}(\gamma_j, Q(s_j|\boldsymbol{y})) \leq h_j(\gamma_j)$ for $\gamma_j > 0$. Therefore, this variant of VMF uses a potentially tighter bound on $\log P(\boldsymbol{y})$ than DSB does. Moreover, since $\log P(\boldsymbol{y}) = -2\phi_{\mathrm{MF}}(\boldsymbol{\gamma}) + \mathrm{D}[Q(\boldsymbol{u}|\boldsymbol{y}) \,\|\, P(\boldsymbol{u}|\boldsymbol{y})]$, this means that DSB minimizes an upper bound to the VMF relative entropy $\mathrm{D}[Q(\boldsymbol{u}|\boldsymbol{y}) \,\|\, P(\boldsymbol{u}|\boldsymbol{y})]$, which can have a simpler structure than the relative entropy itself (see Section 3.4).

[8] A historically more correct terminology would be variational *structured* mean field. In this paper, VMF approximations are those that make use of (6) with restricted families for $Q(\boldsymbol{u}|\boldsymbol{y})$. In general, the term VMF seems reserved for *factorization* restrictions only. In recent papers, structured VMF has been termed "variational Bayes", ignorant of the fact that methods like DSB or EP are variational Bayesian approximations just as well, but are based on different ideas.

[9] In general, VMF approximations do not constitute convex optimization problems. While the bound can be optimized locally, finding a global optimum may still be a hard problem.

A different variant of VMF is discussed in [22], for cases in which the $t_j(s_j)$ have scale mixture decompositions. Once $\boldsymbol{\gamma}$ are introduced alongside $\boldsymbol{u}$, the VMF characterization (6) can be written as maximum over $Q(\boldsymbol{u}, \boldsymbol{\gamma}|\boldsymbol{y})$ just as well, where $P(\boldsymbol{y}, \boldsymbol{u})$ is replaced by $P(\boldsymbol{y}|\boldsymbol{u})N(\boldsymbol{s}|\boldsymbol{0}, \sigma^2\boldsymbol{\Gamma})(f_j(\gamma_j))$ (recall the scale mixture decomposition of $t_j(s_j)$ from Section 3.1). In this case, tractability is obtained by making a *factorization* assumption: $Q(\boldsymbol{u}, \boldsymbol{\gamma}|\boldsymbol{y}) = Q(\boldsymbol{u}|\boldsymbol{y})Q(\boldsymbol{\gamma}|\boldsymbol{y})$. Once more, this free-form assumption implies further simplifications: $Q(\boldsymbol{u}|\boldsymbol{y})$ can be restricted to lie in the Gaussian family (4), with $\boldsymbol{b} = \boldsymbol{0}$. Moreover, it is shown in [22] that single site updating for this VMF variant is precisely equivalent to coordinate descent for DSB. Therefore, this second VMF variant is nothing new, but simply a different way of motivating the DSB variational problem.

3.3. Expectation propagation

The expectation propagation (EP) algorithm [28, 29] is based on the idea of moment matching. It builds on the assumed density filtering (ADF) principle [30]. Suppose a posterior of the form (1) is to be approximated by a Gaussian $Q(\boldsymbol{u}|\boldsymbol{y})$, where (different from the SLM discussed above) the Gaussian factor in $P(\boldsymbol{u}|\boldsymbol{y})$ is normalizable. In exact Bayesian filtering, we start with a Gaussian $Q_0(\boldsymbol{u})$ proportional to this factor, then include one potential $t_j(s_j)$ after the other, stepping from $Q_0(\boldsymbol{u})$ to $\propto Q_0(\boldsymbol{u})t_1(s_1)$, and so on. Even after the first inclusion, the filtering distribution is not Gaussian anymore, and computations become intractable rapidly. In ADF, after each Bayesian inclusion, we project the intermediate distribution back onto a Gaussian, preserving mean and covariance. This is equivalent to finding the solution $Q_1(\boldsymbol{u})$ of $\min_{Q \text{ Gaussian}} \mathrm{D}[\hat{P}_1(\boldsymbol{u}) \,\|\, Q(\boldsymbol{u})]$, where $\hat{P}_1(\boldsymbol{u}) \propto Q_0(\boldsymbol{u})t_1(s_1)$ (see Appendix). Moreover, to step from Q_0 to Q_1, we multiply the present state $Q_0(\boldsymbol{u})$ by $\tilde{t}_1(\boldsymbol{u}) \propto Q_1(\boldsymbol{u})/Q_0(\boldsymbol{u})$. In fact, since t_1 depends on s_1 only, we may choose $\tilde{t}_1(\boldsymbol{u}) = \tilde{t}_1(s_1) := e^{\sigma^{-2}(b_1 s_1 - s_1^2/(2\gamma_1))}$, where b_1, γ_1 can be computed from $Q_0(s_1)$ and $t_1(s_1)$ only (see Appendix). The ADF update is a local computation for models with local potentials $t_j(s_j)$. Moreover, the Gaussian approximation $Q(\boldsymbol{u}|\boldsymbol{y}) = Q_q(\boldsymbol{u})$ lies within the family (4).

Once all potentials $t_j(s_j)$ have been included, ADF has to be terminated. EP extends ADF by the capability of re-visiting previous projections, in order to obtain a full-fledged variational inference technique. If ADF updates were continued after the first round, we would overcount the influence of non-Gaussian potentials. An ADF update at $t_j(s_j)$ is done for a state $Q(\boldsymbol{u}|\boldsymbol{y})$ where $b_j = 1/\gamma_j = 0$. In order to re-create this situation, an EP update starts with dividing out the Gaussian term $\tilde{t}_j(s_j) := e^{\sigma^{-2}(b_j s_j - s_j^2/(2\gamma_j))}$, followed by an ADF update. In other words, the *cavity distribution* $Q_{\setminus j}(s_j|\boldsymbol{y}) \propto Q(s_j|\boldsymbol{y})\tilde{t}_j(s_j)^{-1}$ is used as basis for computing $\hat{P}_j(s_j)$, in place of the marginal $Q(s_j|\boldsymbol{y})$. The EP algorithm is of the single site updating type (see Section 3.1), running EP updates in some ordering until convergence. At a stationary point, *moment consistency* is attained: all $\hat{P}_j(\boldsymbol{u})$ share their Gaussian moments with $Q(\boldsymbol{u}|\boldsymbol{y})$. This condition should be compared with the first VMF variant in Section 3.2, where $\mathrm{D}[Q(\boldsymbol{u}|\boldsymbol{y}) \,\|\, P(\boldsymbol{u}|\boldsymbol{y})]$ is minimized over Gaussians. Following the information-theoretic interpretation of relative entropy $\mathrm{D}[P_a \,\|\, P_b]$ (coding loss if P_b is used instead of the true P_a), the ordering of arguments is the wrong way around. The minimization of $\mathrm{D}[P(\boldsymbol{u}|\boldsymbol{y}) \,\|\, Q(\boldsymbol{u}|\boldsymbol{y})]$ is of course intractable: it is equivalent to finding the *exact* posterior mean and covariance. EP can be seen as using the relative entropy with arguments the right way around, yet replacing the posterior $P(\boldsymbol{u}|\boldsymbol{y})$ by surrogates $\hat{P}_j(\boldsymbol{u})$ ("one step away from Gaussian") for which the divergence can be computed. With EP, γ_j can become negative in general [28]. However, if all $t_j(s_j)$ are log-concave, all γ_j remain positive throughout [12].

What is the variational problem underlying the EP algorithm? Let us introduce additional parameters C_j, defined by moment matching conditions of zero-th order: $\mathrm{E}_{Q_{\setminus j}}[t_j(s_j)] =$

$\mathrm{E}_{Q_{\setminus j}}[C_j\tilde{t}_j(s_j)]$. Replacing $t_j(s_j)$ by $C_j\tilde{t}_j(s_j)$, we obtain the approximation

$$\log P(\boldsymbol{y}) \approx -\phi_{\mathrm{EP}}(\boldsymbol{\gamma}, \boldsymbol{b})/2, \quad \phi_{\mathrm{EP}}(\boldsymbol{\gamma}, \boldsymbol{b}) := -2\log Z_Q - 2\sum_j \log C_j, \tag{8}$$

which is brought into the form previously used by defining $h_j^{\mathrm{EP}}(\gamma_j, b_j, Q(s_j|\boldsymbol{y})) := -2\log C_j$. Every fixed point of the EP algorithm is a stationary point of $\phi_{\mathrm{EP}}(\boldsymbol{\gamma}, \boldsymbol{b})$ [12]. There are important differences to DSB and VMF. First, these stationary points are neither local maxima nor minima, but true saddle points (at least with respect to the unconstrained parameterization in terms of $(\boldsymbol{b}, \boldsymbol{\gamma})$). The criterion $\phi_{\mathrm{EP}}(\boldsymbol{\gamma}, \boldsymbol{b})$ has a more complicated structure than ϕ_{MF} or ϕ_{DSB}, and the EP algorithm does not always converge [28] (a convergence proof has not been given even for log-concave models). While there are provably convergent algorithms [31], they are significantly more expensive to run. Finally, EP seems more susceptible to numerical instability problems than the other techniques discussed here. For the SLM with (log-concave) Laplacian potentials, EP frequently fails on real-world image data [12]. Due to the underdetermined likelihood, cavity variances of $Q_{\setminus j}(s_j|\boldsymbol{y})$ tend to be huge, which leads to runaway numerical errors in the moment computations for $\hat{P}_j(s_j)$, even if these are executed with utmost care. In this case, the EP algorithm is highly sensitive to such errors. A remedy is to use *fractional EP* [32], in that $Q_{\setminus j}(s_j|\boldsymbol{y}) \propto Q(s_j|\boldsymbol{y})\tilde{t}(s_j)^{-\eta}$, $\hat{P}_j(s_j) \propto Q_{\setminus j}(s_j|\boldsymbol{y})t_j(s_j)^{\eta}$ for $\eta < 1$. While this constitutes a variational problem different[10] from standard EP, numerical problems essentially disappear even for $\eta = 0.9$ [12].

3.4. Convexity properties. Scalable double loop algorithms

We discussed a range of continuous-variable variational inference approximations (DSB, VMF, EP) above, along with single site updating algorithms. Implementations of these algorithms tend to be similar, and underlying variational criteria can be brought into similar form. This superficial similarity does not mean that the variational problems can be characterized or solved equally well in practice. We saw that EP is a saddle point rather than an optimization problem, which tends to be numerically more problematic than the others. In this section, we show that for log-concave posteriors (including the SLM introduced above), DSB is a convex optimization problem, and we provide an algorithm to solve it even for very large models. In comparison, VMF is a non-convex problem in general, and no scalable solvers have been proposed so far, let alone for EP.

All criteria share the part $-2\log Z_Q$. Suppose that $\boldsymbol{b}$ is fixed to zero or some other constant, so that $\boldsymbol{\gamma}$ are the sole variational parameters (natural for DSB, can be enforced for VMF, EP). $\boldsymbol{\gamma} \mapsto -2\log Z_Q$ is a convex function for $\boldsymbol{\gamma} \succ \boldsymbol{0}$ [14]. Namely, $\log P(\boldsymbol{y}|\boldsymbol{u}) + \sigma^{-2}(\boldsymbol{b}^T\boldsymbol{u} - \boldsymbol{s}^T\boldsymbol{\Gamma}^{-1}\boldsymbol{s}/2)$ is jointly concave as function of $(\boldsymbol{u}, \boldsymbol{\gamma})$, $\boldsymbol{\gamma} \succ \boldsymbol{0}$, so that $\boldsymbol{\gamma} \mapsto \log Z_Q$ is concave by Prékopa's marginalization theorem [33]. This holds only if $\boldsymbol{b}$ is constant: $\log Z_Q$ is not concave in $(\boldsymbol{b}, \boldsymbol{\gamma})$ in general. It is shown in [14] that for super-Gaussian potentials, $h(\boldsymbol{\gamma})$ is convex iff all sites $t_j(s_j)$ are log-concave: the DSB *variational inference* problem is convex if and only if *MAP estimation* is convex for the same model. An equivalent characterization does not hold for VMF in general, and certainly not for EP.

A scalable algorithm to solve DSB for sparse generalized linear models has recently been proposed [4, 14]. It runs orders of magnitude faster than single site updating, and can be used to successfully address MRI design optimization. In order to understand its benefits, we have to view variational inference from a computational standpoint: *which computations are minimally required in order to solve SLM variational inference problems?* In single site updating, each step

[10] Fractional EP is different from the common practice of *damping*, which does not change the fixed points, but does not help either to alleviate problems in this case.

requires marginal mean and variance of $Q(s_j|\boldsymbol{y}) = N(h_j, \sigma^2\rho_j)$, and each potential has to be visited at least once. For large scale SLMs whose model graph is not a tree, each update requires the solution of a $n \times n$ linear system from scratch (see Section 3.1): these algorithms do not provide scalable sparse inference solutions in general. Can we do better with global steps, say by gradient descent? The computation of $\nabla_{\boldsymbol{\gamma}}\phi_{\rm DSB}$ is dominated by bulk mean and variances computations ("bulk" means that *all* means and variances are required). This is not a problem for the means: all of them are obtained by solving a single $n \times n$ linear system, say by linear conjugate gradients [34]. However, no general method is presently known for approximating all variances in an equally scalable manner, not even if $Q(\boldsymbol{u}|\boldsymbol{y})$ has a sparsely connected graphical model structure such as a nearest-neighbour grid. This fundamental computational difference between bulk mean and variances computation is underlined by theoretical analyses into the convergence of belief propagation in Gaussian Markov random fields [35]. *Any* variational inference algorithm has to compute bulk means and variances eventually, but among different algorithms, those will fare best that require *as few bulk variances computations as possible* until convergence.

Inspired by [25], we can write $\phi_{\rm DSB}(\boldsymbol{\gamma}) = \log|\boldsymbol{A}| + \phi_{\cup}(\boldsymbol{\gamma})$, where $\phi_{\cup}(\boldsymbol{\gamma})$ is a decoupled function. It is the highly coupled term $\log|\boldsymbol{A}|$ that calls for computing variances. Crucially, $\boldsymbol{\gamma}^{-1} \mapsto \log|\boldsymbol{A}|$ is a concave function, so can be upper bounded by affine functions, making use of Legendre/Fenchel duality once more: $\log|\boldsymbol{A}| \le \boldsymbol{z}^T(\boldsymbol{\gamma}^{-1}) - g^*(\boldsymbol{z})$, $\boldsymbol{z} \succ \boldsymbol{0}$ the normal vector, $g^*(\boldsymbol{z})$ the offset. Naturally, this observation leads to a provably convergent *double loop algorithm*, alternating between outer loop updates (computing $\boldsymbol{z}$, $g^*(\boldsymbol{z})$ so to tangentially fit the affine bound at $\boldsymbol{\gamma}$) and inner loop minimizations of the upper bound $\phi_{\cup}(\boldsymbol{\gamma}) + \boldsymbol{z}^T(\boldsymbol{\gamma}^{-1}) - g^*(\boldsymbol{z})$. In the absence of $\log|\boldsymbol{A}|$, these inner minimizations are solved much more efficiently than $\min_{\boldsymbol{\gamma}\succ\boldsymbol{0}} \phi_{\rm DSB}$ itself: they are penalized least squares problems of standard form, which can be solved by IRLS, a variant of Newton-Raphson [4]. Variances computations are not required during inner loops at all. For outer loop updates,

$$\boldsymbol{z} \leftarrow \nabla_{\boldsymbol{\gamma}^{-1}} \log|\boldsymbol{A}| = \mathrm{diag}^{-1}(\boldsymbol{B}\boldsymbol{A}^{-1}\boldsymbol{B}^T) = \sigma^{-2}(\mathrm{Var}_Q[s_j|\boldsymbol{y}]), \quad g^*(\boldsymbol{z}) \leftarrow \boldsymbol{z}^T(\boldsymbol{\gamma}^{-1}) - \log|\boldsymbol{A}|,$$

requiring the bulk marginal variances computation avoided during the inner loop. Since convergence is typically attained after few such outer loop steps, the double loop algorithm is a scalable DSB solver. It can be applied even if $h(\boldsymbol{\gamma})$ is not convex, see [14] for details.

To sum up, the DSB variational relaxation can be solved for very large models by decoupling the critical $\log|\boldsymbol{A}|$ term by way of a double loop algorithm: most of the work is done in standard form decoupled inner loop optimizations. Can the same idea be applied to VMF or EP? A similar direct approach does not work, because their $h_j^{(\cdot)}$ functions depend on $Q(s_j|\boldsymbol{y})$ and cannot simply be upper bounded. An important point for future research is to find scalable algorithms for these variational approximations.

Finally, is it fair to call the double loop algorithm "scalable"? After all, bulk variances still have to be computed a number of times. Moreover, computing many design score values (3) constitutes a closely related problem. While bulk variances cannot at present be approximated nearly as tractably as solving a linear system, this problem is addressed in numerical mathematics. A general idea is to employ a low rank PCA approximation to $\boldsymbol{A}^{-1}$, featuring only the $k \ll n$ smallest eigenvalues and eigenvectors of $\boldsymbol{A}$. Plugging this in, variances or design scores can be approximated efficiently. The optimality of PCA among all rank-k choices in terms of (co)variance explained is well known. Importantly, typical precision matrices $\boldsymbol{A}$ exhibit a roughly linear[11] spectral decay, so that the k smallest eigenvalues of $\boldsymbol{A}$ can be

[11] The linear spectral decay is *good* news, since Lanczos reveals smallest eigenvalues. It is *bad* news for the overall relative accuracy of variance approximations: their dependence on the interior of the spectrum, never penetrated by Lanczos, is significant.

found efficiently by the Lanczos algorithm [36, 4]. The latter scales superlinearly in k and runs up $O(n\,k)$ memory, so only a limited number of iterations can be done. Proposals exist in numerical mathematics for probing deeper into the spectrum of $\boldsymbol{A}$ [37]. For image models with Markov random field structure, alternative low rank approximations have been proposed that circumvent Lanczos computations [38]. Lastly, variances computation can easily be parallelized. Ultimately, if large scale problems are to be addressed with variational relaxations like DSB *without* accurate bulk variances computation in place, successful solvers will have to be robust to errors committed by estimators such as Lanczos. Empirically, at least for SLMs in the context of MR image data, the double loop algorithm behaves well in that respect [14]. Developing a better understanding of the impact of Gaussian variances errors on variational inference results is an important goal for future research.

3.5. Sparse estimation and sparse inference

We have seen that sparse estimation and sparse inference are different problems. They address different goals (for example, image reconstruction from fixed data, versus design optimization) and come with different algorithmic challenges (for example, bulk Gaussian variances computation is required for variational sparse inference, but not for sparse estimation). Why are they so frequently confused? First, they are based on the same underlying models and prior distributions. Second, in the context of variational approximations, their optimization problems may look deceivingly similar. Recall that SBL is a sparse estimation method, yet is motivated as inference approximation in [11]. By tempering with the $h_j(\gamma_j)$ functions in (5), DSB can be turned into SBL, or into any number of other variants. The highly sparse *estimation* algorithm of [25], solving an instance of SBL, is formally similar to the Laplacian SLM *inference* method of [4]: $h_j(\gamma_j) = \tau_j^2\gamma_j$ in the latter becomes $h_j(\gamma_j) = \log\gamma_j$ in the former: outcomes are drastically different, as are computational demands.

For sparse estimation, the goal is to *eliminate* irrelevant variables. In the context of "SBL-like" methods, this means driving many γ_j to exactly zero. Since $\mathrm{Var}_Q[s_j|\boldsymbol{y}] \le \sigma^2\gamma_j$ (Section 3), $\gamma_j = 0$ leads to $s_j = 0$ almost surely under $Q(\boldsymbol{u}|\boldsymbol{y})$. The latter is a *highly degenerate* distribution, representing correlations and mass only for the small number of non-eliminated variables: certainly not a sensible approximation to the true posterior. Bayesian sequential experimental design based on sparse estimation rather than inference [39] fails for real-world signals [9]: the "posterior" precludes any exploration beyond what has already been extracted from the data, and the procedure gets stuck. Sparse estimation is typically computationally simpler than sparse inference. MAP estimation is a convex quadratic problem for the SLM discussed here, without any approximations. In the double loop algorithm of [25], outer loop updates are simple to do using low rank formulae, since they inherit the exact sparsity of $\boldsymbol{\gamma}$: bulk variances computation is not a hard problem for highly degenerate $Q(\boldsymbol{u}|\boldsymbol{y})$.

Sparse estimation and sparse inference are different problems, but what is their theoretical and algorithmic overlap? Given that sparse estimation receives much more attention, advancing this understanding may lead to benefits for sparse inference. For example, the inner loop problem of the double loop DSB algorithm [4] has the form of a commonly used smooth approximation to MAP estimation for the same model. Variational sparse inference methods have to operate without the computational benefit of exact sparsity in $\boldsymbol{\gamma}$, yet they can still benefit from smoothed estimation technology.

4. Discussion

We have contrasted sparse estimation problems (undersampled reconstruction, denoising, decoding of compressed images) with problems of sparse Bayesian inference (measurement design optimization), where the posterior distribution is approximated beyond locating its mode. We have reviewed common variational relaxations of sparse Bayesian inference, related their criteria and discussed algorithmic differences. Adopting a computation-centered view of these techniques, relating them on the same sparse linear model, we aim to clear up currently prevailing confusions (relying on work by David Wipf and colleagues [40, 22, 25]), to point out major avenues for future research, and to show how sparse estimation and inference can benefit from each other.

Acknowledgments
Some of this work has been done jointly with Hannes Nickisch, Max Planck Institute for Biological Cybernetics, Tübingen. We would like to thank David Wipf and Florian Steinke for helpful discussions. This research is partly funded by the Cluster of Excellence on Multimodal Computing and Interaction at Saarland University.

Appendix

In Section 3.2, a variant of variational mean field restricts the maximum in (6) to Gaussian distributions $Q(\boldsymbol{u}|\boldsymbol{y}) = N(\boldsymbol{\mu}, \boldsymbol{\Sigma})$. If $Q(\boldsymbol{u}|\boldsymbol{y})$ is a stationary point, we show that $\boldsymbol{\Sigma}^{-1} = \sigma^{-2}\boldsymbol{X}^T\boldsymbol{X} + \boldsymbol{B}^T\boldsymbol{D}\boldsymbol{B}$ for a diagonal matrix $\boldsymbol{D}$, to that $Q(\boldsymbol{u}|\boldsymbol{y})$ lies in (4), as long as $\boldsymbol{\gamma}$ may have negative entries (if $\boldsymbol{B}$ has full rank n, any mean $\boldsymbol{\mu}$ can be achieved within (4) by varying $\boldsymbol{b}$ alone). Writing $\log P(\boldsymbol{y}|\boldsymbol{u}) = \frac{1}{2}\sigma^{-2}(\boldsymbol{c}^T\boldsymbol{u} - \boldsymbol{u}^T\boldsymbol{P}\boldsymbol{u}) + C$ with $\boldsymbol{P} = \boldsymbol{X}^T\boldsymbol{X}$, and $\nu_j = 2\mathrm{E}_Q[-\log t_j(s_j)]$, then twice the criterion to minimize is $\log|\boldsymbol{\Sigma}| - \sigma^{-2}(\operatorname{tr}\boldsymbol{P}(\boldsymbol{\Sigma} + \boldsymbol{\mu}\boldsymbol{\mu}^T) - \boldsymbol{c}^T\boldsymbol{\mu}) - \boldsymbol{1}^T\boldsymbol{\nu}$. Considering the mean $\boldsymbol{\mu}$ to be fixed, the relevant part is $\mathcal{F} = \log|\boldsymbol{\Sigma}| - \sigma^{-2}\operatorname{tr}\boldsymbol{P}\boldsymbol{\Sigma} - \boldsymbol{1}^T\boldsymbol{\nu}$. Here, ν_j depends on $Q(s_j|\boldsymbol{y})$ only, whose means are fixed, and whose variances are $\boldsymbol{\rho} = \operatorname{diag}^{-1}(\boldsymbol{B}\boldsymbol{\Sigma}\boldsymbol{B}^T)$. If $d_j = \partial\nu_j/(\partial\rho_j)$, $\boldsymbol{D} = \operatorname{diag}\boldsymbol{d}$, then $d\mathcal{F} = \operatorname{tr}(\nabla_{\boldsymbol{\Sigma}}\mathcal{F})(d\boldsymbol{\Sigma}) = \operatorname{tr}\boldsymbol{\Sigma}^{-1}(d\boldsymbol{\Sigma}) - \sigma^{-2}\operatorname{tr}\boldsymbol{P}(d\boldsymbol{\Sigma}) - \operatorname{tr}\boldsymbol{D}\boldsymbol{B}(d\boldsymbol{\Sigma})\boldsymbol{B}^T$. Solving $\nabla_{\boldsymbol{\Sigma}}\mathcal{F} = \boldsymbol{\Sigma}^{-1} - \sigma^{-2}\boldsymbol{P} - \boldsymbol{B}^T\boldsymbol{D}\boldsymbol{B} = \boldsymbol{0}$ establishes the claim.

In Section 3.3, we claim that moment matching of $\hat{P}_1(\boldsymbol{u})$ is equivalent to minimizing $\mathrm{D}[\hat{P}_1(\boldsymbol{u})\,\|\,Q(\boldsymbol{u})]$ over all Gaussians $Q(\boldsymbol{u})$. There is a unique Gaussian $Q_1(\boldsymbol{u})$ sharing mean and covariance with $\hat{P}_1(\boldsymbol{u})$, with which $\mathrm{D}[\hat{P}_1(\boldsymbol{u})\,\|\,Q(\boldsymbol{u})] = \mathrm{D}[\hat{P}_1(\boldsymbol{u})\,\|\,Q_1(\boldsymbol{u})] + \mathrm{E}_{\hat{P}_1}[\log Q_1(\boldsymbol{u}) - \log Q(\boldsymbol{u})]$. Here, $\log Q_1(\boldsymbol{u}) - \log Q(\boldsymbol{u})$ is a quadratic function in $\boldsymbol{u}$. Since $\hat{P}_1(\boldsymbol{u})$ and $Q_1(\boldsymbol{u})$ have the same moments up to second order, $\mathrm{E}_{\hat{P}_1}[\dots]$ can be replaced by $\mathrm{E}_{Q_1(\boldsymbol{u})}[\dots]$, and the second term becomes $\mathrm{D}[Q_1(\boldsymbol{u})\,\|\,Q(\boldsymbol{u})] \geq 0$. Therefore, $\mathrm{D}[\hat{P}_1(\boldsymbol{u})\,\|\,Q(\boldsymbol{u})]$ is minimized uniquely by $Q(\boldsymbol{u}) = Q_1(\boldsymbol{u})$.

Moreover, $\hat{P}_1(\boldsymbol{u}) = \hat{P}_1(s_1)Q_0(\boldsymbol{u}|s_1)$ with $\hat{P}_1(s_1) \propto t_1(s_1)Q_0(s_1)$. By the chain rule of relative entropy, $\mathrm{D}[\hat{P}_1(\boldsymbol{u})\,\|\,Q(\boldsymbol{u})] = \mathrm{D}[\hat{P}_1(s_1)\,\|\,Q(s_1)] + \mathrm{E}_{\hat{P}_1}[\mathrm{D}[Q_0(\boldsymbol{u}|s_1)\,\|\,Q(\boldsymbol{u}|s_1)]]$, to that $Q_1(\boldsymbol{u}|s_1) = Q_0(\boldsymbol{u}|s_1)$ for the minimizer (zeroing the second term), and the marginal $Q_1(s_1)$ matches moments of $\hat{P}_1(s_1)$, which can be obtained as $Q_1(s_1) \propto e^{\sigma^{-2}(b_j s_j - s_j^2/(2\gamma_j))}Q_0(s_1)$, where b_j, γ_j depend on $Q_0(s_1)$ and $t_1(s_1)$ only.

References

[1] Simoncelli E 1999 *Wavelet Applications in Signal and Image Processing VII: Proc. of SPIE* vol 3813 ed Unser M A, Aldroubi A and Laine A F pp 188–95
[2] Lauterbur P 1973 *Nature* **242** 190–1
[3] Garroway A, Grannell P and Mansfield P 1974 *J. Phys. C: Solid State Phys.* **7** L457–62
[4] Seeger M, Nickisch H, Pohmann R and Schölkopf B 2009 *Advances in Neural Information Processing Systems* vol 21 ed Koller D, Schuurmans D, Bengio Y and Bottou L (MIT Press) pp 1441–8
[5] Donoho D 2006 *IEEE Trans. Inform. Theo.* **52** 1289–306
[6] Candès E, Romberg J and Tao T 2006 *IEEE Trans. Inform. Theo.* **52** 489–509
[7] Weaver J, Xu Y, Healy D and Cromwell L 1991 *Magn. Reson. Med.* **21** 288–95
[8] Lustig M, Donoho D and Pauly J 2007 *Magn. Reson. Med.* **85** 1182–95
[9] Seeger M and Nickisch H 2008 *Int. Conf. on Machine Learning 25* ed McCallum A, Roweis S and Silva R (Omni Press)
[10] Chang H, Weiss Y and Freeman W 2009 Informative sensing Tech. Rep. 0901.4275v1 [cs.IT] ArXiv

[11] Tipping M 2001 *J. Mach. Learn. Res.* **1** 211–44
[12] Seeger M 2008 *J. Mach. Learn. Res.* **9** 759–813
[13] Nickisch H and Seeger M 2009 *Int. Conf. on Machine Learning 26* ed Bottou L and Littman M (Omni Press) pp 761–8
[14] Seeger M and Nickisch H 2008 Large scale variational inference and experimental design for sparse generalized linear models Tech. Rep. 175 MPI Biological Cybernetics
[15] Tibshirani R 1996 *J. Roy. Stat. Soc.* B **58** 267–88
[16] Berger J O 1985 *Statistical Decision Theory and Bayesian Analysis* 2nd ed (Springer)
[17] Park T and Casella G 2005 The Bayesian Lasso Tech. rep. University of Florida
[18] Chaloner K and Verdinelli I 1995 *Stat. Sci.* **10** 273–304
[19] Fedorov V 1972 *Theory of Optimal Experiments* (Academic Press)
[20] MacKay D 1991 *Neural Comput.* **4** 589–603
[21] Cover T and Thomas J 1991 *Elements of Information Theory* 1st ed (John Wiley & Sons)
[22] Palmer J, Wipf D, Kreutz-Delgado K and Rao B 2006 *Advances in Neural Information Processing Systems* vol 18 ed Weiss Y, Schölkopf B and Platt J (MIT Press)
[23] Boyd S and Vandenberghe L 2002 *Convex Optimization* (Cambridge University Press)
[24] Jaakkola T and Jordan M 2000 *Stat. Comput.* **10** 25–37
[25] Wipf D and Nagarajan S 2008 *Advances in Neural Information Processing Systems* vol 20 ed Platt J, Koller D, Singer Y and Roweis S (MIT Press)
[26] Girolami M 2001 *Neural Comput.* **13** 2517–32
[27] Wainwright M J and Jordan M I 2008 *Found. Trends Mach. Learn.* **1** 1–305
[28] Minka T 2001 *Uncertainty in Artificial Intelligence 17* ed Breese J and Koller D (Morgan Kaufmann)
[29] Opper M and Winther O 2000 *Neural Comput.* **12** 2655–84
[30] Kushner H and Budhiraja A 2000 *IEEE Trans. Automat. Contr.* **45** 580–5
[31] Opper M and Winther O 2005 *J. Mach. Learn. Res.* **6** 2177–204
[32] Minka T 2004 Power EP Tech. rep. Microsoft Research, Cambridge
[33] Bogachev V 1998 *Gaussian Measures* Mathematical Surveys and Monographs (American Mathematical Society)
[34] Golub G and Van Loan C 1996 *Matrix Computations* 3rd ed (Johns Hopkins University Press)
[35] Malioutov D, Johnson J and Willsky A 2006 *J. Mach. Learn. Res.* **7** 2031–64
[36] Schneider M and Willsky A 2001 *SIAM J. Sci. Comput.* **22** 1840–64
[37] Bekas C, Kokiopoulou E and Saad Y 2008 *SIAM J. Mat. Annal. Appl.* **30** 397–418
[38] Malioutov D, Johnson J, Choi M and Willsky A 2008 *IEEE Trans. Signal Proces.* **56** 4621–34
[39] Ji S and Carin L 2007 *Int. Conf. on Machine Learning 24* ed Ghahramani Z (Omni Press)
[40] Wipf D, Palmer J and Rao B 2004 *Advances in Neural Information Processing Systems* vol 16 ed Thrun S, Saul L and Schölkopf B (MIT Press)

Journal of Physics: Conference Series **197** (2009) 012002 doi:10.1088/1742-6596/197/1/012002

Inference algorithms and learning theory for Bayesian sparse factor analysis

Magnus Rattray[1], **Oliver Stegle**[2,3], **Kevin Sharp**[1] **and John Winn**[4]

[1]School of Computer Science, University of Manchester, Manchester M13 9PL, UK
[2]Max-Planck-Institute for Biological Cybernetics, Tübingen, Germany
[3]Max-Planck-Institute for Developmental Biology, Tübingen, Germany
[4]Microsoft Research Cambridge, Roger Needham Building, Cambridge, CB3 0FB, UK

E-mail: magnus.rattray@manchester.ac.uk

Abstract. Bayesian sparse factor analysis has many applications; for example, it has been applied to the problem of inferring a sparse regulatory network from gene expression data. We describe a number of inference algorithms for Bayesian sparse factor analysis using a slab and spike mixture prior. These include well-established Markov chain Monte Carlo (MCMC) and variational Bayes (VB) algorithms as well as a novel hybrid of VB and Expectation Propagation (EP). For the case of a single latent factor we derive a theory for learning performance using the replica method. We compare the MCMC and VB/EP algorithm results with simulated data to the theoretical prediction. The results for MCMC agree closely with the theory as expected. Results for VB/EP are slightly sub-optimal but show that the new algorithm is effective for sparse inference. In large-scale problems MCMC is infeasible due to computational limitations and the VB/EP algorithm then provides a very useful computationally efficient alternative.

1. Introduction

Factor analysis is a classical statistical approach for discovering latent structure in high-dimensional data. Sparse variants of factor analysis have been applied to the problem of uncovering latent variables that influence gene expression through a sparse regulatory network [1, 2]. Bayesian approaches to sparse factor analysis use sparsity-inducing priors to infer sparse posterior distributions over the factor loading matrix [3].

In this paper we describe a number of algorithms for Bayesian inference in sparse factor analysis models. As well as describing well-established Markov chain Monte Carlo (MCMC) and variational Bayes (VB) algorithms we also describe a new message passing algorithm that can be considered a hybrid between VB and Expectation Propagation (VB/EP). We compare the performance of these algorithms to the theoretical performance predicted by a replica analysis for the single factor case with isotropic noise which corresponds to sparse probabilistic principal component analysis (PCA).

2. Bayesian sparse factor analysis

The basic factor analysis model for a data vector $\boldsymbol{y}$ is given by,

$$\boldsymbol{y} \,|\, \boldsymbol{W}, \boldsymbol{x} \sim \mathcal{N}(\boldsymbol{W}\boldsymbol{x} + \boldsymbol{\mu}, \boldsymbol{\Psi}) \;, \qquad \boldsymbol{x} \sim \mathcal{N}(\boldsymbol{0}, \boldsymbol{I}) \;,$$

where $\boldsymbol{\Psi}$ is a diagonal noise covariance matrix. Such models are applied in many settings but in the case of a transcriptional regulation model we might interpret the data $\boldsymbol{y} = [y_j]$ to represent the logged expression level for genes $j = 1 \ldots p$ while the latent variables (factors) $\boldsymbol{x} = [x_k]$ represents the levels of regulatory proteins $k = 1 \ldots K$ such as transcription factors (TFs). The factor loading matrix $\boldsymbol{W} = [w_{jk}]$ then represents the matrix of regulatory interactions between TFs and genes. We expect this to be sparse in the sense that each gene should be regulated by few TFs. Protein concentration is difficult to measure in a high-throughput manner and TFs are often modified and regulated after transcription so that their expression level may be a poor proxy for the concentration of active protein in the nucleus. Therefore we do not consider them to be observed but instead we treat them as latent variables which have to be integrated out to derive the data likelihood,

$$\boldsymbol{y} \,|\, \boldsymbol{W} \sim \mathcal{N}(\boldsymbol{\mu}, \boldsymbol{\Psi} + \boldsymbol{W}\boldsymbol{W}^{\mathrm{T}}) \, .$$

To simplify the discussion we will assume zero-mean data ($\boldsymbol{\mu} = \boldsymbol{0}$) and we will not explicitly discuss inference of the covariance matrix $\boldsymbol{\Psi}$ although it would typically be inferred along with $\boldsymbol{W}$ in the algorithms that we describe below.

To infer a sparse matrix $\boldsymbol{W}$ we impose the following sparsity-inducing prior on the matrix elements

$$p(\boldsymbol{W} \,|\, \boldsymbol{C}, \lambda) = \prod_{j=1}^{p} \prod_{k=1}^{K} (1 - C_{jk})\delta(w_{jk}) + C_{jk}\mathcal{N}(w_{jk} \,|\, 0, \lambda^{-1}) \, . \tag{1}$$

Here the hyper-parameter $\boldsymbol{C} = [C_{jk}]$ encodes prior knowledge about the probability that there is a regulatory link in the network. The hyper-parameter λ can be learned or more usually is set to a small and uninformative value.

Given a dataset $\boldsymbol{Y} = \{\boldsymbol{y}_1, \boldsymbol{y}_2, \cdots \boldsymbol{y}_n\}$, Bayesian inference can be used to determine the posterior distribution over the loading matrix,

$$p(\boldsymbol{W} \,|\, \boldsymbol{Y}, \boldsymbol{C}, \lambda) \propto p(\boldsymbol{Y} \,|\, \boldsymbol{W}) p(\boldsymbol{W} \,|\, \boldsymbol{C}, \lambda)$$

and hence infer the sparse regulatory network. However, the normalisation of this probability cannot be computed in closed form and approximate inference algorithms are therefore required.

2.1. Markov chain Monte Carlo (MCMC)

The traditional method for carrying out Bayesian inference is to use MCMC. A Markov chain is constructed under which the intractable distribution of interest is invariant. Once convergence is attained, the distribution is approximated by means of a finite set of samples of the states visited.

Gibbs sampling [4] is a variant where each variable is iteratively sampled from its distribution conditioned on the current values of all the others. To construct a Gibbs sampler for this model, a standard way of dealing with the posterior over $\boldsymbol{W}$ induced by the sparsity-inducing mixture prior, is to introduce a binary matrix of indicator variables $\boldsymbol{Z}$ so that:

$$w_{jk} \,|\, z_{jk} = 0 \sim \delta(w_{jk}) \, , \qquad w_{jk} \,|\, z_{jk} = 1 \sim \mathcal{N}(w_{jk} \,|\, 0, \lambda^{-1}) \, ,$$

with independent Bernoulli priors placed over the elements of this matrix:

$$p(\boldsymbol{Z} \,|\, \boldsymbol{C}) = \prod_{j=1}^{p} \prod_{k=1}^{K} (1 - C_{jk})^{(1-z_{jk})} C_{jk}^{z_{jk}} \, . \tag{2}$$

However, although allowing calculation of convenient forms for the conditional distributions of the z_{jk} and w_{jk}, a Gibbs sampler so constructed would mix poorly owing to the high correlation of these variables. Possible refinements that avoid this impasse are either a *collapsed sampler* or a *soft spike and slab sampler*.

In a *collapsed sampler* the z_{jk} are sampled from their conditional distribution from which the w_{jk} have been marginalised. This may be viewed as a way of sampling from their joint distribution conditional on the current values of the other variables $p(\boldsymbol{Z}, \boldsymbol{W} \mid \boldsymbol{\Psi}, \boldsymbol{X}, \boldsymbol{Y})$ (where $\boldsymbol{X} = \{\boldsymbol{x}_1, \boldsymbol{x}_2, \ldots \boldsymbol{x}_n\}$) by first sampling the elements of $\boldsymbol{Z}$ from $p(\boldsymbol{Z} \mid \boldsymbol{\Psi}, \boldsymbol{X}, \boldsymbol{Y})$ followed by those of $\boldsymbol{W}$ from $p(\boldsymbol{W} \mid \boldsymbol{\Psi}, \boldsymbol{X}, \boldsymbol{Y}, \boldsymbol{Z},)$. Provided no other variables are sampled between these steps the posterior remains an invariant distribution of the markov chain. This idea was used by [1] who exploited the conditional independence of the $\boldsymbol{z}_j$ to sample these variables as binary vectors. However, when no constraint is placed on the maximum number of ones in such a vector, normalisation of the resulting multinomial distribution is a combinatorial problem. We avoid this in the same manner as [2] by sampling each z_{jk} independently. Despite this, however, each such step requires inversion of an $s \times s$ matrix where s is the number of ones in the current state of the vector $\boldsymbol{z}_j$

A *soft spike sampler* [5] is a relaxation of the sparsity-inducing mixture prior for the weights, w_{jk}, that approximates $\delta(w_{jk})$ with a narrow Gaussian. Intuitively, the idea is that 'small' values of the w_{jk} are inferred to be 0. The relative widths of the narrow and broader Gaussians determine a trade-off between accuracy and efficient mixing of the chains. Although an approximation, this allows for much cheaper sampling of the z_{jk} as their conditional distributions depend on only the single corresponding w_{jk}, so no costly matrix inversions are required. The sampling steps are essentially the same as for the collapsed sampler, the principal difference being only in the form of the conditional distributions of $\boldsymbol{w}_j$ and z_{jk}.

2.2. Mean-field variational Bayes (VB)

A typically faster alternative to MCMC are deterministic approximation algorithms. Variational Bayes (VB) [6] is a mean-field approximation that can be motivated from statistical physics. The basic idea is to approximate the true posterior $p(\mathbf{W}, \mathbf{X}, \mathbf{Z} \mid \mathbf{Y}, \mathbf{C})$ by a simpler, factorised distribution, $q(\mathbf{W}, \mathbf{X}, \mathbf{Z}) = \prod_{j=1}^{p} q(\mathbf{w}_j) \prod_{i=1}^{n} q(\mathbf{x}_i) \prod_{jk} q(z_{jk})$. Here, we explicitly represented the binary indicator variables z_{jk}, choosing between the two mixture components of the sparsity prior (recall equation 2).

Individual variational factors $q(\cdot)$ are updated iteratively. The corresponding update rules can be derived by minimising the VB KL-divergence

$$\mathrm{KL}_{\mathrm{VB}}[q \,||\, p] = \int_{\mathbf{\Theta}} q(\mathbf{\Theta}) \log \frac{q(\mathbf{\Theta})}{p(\mathbf{\Theta})} \, \mathrm{d}\mathbf{\Theta},$$

where $\mathbf{\Theta}$ denotes the set of all model parameters. Free form variational updates, without specifying the functional form of the approximate factors follow then as

$$q(\cdot) \propto \exp\{\langle \ln p(\mathbf{W}, \mathbf{X}, \mathbf{Y}, \mathbf{C}\rangle_{q\backslash}\},$$

where $\langle\rangle_{q\backslash}$ denotes the expectation value with respect to all approximate factors except for the one that is refined.

As an example, we discuss the update for a single weight vector $\mathbf{w}_j$, which follows as

$$\begin{aligned} q(\mathbf{w}_j) &\propto \exp\{\langle \log p(\mathbf{Y}, \mathbf{C}, \mathbf{W}, \mathbf{X}, \mathbf{Z}, \mathbf{\Psi})\rangle_{q\backslash \mathbf{w}_j}\} \\ &\propto \exp\{\langle \log p(\mathbf{y}_j \mid \mathbf{w}_j, \mathbf{X}, \mathbf{\Psi}) + \log p(\mathbf{w}_j \mid \mathbf{z}_j)\rangle_{q\backslash \mathbf{w}_j}\} \qquad (3) \\ &\propto \underbrace{\exp\{\langle \log p(\mathbf{y}_j \mid \mathbf{w}_j, \mathbf{X}, \mathbf{\Psi})\rangle_{q\backslash \mathbf{w}_j}\}}_{M_{\mathbf{W}\cdot\mathbf{X}\to\mathbf{w}_j}} \underbrace{\exp\{\langle \log p(\mathbf{w}_j \mid \mathbf{z}_j)\rangle_{q\backslash \mathbf{w}_j}\}}_{M_{\mathbf{W}\,|\,\mathbf{Z}\to\mathbf{w}_j}}. \qquad (4) \end{aligned}$$

The resulting Gaussian overall approximate factor, $q(\mathbf{w}_j)$, can be written as a product of two unnormalised Gaussian terms. The first term represents the evidence coming from the data likelihood, and the second term can be identified with the contribution from the sparsity prior. It is instructive to interpret $M_{\mathbf{W}\cdot\mathbf{X}\to\mathbf{w}_j}$ as the message sent from the product factor $f_{\mathbf{W}\cdot\mathbf{X}}$ to the weights variables $\mathbf{w}_j$. The parameters of this Gaussian, $M_{\mathbf{W}\cdot\mathbf{X}\to\mathbf{w}_j} \propto \mathcal{N}\left(\mathbf{w}_j \,\middle|\, \tilde{\mathbf{m}}_{\mathbf{W}\cdot\mathbf{X}\to\mathbf{w}_j}, \tilde{\boldsymbol{\Sigma}}_{\mathbf{W}\cdot\mathbf{X}\to\mathbf{w}_j}\right)$, can be read off from equation (4).

$$\tilde{\boldsymbol{\Sigma}}_{\mathbf{W}\cdot\mathbf{X}\to\mathbf{w}_j} = \left(\left\langle\frac{1}{\Psi_{j,j}}\right\rangle \sum_{i=1}^{n} \left\langle \mathbf{x}_i \mathbf{x}_i^{\mathrm{T}} \right\rangle\right)^{-1} \quad (5)$$

$$\tilde{\mathbf{m}}_{\mathbf{W}\cdot\mathbf{X}\to\mathbf{w}_j} = \tilde{\boldsymbol{\Sigma}}_{\mathbf{w}_j} \left(\left\langle\frac{1}{\Psi_{j,j}}\right\rangle \sum_{i=1}^{n} \langle \mathbf{x}_i\rangle \langle \mathbf{y}_i\rangle\right). \quad (6)$$

Using the definition of this message, the variational update in equation (3) follows as

$$q(\mathbf{w}_j) \propto \exp\Big\{ -\frac{1}{2}\left(\mathbf{w}_j - \tilde{\mathbf{m}}_{\mathbf{W}\cdot\mathbf{X}\to\mathbf{w}_j}\right)^T \tilde{\boldsymbol{\Sigma}}^{-1}_{\mathbf{W}\cdot\mathbf{X}\to\mathbf{w}_j} \left(\mathbf{w}_j - \tilde{\mathbf{m}}_{\mathbf{W}\cdot\mathbf{X}\to\mathbf{w}_j}\right) - \frac{1}{2}\mathbf{w}_j^{\mathrm{T}} \mathrm{diag}\left(\left\{\sum_{c=0}^{1} q(z_{jk}=c)\lambda_c\right\}_k\right)\mathbf{w}_j\Big\}. \quad (7)$$

Update rules for the responsibilities, $q(z_{jk}=1) = \tilde{C}_{jk}$, can be obtained in the same vein using

$$\tilde{C}_{jk} \propto C_{jk} \exp\left\{\left\langle \log \mathcal{N}\left(w_{jk} \,\middle|\, 0, \lambda_1^{-1}\right)\right\rangle_{q\backslash z_{jk}}\right\}$$

$$(1-\tilde{C}_{jk}) \propto (1-C_{jk}) \exp\left\{\left\langle \log \mathcal{N}\left(w_{jk} \,\middle|\, 0, \lambda_0^{-1}\right)\right\rangle_{q\backslash z_{jk}}\right\}. \quad (8)$$

Note that the mixture component corresponding to an inactive weight, $C_{jk} = 0$, is not a delta spike but has been relaxed to a Gaussian with small variance $\lambda_0^{-1} \ll \lambda_1^{-1}$.

2.3. *VB/EP hybrid*

The accuracy of the pure mean-field solution, treating indicators $\mathbf{Z}$ and weights $\mathbf{W}$ as factorised variables can be improved, by considering a hybrid algorithm. This algorithms combines the mean-field learning presented in the previous section with Expectation Propagation [7] (EP), an alternative variational approximation based on a KL divergence with swapped arguments

$$\mathrm{KL}_{\mathrm{EP}}[p \,||\, q] = \int_{\boldsymbol{\Theta}} p(\boldsymbol{\Theta}) \log \frac{p(\boldsymbol{\Theta})}{q(\boldsymbol{\Theta})} \, \mathrm{d}\boldsymbol{\Theta}.$$

Comparing VB and EP, there is no clear-cut answer as to which approximation is superior, although for a number of problems EP was shown to be more accurate [8, 9]. A drawback of EP is that it is more difficult to apply, can lead to improper messages, and for some models is not tractable at all. In fact, full EP inference in the considered sparse factor-analysis model is not feasible. For EP we need the moments of the product factor $f_{\mathbf{X}\cdot\mathbf{W}}$, which are not available in closed form. Note that for observed factor activations $\mathbf{X}$, the factor analyser reduces to sparse linear regression and inference with EP is possible [10].

The idea of the hybrid scheme is to solve the problem of obtaining a posterior for weights and indicators in EP, while keeping the remainder of the inference within the mean-field framework. The posterior distribution of weights and indicators given the incoming message $M_{\mathbf{W}\cdot\mathbf{X}\to\mathbf{w}_j}$ is

$$P\left(\mathbf{w}_j, \mathbf{z}_j \,|\, M_{\mathbf{W}\cdot\mathbf{X}\to\mathbf{w}_j}\right) \propto \mathcal{N}\left(\mathbf{w}_j \,\middle|\, \tilde{\mathbf{m}}_{\mathbf{W}\cdot\mathbf{X}\to\mathbf{w}_j}, \tilde{\boldsymbol{\Sigma}}_{\mathbf{W}\cdot\mathbf{X}\to\mathbf{w}_j}\right) \prod_{k=1}^{K} p(w_{jk} \,|\, z_{jk}) p(z_{jk} \,|\, C_{jk}). \quad (9)$$

As for VB, we choose an approximate form

$$q(\mathbf{w}_j, \mathbf{z}_j) \;= s \cdot \mathcal{N}\left(\mathbf{w}_j \,\middle|\, \tilde{\mathbf{m}}_{\mathbf{W}\cdot\mathbf{X}\to\mathbf{w}_j}, \tilde{\mathbf{\Sigma}}_{\mathbf{W}\cdot\mathbf{X}\to\mathbf{w}_j}\right) \prod_{k=1}^{K} q(w_{jk}) q(z_{jk}), \tag{10}$$

where the factors $q(w_{jk})q(z_{jk})$ are meant to approximate $p(w_{jk} \,|\, z_{jk})p(z_{jk} \,|\, C_{jk})$. The explicit scale of the approximation, s, will be dropped in the following. Choosing factor distributions that match the VB approximation, q-distributions for weights are Gaussian, $q(w_{jk}) = \mathcal{N}\left(w_{jk} \,\middle|\, \tilde{\mu}_{w_{jk}}, \tilde{\sigma}^2_{w_{jk}}\right)$, and factors of indicators are Bernoulli distributed, $q(z_{jk}) = \text{Bernoulli}(z_{jk} \,|\, \tilde{C}_{jk})$. While the overall approximation in equation (10) is fully factorised over indicators z_{jk}, it is multivariate Gaussian in the weights $\mathbf{w}_j$. Writing out the product of the Gaussian prior and the individual Gaussian factors $q(w_{jk})$ yields

$$q(\mathbf{w}_j, \mathbf{z}_j) \propto \mathcal{N}\left(\mathbf{w}_j \,\middle|\, \tilde{\mathbf{m}}_{\mathbf{w}_j}, \tilde{\mathbf{\Sigma}}_{\mathbf{w}_j}\right) \prod_{k=1}^{K} q(z_{jk}). \tag{11}$$

Defining $\tilde{\boldsymbol{\mu}} = \left(\tilde{\mu}_{w_{j1}}, \ldots, \tilde{\mu}_{w_{jK}}\right)$ and $\tilde{\mathbf{\Sigma}} = \text{diag}(1/\tilde{\sigma}^2_{w_{j1}}, \ldots, 1/\tilde{\sigma}^2_{w_{jK}})$, the covariance and the mean of this Gaussian follow as

$$\tilde{\mathbf{\Sigma}}_{\mathbf{w}_j} = \left(\tilde{\mathbf{\Sigma}}^{-1}_{\mathbf{W}\cdot\mathbf{X}\to\mathbf{w}_j} + \tilde{\mathbf{\Sigma}}\right)^{-1} \qquad \tilde{\mathbf{m}}_{\mathbf{w}_j} = \tilde{\mathbf{\Sigma}}_{\mathbf{w}_j}\left[\tilde{\mathbf{\Sigma}}^{-1}_{\mathbf{W}\cdot\mathbf{X}\to\mathbf{w}_j} \tilde{\mathbf{m}}_{\mathbf{W}\cdot\mathbf{X}\to\mathbf{w}_j} + \tilde{\mathbf{\Sigma}}^{-1}\tilde{\boldsymbol{\mu}}\right]. \tag{12}$$

The idea of EP is to iteratively refine individual pairs of factors for indicators and weights, leaving all other factors fixed. To update the ith pair, $q(w_{ji})q(z_{ji})$, the local KL divergence is

$$\begin{aligned} \text{KL}\Big[\quad & \mathcal{N}\left(\mathbf{w}_j \,\middle|\, \tilde{\mathbf{m}}_{\mathbf{W}\cdot\mathbf{X}\to\mathbf{w}_j}, \tilde{\mathbf{\Sigma}}_{\mathbf{W}\cdot\mathbf{X}\to\mathbf{w}_j}\right) \prod_{k\neq i} q(w_{jk})q(z_{jk}) \overbrace{p(w_{ji} \,|\, z_{ji})p(z_{ji})}^{\text{exact factor}} \Big\| \\ & \mathcal{N}\left(\mathbf{w}_j \,\middle|\, \tilde{\mathbf{m}}_{\mathbf{W}\cdot\mathbf{X}\to\mathbf{w}_j}, \tilde{\mathbf{\Sigma}}_{\mathbf{W}\cdot\mathbf{X}\to\mathbf{w}_j}\right) \prod_{k\neq i} q(w_{jk})q(z_{jk}) \quad \underbrace{q(w_{ji})q(z_{ji})}_{\text{approximation}} \quad \Big]. \end{aligned} \tag{13}$$

As the arguments of the KL divergence differ only in that ith factor, all other dimensions are marginalised out. This motivates the definition of a cavity distribution

$$\begin{aligned} q_{\setminus i}(w_{ji}) \;&=\; \int_{\mathbf{w}_{j\setminus i}} \mathcal{N}\left(\mathbf{w}_j \,\middle|\, \tilde{\mathbf{m}}_{\mathbf{W}\cdot\mathbf{X}\to\mathbf{w}_j}, \tilde{\mathbf{\Sigma}}_{\mathbf{W}\cdot\mathbf{X}\to\mathbf{w}_j}\right) \prod_{k\neq i} q(w_{jk})\, \mathrm{d}\mathbf{w}_{j\setminus i} \\ &=\; \mathcal{N}\left(w_{ji} \,\middle|\, \tilde{\mu}_{\setminus i}, \tilde{\sigma}^2_{\setminus i}\right). \end{aligned} \tag{14}$$

The cavity distribution $q_{\setminus i}(w_{ji})$ can be calculated efficiently from the current full approximation (equation (11)), dividing out the contribution of the ith factor (see for example [11], chapter 3).

Using this definition, the KL-divergence in equation (13) can be expressed in a compact form

$$\text{KL}\left[q_{\setminus i}(w_{ji}) \overbrace{p(w_{ji} \,|\, z_{ji})p(z_{ji})}^{\text{exact factor}} \,\middle\|\, q_{\setminus i}(w_{ji}) \underbrace{q(w_{ji} \,|\, \tilde{\mu}_{w_{ji}}, \tilde{\sigma}^2_{w_{ji}}) q(z_{ji} \,|\, \tilde{C}_{ji})}_{\text{approximation}}\right]. \tag{15}$$

Minimising equation (15) with respect to the parameters of the Gaussian factor $q(w_{ji})$ leads to moment-matching conditions [12]. The new parameters of the approximation $q(w_{ji})$ are set such that the moments of both arguments of the KL divergence match. The task hence reduces to calculating a set of moments under the exact factor

$$\begin{aligned} F_C &= \int_{w_{ji}} q_{\setminus i}(w_{ji}) \sum_{c=\{0,1\}} p(w_{ji} \mid z_{ji}=c) p(z_{ji}=c) \, \mathrm{d}w_{ji} \\ F_\mu &= \frac{1}{F_C} \int_{w_{ji}} q_{\setminus i}(w_{ji}) \sum_{c=\{0,1\}} p(w_{ji} \mid z_{ji}=c) p(z_{ji}=c) w_{ji} \, \mathrm{d}w_{ji} \\ F_{\sigma^2} + F_\mu^2 &= \frac{1}{F_C} \int_{w_{ji}} q_{\setminus i}(w_{ji}) \sum_{c=\{0,1\}} p(w_{ji} \mid z_{ji}=c) p(z_{ji}=c) w_{ji}^2 \, \mathrm{d}w_{ji} . \end{aligned} \tag{16}$$

Analytic expressions for these moments can be derived considering the moment generating function [13].

In the same vein, optimisation of equation (15) with respect to $\tilde{C}_{ji}$ leads to updates of the posterior over the indicator variables

$$\begin{aligned} \tilde{C}_{ji} &\propto C_{ji} \int_{w_{ji}} q_{\setminus i}(w_{ji}) \mathcal{N}\left(w_{ji} \,\middle|\, 0, \lambda_1^{-1}\right) \mathrm{d}w_{ji} \\ (1 - \tilde{C}_{ji}) &\propto (1 - C_{ji}) \int_{w_{ji}} q_{\setminus i}(w_{ji}) \mathcal{N}\left(w_{ji} \,\middle|\, 0, \lambda_0^{-1}\right) \mathrm{d}w_{ji} . \end{aligned} \tag{17}$$

3. Replica theory for a single latent factor

The replica method from statistical mechanics can be used to derive the performance of learning for an idealized situation in which data are produced by a sparse factor analysis generative model. We consider the simplest case of a single latent factor and set the data covariance $\boldsymbol{\Psi} = \boldsymbol{I}$ which corresponds to the probabilistic PCA model [14],

$$\boldsymbol{y} = \boldsymbol{w} x + \boldsymbol{\epsilon} \, , \tag{18}$$

where $x \sim \mathcal{N}(0,1)$ and $\boldsymbol{\epsilon} \sim \mathcal{N}(\boldsymbol{0}, \boldsymbol{I})$. Integrating out the latent factor gives the data density under the model,

$$\boldsymbol{y} \mid \boldsymbol{w} \sim \mathcal{N}(\boldsymbol{0}, \boldsymbol{I} + \boldsymbol{w}\boldsymbol{w}^{\mathrm{T}}) \, , \tag{19}$$

where the model parameters are now in a vector $\boldsymbol{w} = [w_j]$ since there is only a single latent factor. The log-likelihood for dataset $\boldsymbol{Y}$ is given by,

$$\begin{aligned} \ln p(\boldsymbol{Y} \mid \boldsymbol{w}) &= -n \ln \sqrt{(2\pi)^p |\boldsymbol{I} + \boldsymbol{w}\boldsymbol{w}^{\mathrm{T}}|} - \frac{1}{2} \sum_{i=1}^{n} \boldsymbol{y}_i^{\mathrm{T}} (\boldsymbol{I} + \boldsymbol{w}\boldsymbol{w}^{\mathrm{T}})^{-1} \boldsymbol{y}_i \\ &= -\frac{np}{2} \ln(2\pi) - \frac{n}{2} \ln(1 + ||\boldsymbol{w}||^2) - \frac{1}{2} \sum_{i=1}^{n} \boldsymbol{y}_i^{\mathrm{T}} \boldsymbol{y}_i + \frac{1}{2} \sum_{i=1}^{n} \frac{(\boldsymbol{y}_i^{\mathrm{T}} \boldsymbol{w})^2}{1 + ||\boldsymbol{w}||^2} \, . \end{aligned}$$

We use an equal sparsity hyper-parameter $C_j = C$ for each parameter vector component $j = 1, \ldots, p$ in the mixture prior $p(\boldsymbol{w} \mid C, \lambda)$ (recall equation (1)).

The marginal likelihood $p(\boldsymbol{Y} \mid C, \lambda)$ obtained by integrating out the model parameters $\boldsymbol{w}$ is analogous to a partition function in Statistical Mechanics. The parameter-dependent terms in the log-likelihood can be written,

$$E(\boldsymbol{w}) = \frac{n}{2} \ln(1 + ||\boldsymbol{w}||^2) - \frac{1}{2} \sum_{i=1}^{n} \frac{(\boldsymbol{y}_i^{\mathrm{T}} \boldsymbol{w})^2}{1 + ||\boldsymbol{w}||^2} \, . \tag{20}$$

Then $p(\boldsymbol{Y} \mid C, \lambda) \propto Z$ which is defined,

$$Z = \int \exp\left(-E(\boldsymbol{w})\right) p(\boldsymbol{w} \mid C, \lambda)\, \mathrm{d}\boldsymbol{w} \; .$$

To study learning performance we assume that the data are produced by a similar "teacher" distribution to the model in equation (19). The teacher parameters $\boldsymbol{w}^t$ are also generated from a similar prior $p(\boldsymbol{w}^t \mid C_t, \lambda_t)$ except that the hyper-parameters λ_t and C_t may differ from those of the model.

The replica calculation has similarities to previous work on non-sparse PCA [15], diluted neural networks [16] and a sparse Bayesian classifier [17]. The replica method makes use of the identity

$$\langle \ln Z \rangle_{\boldsymbol{Y},\boldsymbol{w}^t} = \lim_{m \to 0} \frac{\partial \ln \langle Z^m \rangle_{\boldsymbol{Y},\boldsymbol{w}^t}}{\partial m} \; . \tag{21}$$

The left-hand side shows the average we wish to compute but the calculation is intractable. However, we can compute the average over the right-hand side for integer m, then make an analytical continuation to real m and take the limit (see e.g. [18]). Before that we also take the limit $p \to \infty$ in order to use the saddle point method. We find,

$$p^{-1} \langle \ln Z \rangle_{\boldsymbol{Y},\boldsymbol{w}^t} = \operatorname*{Ext}_{r,k,q,\hat{r},\hat{k},\hat{q}} \; \alpha G_0(r,q,k) - r\hat{r} + k\hat{k} - \frac{1}{2} q\hat{q} + G_1(\hat{r},\hat{q},\hat{k}) \tag{22}$$

where,

$$G_0(r,q,k) \;=\; \frac{1}{2}\left(\frac{q+r^2}{1+q} - \ln(1+q)\right) , \tag{23}$$

$$G_1(\hat{r},\hat{q},\hat{k}) \;=\; \left\langle \int_{-\infty}^{\infty} \frac{\mathrm{d}\eta}{\sqrt{2\pi}} \mathrm{e}^{\frac{-\eta^2}{2}} \ln \langle \mathrm{e}^{p\hat{r}ww^t - p(\hat{k}+\frac{1}{2}\hat{q})w^2 + \eta\sqrt{p\hat{q}}w} \rangle_{w|\theta} \right\rangle_{w^t} . \tag{24}$$

The angled brackets denote averages with respect to individual components of the vectors $\boldsymbol{w}$ and $\boldsymbol{w}^t$. Notice that although we have taken the limit $p \to \infty$ it still appears in the terms multiplied by w or w^t. This is because individual components of the parameter vectors should be scaled to be $O(1/\sqrt{p})$ so that the length of these vectors remains $O(1)$ as $p \to \infty$. This is achieved by choosing an appropriate scale for the hyper-parameters.

The order parameters r, q and k obtained by solving the saddle point equations represent the following quantities,

$$q = ||\langle \boldsymbol{w} \rangle_{\text{post}}||^2 = ||\boldsymbol{w}^{\text{PM}}||^2 , \quad k = \langle ||\boldsymbol{w}||^2 \rangle_{\text{post}} , \quad r = \langle \boldsymbol{w} \cdot \boldsymbol{w}^t \rangle_{\text{post}} = \boldsymbol{w}^{\text{PM}} \cdot \boldsymbol{w}^t , \tag{25}$$

where averages are over the posterior distribution and $\boldsymbol{w}^{\text{PM}}$ denotes the posterior mean (PM) parameter. These order parameters can be used to assess learning performance with respect to the underlying data generating process.

A useful measure of performance is the cosine-angle, which we will refer to as the overlap, between the parameter estimate and the true parameter. For the posterior mean parameter estimate this can be written in terms of the order parameters as,

$$\rho^{\text{PM}} = \frac{r}{\sqrt{Tq}}$$

where $T = \langle ||\boldsymbol{w}^t||^2 \rangle = C_t/\lambda_t$.

Another relevant measure of performance is the mean log-probability of a test data point under the model with the posterior mean parameter. This can also be written in terms of the order parameters,

$$\mathcal{L}^{\rm PM} = \langle \ln p(\boldsymbol{y}^{\rm test} \,|\, \boldsymbol{w}^{\rm PM}) \rangle_{\boldsymbol{y}^{\rm test}} = \frac{1}{2}\left(\frac{q+r^2}{1+q} - \ln(1+q)\right) + \kappa \,, \tag{26}$$

where $\kappa = -p\ln(2\pi)/2 - (1+T)/2$ is a constant term that does not depend on the model parameters.

For Bayesian learning a better prediction of test data may be obtained by averaging over the posterior distribution. In this case we have to average the log of the predictive distribution over test data,

$$\mathcal{L}^{\rm bayes} = \langle \ln\langle p(\boldsymbol{y}^{\rm test} \,|\, \boldsymbol{w})\rangle_{\rm post}\rangle_{\boldsymbol{y}^{\rm test}} = \left\langle \ln\left\langle \frac{1}{\sqrt{1+||\boldsymbol{w}||^2}} \exp\left(\frac{(\boldsymbol{y}^{\rm T}\boldsymbol{w})^2}{2(1+||\boldsymbol{w}||^2)}\right)\right\rangle_{\rm post}\right\rangle_{\boldsymbol{y}|\boldsymbol{w}^t} + \kappa \,. \tag{27}$$

For large p we assume that the central limit theorem can be applied to the sum $\boldsymbol{y}^{\rm T}\boldsymbol{w} = \sum_j y_j w_j$ in which case it has a Gaussian distribution for large p with mean and variance,

$$\begin{aligned} E_{\rm post}[\boldsymbol{y}^{\rm T}\boldsymbol{w}] &= rx + \boldsymbol{\epsilon}^{\rm T}\boldsymbol{w}^{\rm PM} \,, \\ {\rm Var}_{\rm post}[\boldsymbol{y}^{\rm T}\boldsymbol{w}] &= \langle \boldsymbol{w}^{\rm T}\boldsymbol{\epsilon}\boldsymbol{\epsilon}^{\rm T}\boldsymbol{w}\rangle_{\rm post} - \langle \boldsymbol{w}^{\rm T}\rangle_{\rm post}\boldsymbol{\epsilon}\boldsymbol{\epsilon}^{\rm T}\langle \boldsymbol{w}\rangle_{\rm post} + O(1/p) \\ &= k - q + O(1/p) \,, \end{aligned}$$

where x and $\boldsymbol{\epsilon}$ are defined after (18). The variance is self-averaging for large p and this allows us to take the average with respect to the posterior distribution. We write $\boldsymbol{y}^{\rm T}\boldsymbol{w} = rx + \boldsymbol{\epsilon}^{\rm T}\boldsymbol{w}^{\rm PM} + \nu\sqrt{k-q}$ with $\nu \sim \mathcal{N}(0,1)$. The other terms in the log-likelihood only involve $||\boldsymbol{w}||^2$, which is also self-averaging for large p, and we therefore obtain,

$$\begin{aligned} \mathcal{L}^{\rm bayes} &= \left\langle \ln\left\langle \frac{1}{\sqrt{1+k}} \exp\left(\frac{(rx + \boldsymbol{\epsilon}^{\rm T}\boldsymbol{w}^{\rm PM} + \nu\sqrt{k-q})^2}{2(1+k)}\right)\right\rangle_{\nu}\right\rangle_{z,\boldsymbol{\epsilon}} + \kappa \\ &= \frac{1}{2}\left(\frac{r^2+q}{1+q} - \ln(1+q)\right) + \kappa = \mathcal{L}^{\rm PM} \quad \text{as } p \to \infty \,. \end{aligned} \tag{28}$$

So we see that the performance of full Bayesian inference is equivalent to using the posterior mean parameter in the large p limit considered here.

4. Simulation results

In figure 1 we compare the replica theory with results from MCMC and the VB/EP algorithm applied to simulated data. MCMC can be considered a gold standard given sufficient computation time and we see that the results agree closely with the theoretical performance. As expected, optimal performance is achieved when the sparsity of the model matches the sparsity in the data ($C = C_t = 0.1$). The results for VB/EP show lower accuracy but still demonstrate that sparse inference provides a significant benefit over non-sparse PCA ($C = 1$ is the non-sparse result). Also, we observe that the optimal performance of VB/EP is obtained when the model and data sparsity are matched. Similar results are obtained for the predictive likelihood.

5. Conclusion

We have presented a number of MCMC and deterministic inference algorithms for Bayesian sparse factor analysis, including a novel VB/EP hybrid algorithm. We compared the empirical

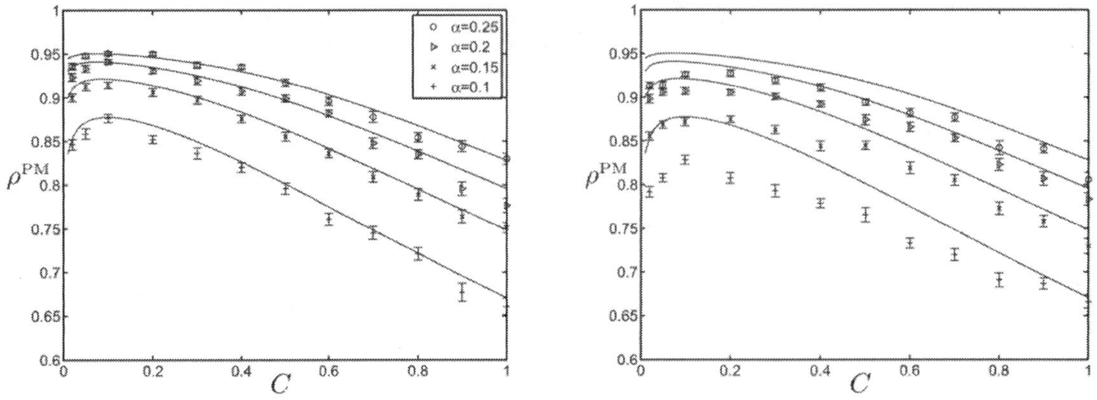

(a) MCMC versus theory

(b) VB/EP versus theory

Figure 1. The theory for sparse PCA is compared to (a) MCMC results and (b) VB/EP results for data with 90% sparsity ($C_t = 0.1$). We show the overlap $\rho^{\rm PM}$ between the true underlying vector and the posterior mean parameter as a function of the sparsity hyper-parameter C with $C = 1$ corresponding to standard non-sparse PCA. Results are shown for $n = 100$ with different values of $\alpha = n/p$ and $\lambda = \lambda_t = 0.01$. For the MCMC results we generated 500 Gibbs samples after discarding 500 as burn-in. Results are averaged over 50 replicate datasets.

performance of MCMC and the VB/EP algorithm with results from a replica theory of learning performance. The MCMC results confirmed the accuracy of the theory and results for the VB/EP algorithm are encouraging. Our ongoing work shows that this algorithm can provide very significant gains over MCMC in large-scale applications where MCMC has no hope of converging in a reasonable time.

Acknowledgments

We thank Michalis Titsias for useful discussions. OS gratefully acknowledges financial support from the Cambridge Gates Trust. MR thanks Martin Weigt and Andrea Pagnani for hosting visits to ISI Turin where the theoretical work was initiated.

References

[1] Sabatti C and James G M 2006 *Bioinformatics* **22** 739–46
[2] Pournara I and Wernisch L 2007 *BMC Bioinformatics* **8** 61
[3] West M 2003 *Bayesian Statistics* vol 7 (Oxford University Press) pp 723–32
[4] Geman S and Geman D 1984 *IEEE Trans. Pattern. Anal.* **6** 721741
[5] George E and McCulloch R 1993 *J. Am. Stat. Assoc.* **88** 881–9
[6] Jordan M, Ghahramani Z, Jaakkola T S and Saul L K 1999 *Mach. Learn.* **37** 183–233
[7] Minka T P 2001 *UAI '01: Proc. of the 17th Conf. in Uncertainty in Artificial Intelligence* vol 17 (San Fransisco: Morgan Kaufmann Publishers) pp 362–9
[8] Nickisch H and Rasmussen C E 2008 *J. Mach. Learn. Res.* **9** 2035–78
[9] Frey B J, Patrascu R, Jaakkola T and Moran J 2000 *Advances in Neural Information Processing Systems* vol 12 pp 493–9
[10] Seeger M W 2008 *J. Mach. Learn. Res.* **9** 759–813
[11] Rasmussen C E and Williams C K I 2006 *Gaussian Processes for Machine Learning* (MIT Press)
[12] Minka T P 2001 *A family of algorithms for approximate Bayesian inference* Ph.D. thesis Massachusetts Institute of Technology
[13] DeGroot M and Schervish M 2002 *Probability and Statistics* (Addison-Wesley Boston)
[14] Tipping M E and Bishop C M 1999 *J. R. Stat. Soc.* B **61** 611–22
[15] Reimann P, Van den Broeck C and Bex G J 1996 *J. Phys. A: Math. Gen.* **29** 3521–35

[16] Kuhlmann P and Müller K R 1994 *J. Phys. A: Math. Gen.* **27** 3759–74
[17] Uda S and Kabashima Y 2005 *J. Phys. Soc. Japan* **74** 2233–42
[18] Engel A and Van den Broeck C 2001 *Statistical Mechanics of Learning* (Cambridge University Press)

International Workshop on Statistical-Mechanical Informatics 2009 (IW-SMI 2009) IOP Publishing
Journal of Physics: Conference Series **197** (2009) 012003 doi:10.1088/1742-6596/197/1/012003

A statistical-mechanical study of evolution of robustness: An approach from two-temperature models

Ayaka Sakata[1], **Koji Hukushima**[1] **and Kunihiko Kaneko**[1,2]

[1]Graduate school of Arts and Sciences, The University of Tokyo, Komaba, Meguro-ku, Tokyo 153-8902, Japan

[2]Complex Systems Biology Project, Exploratory Research for Advanced Technology (ERATO), Japan Science and Technology Agency (JST), Tokyo, Japan

E-mail: ayaka@huku.c.u-tokyo.ac.jp, hukusima@phys.c.u-tokyo.ac.jp, kaneko@complex.c.u-tokyo.ac.jp

Abstract. Recently we proposed a simple statistical-mechanical model of interacting spins for studying evolutional process in biological system. In our model, a configuration of spins and their interactions $\boldsymbol{J}$ represent a phenotype and genotype, respectively. The phenotype dynamics are given by a stochastic process with temperature T_S under a fixed Hamiltonian with $\boldsymbol{J}$. Meanwhile, the evolution of $\boldsymbol{J}$ is also stochastic with temperature T_J and follows mutations introduced into $\boldsymbol{J}$ and selection based on a fitness given by equilibrium spin configurations of a given set of target spins. Using Monte Carlo simulations, it is found that frustration around the target spins is strongly suppressed for the interactions $\boldsymbol{J}$ evolved in the intermediate T_S temperature region. The evolved $\boldsymbol{J}$'s give the funnel-like dynamics and show robustness to mutations. This model can be regarded as an extension of partial annealing model, in which the interactions between spins are dynamical variables with characteristic time scale widely separated from that of the spins. We also discuss the partially annealed Sherrington-Kirkpatrick model by particularly paying attention to the frustration.

1. Introduction

The function of a biological unit is generally determined by a biological state, phenotype, that is shaped by a dynamical process that yields a specific pattern or structure. The dynamics themselves are governed by a genetic sequence. In evolution governed by a fixed condition, a phenotype that gives a higher value for the fitness function has a high chance of survival. The genes that produce such a phenotype are transferred to the next generation, and thus, specific genetic sequences are selected. However, it should be noted that the dynamical process producing a phenotype is subject to noise. For example, genetic information determines the amino acid sequence for a protein, while the tertiary structure responsible for its functions is generated only by folding dynamics. The genotype-phenotype mapping is formed by this folding process, and this mapping is stochastic because of the thermal noise in the folding process[1]. Related folding dynamics also occur in t-RNA[2]. Hence, phenotypes shaped from a genotype through the dynamical process are not necessarily identical, and therefore, they form a distribution.

The phenotype for the function is expected to be insensitive to noise encountered during the dynamical process[3]. However, such robustness is not a general property of dynamics. For example, the complex folding process of a hetero-polymer from an arbitrary random sequence might not have such robustness. In this sense, the robustness could be a result of evolution. To have robustness to noise in such dynamics, it is ideal to adopt dynamics in which the target phenotype is reached smoothly and globally from a variety of initial configurations and is maintained thereafter. In fact, the existence of such global attraction in the protein folding process was proposed as a consistency principle by Go [4] and as "funnel" landscape by Onuchic et al. [5, 6], while similar global attraction dynamics have been discovered recently in gene regulatory networks [7] and developmental dynamics [8]. In spite of the ubiquity of such funnel-like structures for phenotype dynamics, little is understood about how these structures are shaped by the evolutionary process [1].

In addition to being robust to noise, a biological system has to remain relatively robust to mutations in the genetic sequence that occur through evolution; the phenotype has to be rather insensitive to changes in the genetic sequence. Are these two types of robustness correlated? Indeed, a possible relationship between robustness to noise and robustness to mutation has been discussed [9, 10, 11, 12, 13, 14, 15], following the pioneering study by Waddington[16] on the evolution-development relationship, which is referred to as canalization and genetic assimilation. However, a theoretical understanding of the evolution of robustness is still insufficient.

In order to understand the concepts in the evolution of robustness, we introduce a simple statistical-mechanical spin model that consists of N Ising spins interacting globally[17, 18]. Indeed, this type of model has also been adopted by Saito et al. [1], who utilized it in the study of the evolution of protein folding dynamics. Even though the spin model is abstract, it can account for the basic structures required to study the evolution of genotype and phenotype. In comparison with the gene transcription network model utilized in the study of the evolution of robustness [9, 10, 13, 14], the present spin model is computationally efficient in Monte Carlo simulations and the methods developed in statistical mechanics of spin systems can be applied to study the general questions on evolution. Our model mainly discussed in this paper can be regarded as a variant of partial annealing models with two temperatures[19, 20]: one for the spins and the other for the interactions. The original partial annealing models can be analyzed by replica method with a replica number fixed to be ratio of two temperatures, while the corresponding replica analysis for our model is straightforward unfortunately. We thus reveal novel features of our model by using Monte Carlo simulations and compare the results with those of a partial annealing model.

This paper is organized as follows. In Sec. 2, we explain the model setup that captures the essential features of the evolution. In Sec. 3, the numerical results for the model are reviewed. This part is largely based on our previous papers[17, 18]. Then, we show that a robust system is realized at an intermediate temperature. A key for the robustness is frustration of the interactions. In Sec. 4, we discuss the frustration of partially annealed Sherrington-Kirkpatrick (SK) model in comparison with our findings. Finally, Sec. 5 is devoted to the conclusions and prospects for further development.

2. A spin model for evolution of robustness

We introduce a statistical-mechanical spin model in which the phenotype and genotype are represented by configurations of spin variables S_i and an interaction matrix J_{ij}, respectively, with $i, j = 1, \cdots, N$. The spins S_i and J_{ij} can take one of only two values ± 1, and the interaction matrix is assumed to be symmetric, i.e., $J_{ij} = J_{ji}$. A set of configurations is denoted by $\boldsymbol{S}$ for the phenotype and by $\boldsymbol{J}$ for the genotype. The dynamics of the phenotype are given by a flip-flop update of each spin with an energy function, which is defined by the Hamiltonian for a given $\boldsymbol{J}$,

$$H(\boldsymbol{S}|\boldsymbol{J}) = -\frac{1}{\sqrt{N}} \sum_{i<j} J_{ij} S_i S_j. \quad (1)$$

We adopt the Glauber dynamics as an update rule of $\boldsymbol{S}$, where the N spins are in contact with their own heat bath at temperature T_S. The Glauber dynamics, satisfying the detailed balance conditions, yields an equilibrium distribution for a given $\boldsymbol{J}$:

$$P(\boldsymbol{S}|\boldsymbol{J},T_S) = \frac{e^{-\beta_S H(\boldsymbol{S}|\boldsymbol{J})}}{Z_S(\boldsymbol{J},T_S)}, \tag{2}$$

where $Z_S(\boldsymbol{J},T_S) = \mathrm{Tr}_{\boldsymbol{S}}\, e^{-\beta_S H(\boldsymbol{S}|\boldsymbol{J})}$ and $\beta_S = T_S^{-1}$. After a relaxation process, the phenotype $\boldsymbol{S}$ follows from the equilibrium distribution, and it is not determined uniquely from the genotype $\boldsymbol{J}$; rather, it is distributed, except at zero temperature. The phenotype fluctuation is computed from the resulting equilibrium probability distribution. Thus, the degree of fluctuation is characterized by the temperature T_S. When the interactions $\boldsymbol{J}$ are quenched variables given by Gaussian distribution, the model is identical to the SK model. In our model, however, they are supposed to be dynamical variables.

Next, we introduce evolutionary dynamics for the genotype $\boldsymbol{J}$. The genotype is transmitted to the next generation with some variation, while genotypes that produce a phenotype with higher fitness are selected. The time scale for genotype change is generally much larger than that of the phenotype dynamics. We assume that the two time scales for the phenotype expression dynamics and the genotypic evolutionary dynamics are widely separated, so that the variables $\boldsymbol{S}$ are well equilibrating within the unit time scale of the slow variable $\boldsymbol{J}$. Then, the fitness should be expressed by a function of the phenotype $\boldsymbol{S}$ that is averaged with respect to the distribution. Here, we define the fitness as

$$\Psi(\boldsymbol{J}|T_S) = \Big\langle \prod_{i<j\in\boldsymbol{t}} \delta(S_i - S_j) \Big\rangle \equiv \big\langle \psi \big\rangle, \tag{3}$$

where $\langle\cdots\rangle$ denotes the expectation value with respect to the equilibrium probability distribution of eq.(2). The set $\boldsymbol{t}$ denotes a subset of $\boldsymbol{S}$ with size t ; the members of $\boldsymbol{t}$ are termed as target spins. We refer to configurations such that all target spins are aligned in parallel as target configurations, which are assumed to give a requested appropriate function. By a gauge transformation on the target spin and the corresponding elements of $\boldsymbol{J}$, a choice of any other form of spin alignment for the fitness function, instead of the "ferromagnetic" configuration, yields the same result [21]. The fitness can be interpreted as the average frequency of finding the target configurations in equilibrium for a given $\boldsymbol{J}$. It should be noted that in our model, only the target spins contribute explicitly to the fitness and the remaining spins have no direct influence on the fitness and the selection of genes. Hence, the spin configuration for a given fitness has redundancy.

The genotype dynamics are a result of mutations and selection, i.e., changes according to the fitness function following random flip-flops of genes. Hence, for a genetic dynamics, we once again adopt the Glauber dynamics by using the fitness instead of the Hamiltonian in the phenotype dynamics, where the genotype $\boldsymbol{J}$ is in contact with a heat bath whose temperature T_J is different from T_S. The dynamics for the genotype are given by a stochastic Markov process with the following stationary distribution:

$$P(\boldsymbol{J}|T_S,T_J) = \frac{e^{\beta_J \Psi(\boldsymbol{J}|T_S)}}{Z_J(T_S,T_J)}, \tag{4}$$

where $Z_J(T_S,T_J) = \mathrm{Tr}_{\boldsymbol{J}}\, e^{\beta_J \Psi(\boldsymbol{J}|T_S)}$ and $\beta_J = T_J^{-1}$. According to the dynamics, genotypes are selected rather uniformly at high values of the temperature T_J, irrespective of the fitness, whereas at low values of T_J, the genotypes with higher fitness values are preferentially selected. In this sense, the temperature T_J represents the selection pressure among mutated genotypes. In actual

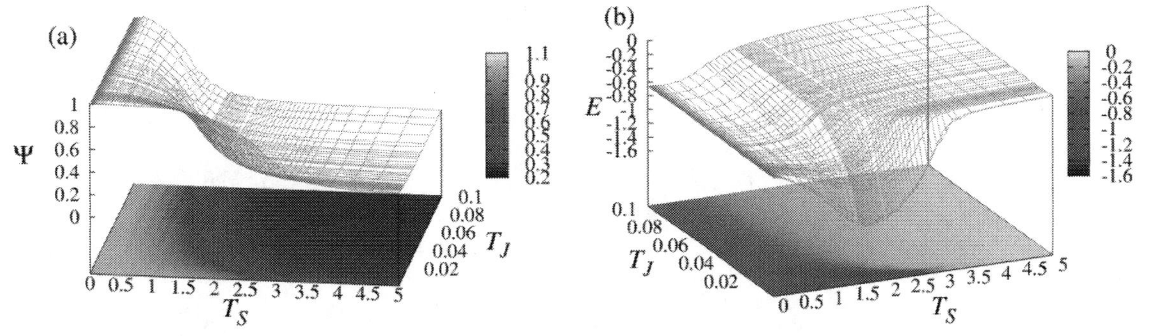

Figure 1. Two-temperature dependence of the fitness $\Psi(\boldsymbol{J})$ in (a) and the energy in (b), respectively, with $N = 15$ and $t = 3$. Figure is quoted from [18].

simulations, a candidate $\boldsymbol{J}'$ for the next generation is set by some flips of a randomly chosen J_{ij} from the current $\boldsymbol{J}$, while the transition probability from $\boldsymbol{J}$ to $\boldsymbol{J}'$ is given by Metropolis rules, $\min(1, \exp(\beta_J(\Psi(\boldsymbol{J}) - \Psi(\boldsymbol{J}'))).$

3. Numerical Results

3.1. Fitness and Energy

We have carried out MC simulations of the model discussed above and studied the dependence of the fitness and energy on T_S and T_J. They are given by

$$\Psi(T_S, T_J) = [\Psi(\boldsymbol{J}|T_S)]_J, \quad E(T_S, T_J) = [\langle H(\boldsymbol{S}|\boldsymbol{J})\rangle]_J, \tag{5}$$

respectively, where $[\cdots]_J$ denotes the average with respect to the equilibrium probability distribution, $P(\boldsymbol{J}|T_S, T_J)$. MC sampling with temperature T_S under the Hamiltonian H and the stochastic selection process governed by the fitness are carried out alternately. In our simulations of the spin dynamics, the exchange Monte Carlo simulation (EMC) [22] is introduced to accelerate the relaxation time to equilibrium and obtain the equilibrium spin distribution efficiently. In this section, we concentrate on the analysis of the equilibrium state. Figure 1 (a) and (b) show the dependence of the fitness and the energy on T_S and T_J, respectively, for $N = 15$ and $t = 3$. For each generation of the genotype dynamics, the fitness and energy are averaged with respect to the equilibrium distribution over 1500 MC steps after discarding the first 1500 MC steps; this number of steps is sufficient for equilibration. The data are averaged over the last 1000 generations. For any T_S, the fitness decreases monotonically with T_J, but the rate of decrease is affected significantly by T_S. The fitness for sufficiently low T_S remains at a high level and decreases only slightly with an increase in T_J, while for a medium value of T_S, the fitness gradually decreases to a lower level as a function of T_J, and eventually, for a sufficiently high value of T_S, it never reaches a high level. This result implies that the structure of the fitness landscape depends on T_S, the temperature at which the system has evolved.

The energy function, on the other hand, shows a significant dependence on T_S. While the energy is represented by a monotonically increasing function of T_S for high T_J, it exhibits non-monotonic behavior for low T_J and has a minimum at $T_S \simeq 2.0$. The configurations that include the target pattern at the energy minimum are obtained around $T_S \simeq 2.0$. This non-monotonicity of the energy corresponds to a negative specific heat in the sense of standard thermodynamics. The configurations of $\boldsymbol{J}$ giving rise to the highest fitness value generally have a huge redundancy. Using a fluctuation induced by T_S, a specific subset of the configurations of $\boldsymbol{J}$ with lower energy is selected among the redundant configurations at around $T_S \simeq 2.0$.

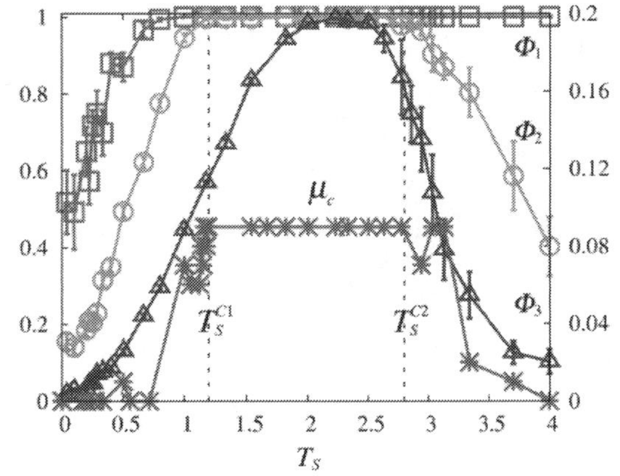

Figure 2. Dependence of local frustrations on T_S,; $\Phi_1(\Box), \Phi_2(\bigcirc), \Phi_3(\triangle)$ represents the left axis and $\mu_c(*)$ represents the right axis by fixing T_J at 0.5×10^{-3}. The data are computed by taking averages over 150 genotypes $\boldsymbol{J}$ evolved at a given temperature T_S. μ_c is a threshold value, beyond which the fitness of the mutated genotype begins to decrease. The transition points T_S^{c1} and T_S^{c2} are estimated as a temperature at which Φ_2 deviates from 1. Figure is quoted from [17].

3.2. Frustration

In the medium-temperature range, both a lower energy and a higher fitness are achieved. What is the structure in $\boldsymbol{J}$ configurations that helps to achieve this? The statistical physics of spin systems tells us that a decrease in energy implies a decrease of the frustration in spin configurations. Triplets of interactions that satisfy $J_{ij}J_{jk}J_{ki} < 0$ are known to yield frustration, which is an obstacle to attaining the unique global energy minimum[21, 23]. Meanwhile, all the interactions satisfying $J_{ij}J_{jk}J_{ki} > 0$ do not have frustration, and the energy of the spin states attains the minimum value. However, the energetically favorable spin configuration cannot be uniquely determined only by the condition $J_{ij}J_{jk}J_{ki} > 0$. The spin configurations that have low energy should be the target configurations when both the decrease in energy and the increase in fitness are simultaneously achieved. In our case, the target spins play a distinct role, and therefore, we need to quantify the frustration while distinguishing between target and non-target spins; this is in contrast to the standard spin-glass study.

The interactions are divided into three categories: those between target spins, $\boldsymbol{J}_{tt}$ ($\{J_{ij} \mid i, j \in \boldsymbol{t}\}$), those between target and non-target spins, $\boldsymbol{J}_{to}$ ($\{J_{ij} \mid i \in \boldsymbol{t}, j \notin \boldsymbol{t}\}$), and those between non-target spins, $\boldsymbol{J}_{oo}$ ($\{J_{ij} \mid i, j \notin \boldsymbol{t}\}$). It can be assumed that the frustration of all categories decreases at intermediate T_S. To confirm this, we should define the conditional frustration for each category of spins. We first define Φ_1 as the frequency of positive coupling among target spins, i.e.,

$$\Phi_1(T_S, T_J) = \frac{2}{t(t-1)} \Big[\sum_{i<j\in \boldsymbol{t}} J_{ij} \Big]_J. \tag{6}$$

The target configurations are energetically preferred under ferromagnetic coupling, i.e., $\Phi_1 = 1$, for which no frustration exists among the target spins. Second, we define Φ_2 as

$$\Phi_2(T_S, T_J) = \frac{2}{t(t-1)(N-t)} \Big[\sum_{i<j\in \boldsymbol{t}} \sum_{k\notin \boldsymbol{t}} J_{ik}J_{jk} \Big]_J, \tag{7}$$

where $t(t-1)(N-t)/2$ is the total number of possible spins: two target spins and one non-target spin. Here, $\Phi_2 = 1$ implies that no frustration exists in the interactions between target and non-target spins, and thus, the target configuration is at an energy minimum even when these interactions are included. Lastly, as a measure of the frustration among non-target spins, Φ_3 is defined as

$$\Phi_3(T_S, T_J) = \frac{1}{C_2^{N-t}} \Big[\sum_{k<l\notin \boldsymbol{t}} \Big(\frac{1}{t}\sum_{i\in \boldsymbol{t}} J_{ik}\Big) J_{kl} \Big(\frac{1}{t}\sum_{j\in \boldsymbol{t}} J_{jl}\Big) \Big]_J, \tag{8}$$

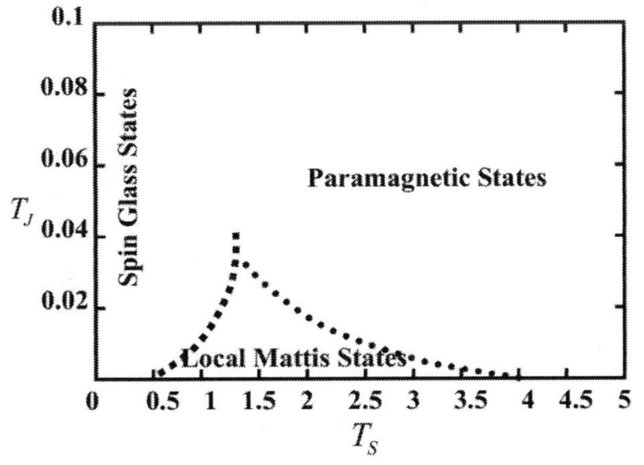

Figure 3. Phase diagram of the evolved $\boldsymbol{J}$s at $N = 15$ and $t = 3$. Three types of evolved $\boldsymbol{J}$ are defined on the basis of the value of the fitness and Φ_2.

where C_2^{N-t} is the total number of possible pairs of non-target spins. If Φ_3 is equal to 1, the frustration is not introduced by the interactions between non-target spins; in other words, there is no frustration globally. Hence, the system with $\Phi_1 = \Phi_2 = \Phi_3 = 1$ is in the Mattis state [24], which can be transformed to ferromagnetic interaction by a gauge transformation.

For the interaction $\boldsymbol{J}$ with evolution under an environment with temperature T_S, we have computed Φ_1, Φ_2, and Φ_3 by performing MC simulations. In Figure 2, we present the T_S dependence of $\Phi_1(T_S, T_J)$, $\Phi_2(T_S, T_J)$, and $\Phi_3(T_S, T_J)$ at a fixed $T_J = 0.5 \times 10^{-3}$. At sufficiently low T_J, the frustration parameters attain the maximum value 1 at the intermediate T_S, where the frustrations are extensively eliminated, while they remain finite at low T_S and high T_S. We define the intermediate temperature region as $T_S^{c1} < T_S < T_S^{c2}$, where the frustration parameter Φ_2 equals 1. These temperatures depend on T_J, and we express them as $T_S^{c1}(T_J)$ and $T_S^{c2}(T_J)$. We see that for low T_J ($\lesssim 0.05$), the phase diagram is split into three phases. The first one is frustrated and adapted phases for $T_S < T_S^{c1}(T_J)$. For $T_S < T_S^{c1}(T_J)$, all Φ_i $(i = 1, 2, 3)$ are less than unity, and hence, the frustration remains for target and non-target spins.

For $T_S \geq T_S^{c1}(T_J)$, Φ_1 equals 1, so that a target configuration is embedded as an energetically favorable state. For a finite system with finite T_J, Φ_j cannot be exactly 1. However, as long as T_J is low, the deviation of Φ_j from 1 at the intermediate temperature is negligible. For $T_S^{c1}(T_J) \leq T_S \leq T_S^{c2}(T_J)$, Φ_2 is also equal to 1, implying that the frustration among spins is not introduced via interactions with a non-target spin. In this temperature range, Φ_3 is not always equal to 1, except for $T_S \sim 2.0$, where the Mattis state arises. When $\Phi_2 = 1$ and $\Phi_3 \neq 1$, the frustration is not completely eliminated from the non-target spin interactions $\boldsymbol{J}_{oo}$; this is in contrast to the Mattis state. We call such a $\boldsymbol{J}$ configuration "local Mattis state"(LMS), and it is characterized by $\Phi_1 = \Phi_2 = 1$ but $\Phi_3 \neq 1$. This implies that the interactions $\boldsymbol{J}$ have no frustration around the target spins, but there is some frustration between non-target spin interactions. The interactions $\boldsymbol{J}$ required to form such a local Mattis state are obtained as a consequence of the evolution around $T_S^{c1}(T_J) \leq T_S \leq T_S^{c2}(T_J)$ for low T_J, where both the fitted target configuration and lower energy are achieved. The T_S range in which the local Mattis state is stabilized becomes narrower with an increase in T_J. The phase diagram of the model is shown in figure 3. For $T_S > T_S^{c2}(T_J)$, the frustration parameter Φ_2 is less than 1, and consequently, the frustration remains, and the fitness Ψ starts to decrease and the energy increases. Thus, neither adaptation nor energy minimization is achieved.

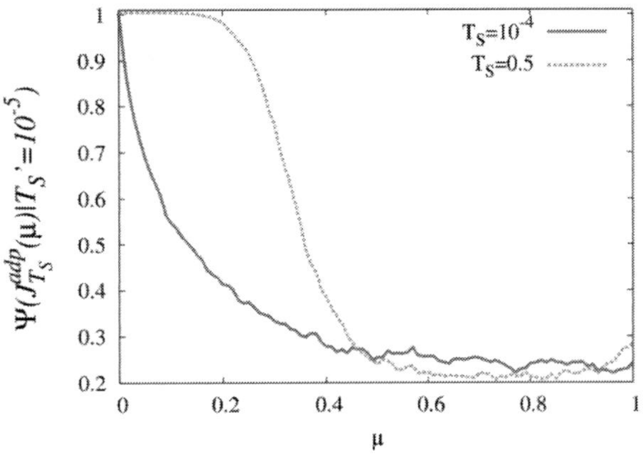

Figure 4. Fitness of the mutated $\boldsymbol{J}$ as a function of the mutation rate μ for $T_S = 10^{-4}$ (solid curve) and $T_S = 2.0$ (dotted curve). T_J is fixed at 0.5×10^{-3} for both cases. For each adapted genotype, the mutated genotypes $\boldsymbol{J}$ are generated 150 times by flipping randomly chosen elements J_{ij} with $\boldsymbol{J}$ governed by the rate μ, and the average is taken over 150 adapted genotypes. Figure is quoted from [17].

3.3. Robustness to mutation

We examine the robustness that represents the stability of $\boldsymbol{J}$'s fitness with respect to changes in the $\boldsymbol{J}$ configuration. ¿From the evolved genotype under a given T_S and T_J, denoted as $\boldsymbol{J}^{adp}_{T_S}$, mutations are imposed by flip-flopping the sign of a certain fraction of randomly chosen matrix elements in $\boldsymbol{J}^{adp}_{T_S}$. The value of the fraction represents the mutation rate μ. We evaluate the fitness of the mutated $\boldsymbol{J}^{adp}_{T_S}(\mu)$ at $T'_S = 10^{-5}(\neq T_S)$, i.e., $\Psi(\boldsymbol{J}^{adp}_{T_S}(\mu)|T'_S = 10^{-5})$, by taking an average over 150 samples of mutated $\boldsymbol{J}^{adp}_{T_S}(\mu)$. Figure 4 shows μ dependence of the fitness for $T_S = 10^{-4}$ and $T_S = 2.0$. For low values of T_S, the fitness of mutated $\boldsymbol{J}^{adp}_{T_S}(\mu)$ exhibits a rapid decrease with respect to the mutation rate, but for T_S between T_S^{c1} and T_S^{c2}, it does not decrease until the mutation rate reaches a specific value. We define $\mu_c(T_S)$ as the threshold point in the mutation rate beyond which the fitness $\Psi(\boldsymbol{J}^{adp}_{T_S}(\mu)|T'_S)$ begins to decrease from unity. Figure 2 shows the dependence of μ_c on T_S, which has a plateau at $T_S^{c1} \leq T_S \leq T_S^{c2}$ where Φ_2 is unity. This range of temperatures that exhibits mutational robustness agrees with the range that gives rise to the LMS. In other words, mutational robustness is realized for a genotype with no frustration around the target spins. Evolution in a mutationally robust genotype $\boldsymbol{J}$ is possible only when the phenotype dynamics is subjected to noise within the range $T_S^{c1} \leq T_S \leq T_S^{c2}$. This mutational robustness is interpreted as a consequence that the fitness landscape becomes non-neutral for $T_S \geq T_S^{c1}$ [18].

4. Frustration of partially annealed SK model

In our model discussed so far, the fitness is defined as a function of equilibrium configurations of the target spins. Instead, when the free energy $\beta_S F(\boldsymbol{J}, T_S) = -\log Z_S(\boldsymbol{J}, T_S)$ is chosen as the fitness, the model system is equivalent to the partial annealing[19, 20] for the SK model, which enables us to study by the replica method. The frustration parameter Φ_3 might be defined as

$$\Phi_3 = \frac{1}{C_3^N} \sum_{i<j<k} J_{ij} J_{jk} J_{ki}. \tag{9}$$

In the case of the SK model, the frustration $[\Phi_3]_J$ can be calculated analytically, as shown in Dotsenko et al.[20]. We explain their method briefly and derive the expression of $[\Phi_3]_J$ under the replica symmetry (RS) ansatz. The product of $\boldsymbol{J}$ can be described by the product of spins by using the fact that $P_0(\boldsymbol{J})$ is Gaussian distribution, as follows,

$$[\Phi_3]_J = \frac{1}{[Z^n]_J} \int D\boldsymbol{J} P_0(\boldsymbol{J}) Z^n(\boldsymbol{J}) \Big(\frac{1}{C_3^N} \sum_{i<j<k} J_{ij} J_{jk} J_{ki} \Big) = \left(\frac{\beta_S}{\sqrt{N} C_3^N} \right)^3 \sum_{\alpha,\beta,\gamma} \sum_{i<j<k} [\langle S_i^\alpha S_j^\alpha S_j^\beta S_k^\beta S_k^\gamma S_i^\gamma \rangle]_J.$$

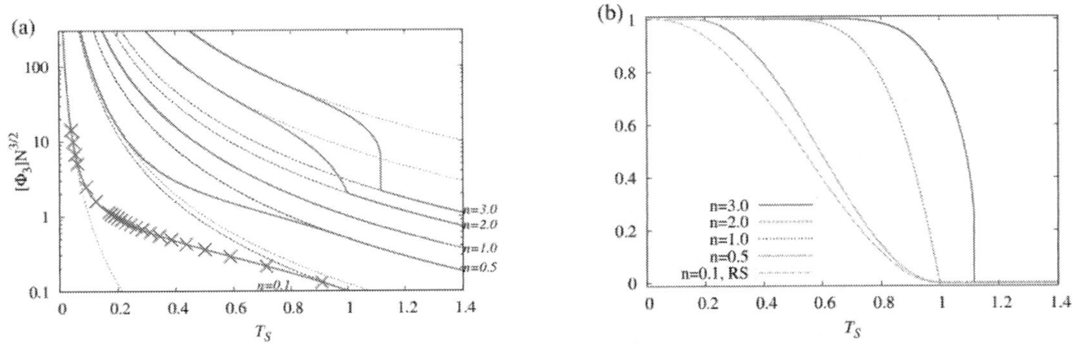

Figure 5. (a): T_S dependence of the frustration $[\Phi_3]_J$ for $n = 3, 2, 1, 0.5$ and 0.3 by solid lines from top to bottom. The dashed line is $n\beta_S^3$ and the dotted line is $\beta_J^3 = (n\beta_S)^3$. The point $\times$ shows the $[\Phi_3]_J$ for the 1 step RSB solution obtained in eq.(A.3).
(b): T_S dependence of the normalized frustration parameter, $([\Phi_3]_J - n\beta_S^3)/(\beta_J^3 - n\beta_S^3)$, for each n shown in (a).

Taking the leading term of N in the summation $\sum_{i<j<k}$, $[\Phi_3]_J$ can be expressed as

$$N^{3/2}[\Phi_3]_J = \left(\frac{\beta_S}{N}\right)^3 \sum_{\alpha,\beta,\gamma} \left[\left\langle \left(\sum_i S_i^\alpha S_i^\gamma \sum_j S_j^\alpha S_j^\beta \sum_k S_k^\beta S_k^\gamma \right) \right\rangle\right]_J = \beta_S^3 \, \mathrm{Tr}_n (I+Q)^3, \quad (10)$$

where Q is $n \times n$ matrix whose diagonal elements take 0 and off-diagonal ones are $q^{\alpha\beta}$ $(\alpha \neq \beta)$, and I is $n \times n$ identical matrix. In the RS case, all of the non-diagonal components of the $n \times n$ matrix Q take the same value q. The frustration $[\Phi_3]_J$ is written as

$$[\Phi_3]_J N^{3/2} = (\beta_S)^3 \left(\lambda_1^3 + (n-1)\lambda_2^3\right), \quad (11)$$

with the eigen values λ_1 and λ_2 of the matrix $(I+Q)$, $\lambda_1 = 1 + (n-1)q$ and $\lambda_2 = (1-q)$. The spin-glass order parameter q under the RS ansatz with finite-n is given as

$$q = \frac{\int Dz \tanh^2(\beta_S\sqrt{q}z)\cosh^n(\beta_S\sqrt{q}z)}{\int Dz \cosh^n(\beta_S\sqrt{q}z)}, \quad (12)$$

where $Dz = dz\; e^{-z^2/2}/\sqrt{2\pi}$. The expression of $[\Phi_3]_J$ for 1 step replica-symmetry breaking is shown in Appendix.

Figure 5(a) shows the dependence on T_S of $[\Phi_3]_J$ in the thermodynamic limit for $n = 3, 2, 1, 0.5$, and 0.3, beginning at the top, respectively. The dashed and dotted lines are $n\beta_S^3$ and β_J^3, respectively. The spin-glass transition temperature is $T_S^c \simeq 1.15$ for $n = 3$ and $T_S^c = 1.0$ for other n. At $T_S > T_S^c$, $[\Phi_3]_J$ behaves as $n\beta_S^3$. At T_S^c, the spin-glass order parameter q begins to take a finite value and $[\Phi_3]_J$ deviates from $n\beta_S^3$, except in the case of $n = 1$ where $[\Phi_3]_J$ always behaves as $n\beta_S^3$ for any q. At sufficiently low T_S, $[\Phi_3]_J$ scales as β_J^3 below a certain temperature that varies with n. In this way, crossover of frustration from $n\beta_S^3$ to $\beta_J^3 = (n\beta_S)^3$ is observed except for $n = 1$. The fact that $[\Phi_3]_J$ is equal to β_J^3 means that the overlap q takes the value 1. Then, $[\Phi_3]_J$ takes the possible maximum value, implying that the frustration is reduced as less as possible.

For $n > 1$, β_J^3 is larger than $n\beta_S^3$ for whole temperature region. Therefore, the spin-glass transition induces the increase of $[\Phi_3]_J$ that enhances the elimination of the frustration. On

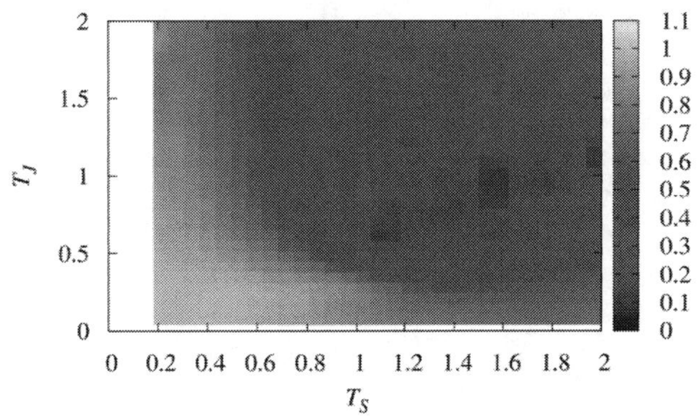

Figure 6. A contour map on T_S and T_J of the normalized frustration parameter. At the temperature region where it takes 1, the frustration parameter scales as β_J^3 and the frustration is completely vanished.

the other hand, for $n \leq 1$ where $n\beta_S^3$ is less than β_J^3, $[\Phi_3]_J$ increases slowly in comparison with the case for $n > 1$ and the crossover to less-frustrated state occurs at very low temperature. In figure 5(b), we show the normalized frustration parameter, $([\Phi_3]_J - n\beta_S^3)/(\beta_J^3 - n\beta_S^3)$, which takes the value within [0,1]. It begins to increase at the spin-glass transition temperature and converges to 1 at low T_S.

To show a temperature region where the frustration sufficiently close to the scaling β_J^3, we plot a contour map on T_S and T_J of the normalized frustration parameter in figure 6. At low T_J, the less-frustrated region is expanded, similar to that found in our model. However, the reduction of frustration is induced by the emergence of spin-glass order in this partially annealed SK model. In fact, the less-frustrated region appears at high T_J and low T_S, where the spin-glass order exists, and this is sharp contrast to that found in our model. This might be due to the fact that the free energy is chosen as the fitness. In order to understand the local Mattis state found in our model, some local structure in the fitness is to be taken into account. This is an issue in the future work.

5. Summary and discussion

We have considered the evolution of a Hamiltonian system to generate a specific configuration for target spins that captures the basic features required to study the evolution. In this study, we adopted a Markov process, which is given by temperature T_J and fitness $\Psi(\boldsymbol{J})$, for evolutional dynamics. By performing numerical simulation, we found that a specific subset of $\boldsymbol{J}$ with low energy and high fitness is evolved at an intermediate T_S and low T_J. From the statistical-mechanical viewpoint, we focused on frustration and found that the interactions $\boldsymbol{J}$ that evolved at the intermediate T_S are less frustrated. We called these $\boldsymbol{J}$ the local Mattis states. In general, the less frustrated $\boldsymbol{J}$ states are robust to mutation. Hence, the robustness of evolving states to mutation is realized at intermediate temperatures $T_S^{c1} \leq T_S \leq T_S^{c2}$. In other words, robustness to thermal noise introduces mutational robustness; this has also been recently discussed for gene regulation network models[13]. The relevance of thermal noise to robust evolution is thus demonstrated.

In this study, we observed transitions at T_S^{c1} and T_S^{c2}. The phase below T_S^{c1} corresponds to the spin-glass phase and that above T_S^{c2} corresponds to the paramagnetic phase in the context of statistical physics; in contrast, the local Mattis phase between the two phases, which could correspond to the funnel landscape, is a novel discovery in this study. Since a framework of statistical physics of spin systems has been adopted in the present study, the theoretical concepts developed therein, such as replica symmetry breaking, may be applicable in the context developed here to understand this transition. A promising direction would be the study of the partial annealing models with the replica theory. Some related systems based on the partial annealing have been studied in various areas; protein folding and its evolution [25], ligand-receptor reaction [26] and so on. In fact, when the fitness is set to the free energy, our model is reduced to the partial annealing version of the SK model. We can find a similarity between our model and its partial annealing model. However, some points, in particular the mechanism for the emergence of local Mattis state, are less clear and remains to be solved in the future.

Acknowledgments
We would like to thank C. H. Nakajima and T. Nakajima for helpful comments and discussions. This work was partially supported by a Grant-in-Aid for Scientific Research (No.18079004) from MEXT and JSPS Fellows (No.20−10778) from JSPS.

Appendix A. 1RSB expression of frustration $[\Phi_3]_J$ for the Sherrington-Kirkpatrick model with finite replica number n

In this appendix, we discuss how the frustration $[\Phi_3]_J$ is modified by taking into account replica symmetry breaking. At the level of one-step replica symmetry breaking (1RSB), the saddle-point equations for the order parameters q_0 and q_1 are given as

$$q_0 = \int Du\Big(\int Dv \cosh^m \Xi\Big)^{n/m}\Big(\frac{\int Dv \cosh^m \Xi \tanh \Xi}{\int Dv \cosh^m \Xi}\Big)^2 \Big/ \int Du\Big(\int Dv \cosh^m \Xi\Big)^{n/m},$$
$$q_1 = \int Du\Big(\int Dv \cosh^m \Xi\Big)^{n/m}\frac{\int Dv \cosh^m \Xi \tanh^2 \Xi}{\int Dv \cosh^m \Xi} \Big/ \int Du\Big(\int Dv \cosh^m \Xi\Big)^{n/m} \qquad \text{(A.1)}$$

with $\Xi = \beta_S(\sqrt{q_0}u + \sqrt{q_1 - q_0}v)$. Here Du and Dv are the Gaussian kernel, $DX = dX\ e^{-X^2/2}/\sqrt{2\pi}$. The value of replica-partition parameter m is determined as a solution of

$$\frac{\beta_S^2}{4}(q_1^2 - q_0^2)m = \int \tilde{D}u \frac{\int Dv \cosh^m \Xi \log\cosh \Xi}{\int Dv \cosh^m \Xi} - \frac{1}{m}\int \tilde{D}u \log\Big(\int Dv \cosh^m \Xi\Big). \qquad \text{(A.2)}$$

With these parameters, the frustration $[\Phi_3]_J$ is written as

$$[\Phi_3]_J^{1RSB} = (\beta_S N)^3\Big\{\lambda_1^3 + (m-1)\lambda_2^3 + \Big(\frac{n}{m}-1\Big)\lambda_3^3 + (m-1)\Big(\frac{n}{m}-1\Big)\lambda_4^3\Big\} \qquad \text{(A.3)}$$

with these eigen values

$$\lambda_1 = 1 + (m-1)q_1 + (n-m)q_0, \quad \lambda_2 = 1 - q_1$$
$$\lambda_3 = 1 + (m-1)q_1 - mq_0, \quad \lambda_4 = 1 - q_1.$$

The frustration for RS solution, eq.(11) can be recovered by setting $q_1 = q_2 = q$.

The frustration parameter $[\Phi_3]_J$ of the 1 step RSB version is shown in Figure 5 for $n = 0.1$. The behavior of $[\Phi_3]_J^{1RSB}$ is almost the same as that of $[\Phi_3]_J^{RS}$. At low T_S with $n = T_S/T_J$ kept, the overlap parameters takes the value $q_1 \simeq q_0$ and actually $[\Phi_3]_J^{RS}$ is recovered. It is hard to detect the breaking of the replica symmetry through the behavior of the frustration.

References

[1] Saito S, Sasai M and Yomo T 1997 *Proc. Natl. Acad. Sci.* USA **94** 11324
[2] Ancel L W and Fontana W 2000 *J. Exp. Zool.* (*Mol. Dev. Evol.*) **288** 242
[3] Alon U, Surette M G, Barkai N and Leibler S 1999 *Nature* **397** 168
[4] Go N 1983 *Ann. Rev. Biophys. Bioeng.* **12** 183

[5] Bryngelson J D and Wolynes P G 1987 *Proc. Natl. Acad. Sci.* USA **84** 84
[6] Onuchic J N and Wolynes P G 2004 *Curr. Opin. Struc. Biol.* **14** 70
[7] Li F, Long T, Lu Y, Ouyang Q and Tang C 2004 *Proc. Natl. Acad. Sci. USA* **101** 4781
[8] Kaneko K, Sato K, Michiue T, Okabayashi K, Ohnuma K, Danno H and Asashima M 2008 *J. Exp. Zool.* B **310** 492
[9] Ciliberti S, Martin O C and Wagner A 2007 *PLoS Comput. Biol.* **3** e15
[10] Wagner A 2005 *Robustness and Evolvability in Living Systems* (New Jersey: Princeton University Press)
[11] Sun J and Deem M W 2007 *Phys. Rev. Lett.* **99** 228107
[12] Kaneko K 2006 *Life: An Introduction to Complex Systems Biology* (Berlin, New York: Springer-Verlag)
[13] Kaneko K 2007 *PLoS ONE* **2** e434
[14] Kaneko K 2008 *Chaos* **18** 026112
[15] Kaneko K and Furusawa C 2006 *J. Theor. Biol.* **240** 78
[16] Waddington C H 1957 *The Strategy of the Genes* (Bristol: George Allen & Unwin LTD)
[17] Sakata A, Hukushima K and Kaneko K 2009 *Phys. Rev. Lett.* **102** 148101
[18] Sakata A, Hukushima K and Kaneko K 2009 *Preprint* arXiv: 0906.0900 (to be published in *Phys. Rev.* E)
[19] Penney R W, Coolen A C C and Sherrington D 1993 *J. Phys. A: Math. Gen.* **26** 3681
[20] Dotsenko V, Frantz S and Mézard M 1994 *J. Phys. A: Math. Gen.* **27** 2351
[21] Nishimori H 2001 *Statistical Physics of Spin Glasses and Information Processing: An Introduction* (New York: Oxford University Press)
[22] Hukushima K and Nemoto K 1996 *J. Phys. Soc. Japan* **65** 1604
[23] Mézard M, Parisi G and Virasoro M A 1987 *Spin Glass Theory and Beyond* (Singapore: World Scientific Publisher)
[24] Mattis D C 1976 *Phys. Rev.* **56** 421
[25] Rabello S, Coolen A C C, Perez-Vicente C J and Fraternali F 2008 *J. Phys. A: Math. Theor.* **41** 285004
[26] Behringer H, Degenhard A and Schmid F 2006 *Phys. Rev. Lett.* **97** 128101

Journal of Physics: Conference Series **197** (2009) 012004 doi:10.1088/1742-6596/197/1/012004

Monte Carlo dynamically weighted importance sampling for spatial models with intractable normalizing constants

Faming Liang

Department of Statistics, Texas A&M University, College Station, TX 77843-3143, USA

E-mail: `fliang@stat.tamu.edu`

Sooyoung Cheon

KU Industry-Academy Cooperation Group Team of Economics and Statistics, Korea University, Jochiwon 339-700, Korea.

E-mail: `stat-csy@hanmail.net`

Abstract. The problem of simulating from distributions with intractable normalizing constants has received much attention in the recent literature. In this paper, we propose a new MCMC algorithm, the so-called Monte Carlo dynamically weighted importance sampler, for tickling this problem. The new algorithm is illustrated with the spatial autologistic models. The novelty of our algorithm is that it allows for the use of Monte Carlo estimates in MCMC simulations, while still leaving the target distribution invariant under the criterion of dynamically weighted importance sampling. Unlike the auxiliary variable MCMC algorithms, the new algorithm removes the need of perfect sampling, and thus can be applied to a wide range of problems for which perfect sampling is not available or very expensive. The new algorithm can also be used for simulating from the incomplete posterior distribution for the missing data problem.

1. Introduction

Spatial models, e.g., the autologistic model, the Potts model, and the autonormal model [1], have been used in modeling of many scientific problems. Examples include image analysis [2], disease mapping [3], genetic analysis [4], among others. A major problem with these models is that the normalizing constant is intractable. The problem can be described as follows. Suppose we have a dataset $\boldsymbol{X}$ generated from a statistical model with the likelihood function

$$f(\boldsymbol{x}|\theta) = \frac{p(\boldsymbol{x},\theta)}{Z(\theta)}, \quad \boldsymbol{x} \in \mathcal{X}, \quad \theta \in \Theta, \tag{1}$$

where θ is the parameter, and $Z(\theta)$ is the normalizing constant which depends on θ and is not available in closed form. Let $\pi(\theta)$ denote the prior density of θ. The posterior distribution of θ given $\boldsymbol{X} = \boldsymbol{x}$ is then given by

$$\pi(\theta|\boldsymbol{x}) \propto \frac{1}{Z(\theta)} p(\boldsymbol{x},\theta)\pi(\theta). \tag{2}$$

Since the closed form of $Z(\theta)$ is not available, inference for the model poses a great challenge on the current statistical methods.

The Metropolis-Hastings (MH) algorithm [5, 6] cannot be directly applied to simulate from $\pi(\theta|\boldsymbol{x})$, because the acceptance probability would involve the unknown ratio $Z(\theta)/Z(\theta')$, where θ' denotes the proposed value. To circumvent this difficulty, various methods have been proposed in the literature. These methods can be roughly divided into two categories, namely, the likelihood approximation-based methods and the auxiliary variable MCMC methods. They are described, respectively, as follows.

- *The likelihood approximation-based methods.* A very first work of this category is the pseudo-likelihood method [1], which approximates the likelihood function by a tractable pseudo-likelihood function. This method is easy to use, but it typically performs less well for the models for which neighboring dependence is strong. Later, Geyer and Thompson [7] proposed to approximate $Z(\theta)$ and thus the likelihood function using Monte Carlo samples generated from a distribution $f(\boldsymbol{x}|\theta_0)$ with a fixed value of θ_0. This approach was refined in [8], where the likelihood is approximated using the Monte Carlo samples generated from a mixture distribution defined on a lattice of points of θ. Liang [9] proposed an alternative Monte Carlo approach to approximate $Z(\theta)$, where $Z(\theta)$ is viewed as a marginal distribution of the unnormalized distribution $p(\boldsymbol{x}, \theta)$ and is estimated by an adaptive kernel density estimator using Monte Carlo draws. A very recent work belonging to this category is the double MH algorithm [10], which does not approximate the likelihood directly, but uses the samples generated with a finite number of MH moves to substitute for perfect samples of $f(\boldsymbol{x}|\theta')$ in a MH update of θ.
- *Auxiliary variable MCMC methods.* The algorithms belonging to in this category include the Møller algorithm [11] and the exchange algorithm[12], which work by including some auxiliary variables in state transitions such that the normalizing constant can be canceled. A serious weakness of the two algorithms is that they require the auxiliary variables to be generated by a perfect sampler [13]. Unfortunately, perfect sampling is very expensive or impossible for many statistical models whose normalizing constant is intractable.

In this paper, we propose a new MCMC algorithm, the so-called Monte Carlo dynamically weighted importance sampler (MCDWIS), for simulating from the distributions with intractable normalizing constants. The MCDWIS removes the need of perfect sampling, and thus can be applied to many statistical models for which perfect sampling is not available or very expensive.

The MCDWIS is a Monte Carlo version of the dynamically weighted importance sampling (DWIS) algorithm [14]. As in DWIS, the state space of the Markov chain is augmented to a population, a collection of weighted samples $(\boldsymbol{\theta}, \boldsymbol{w}) = \{\theta_1, w_1; \ldots; \theta_n, w_n\}$, where n is called the population size. With a slight abuse of notation, (θ_i, w_i) is called an individual state of the population $(\boldsymbol{\theta}, \boldsymbol{w})$. Given the current population $(\boldsymbol{\theta}_t, \boldsymbol{w}_t)$, the population at the tth iteration, one iteration of the MCDWIS, as illustrated in Figure 1, involves two steps (described in detail later):

1. *Monte Carlo Dynamic weighting (MCDW):* Update each individual state of the current population by a MCDW transition.
2. *Population control:* Split or replicate the individual states with large weights and discard the individual states with small weights.

The MCDW step allows for the use of Monte Carlo estimates in MCMC simulations. The bias induced thereby is counterbalanced by giving different weights to the new samples produced. Hence, a Monte Carlo estimate of $Z(\theta)/Z(\theta')$ can be incorporated into the MCDWIS simulations of $\pi(\theta|\boldsymbol{x})$, while leaving $\pi(\theta|\boldsymbol{x})$ invariant with respect to the dynamic importance weights. Note that the conventional MCMC algorithms, e.g., the MH algorithm, do not allow for the use

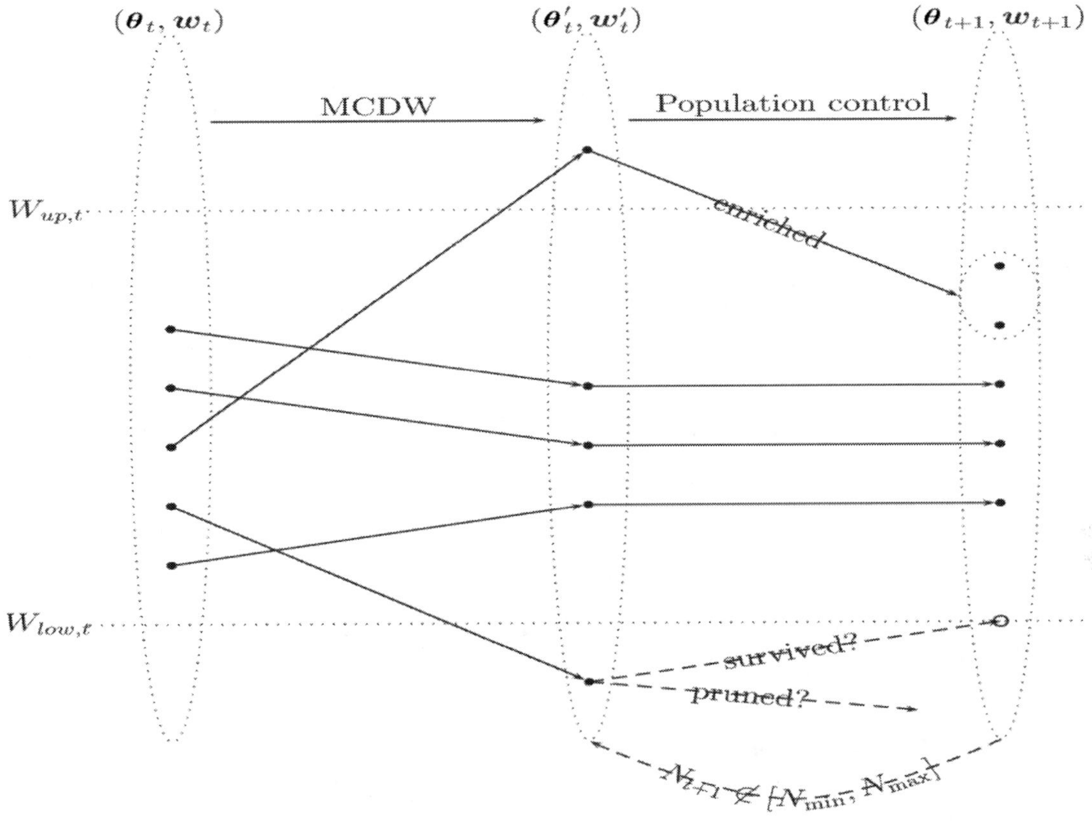

Figure 1. A diagram of the MCDWIS algorithm.

of Monte Carlo estimates in simulations. Otherwise, the detailed balance condition will be violated. Inclusion of the population control step is also quite crucial to the MCDWIS, which has the weights and population size controlled to suitable ranges. Otherwise, the MCDWIS will perform like a multi-chain dynamic weighting algorithm, and the resulting estimates would be highly variable as in a single chain dynamic weighting algorithm. Refer to [15, 16] for detailed descriptions and theoretical properties of the dynamic weighting algorithm.

The remainder of this paper is organized as follows. In Section 2, we provide a brief review for the theory of dynamically weighted importance sampling. In Section 3, we describe the MCDWIS and justify it with the theory introduced in Section 2. In Section 4, we illustrate the MCDWIS with spatial autologistic models. In Section 5, we conclude the paper with a brief discussion.

2. A theory of dynamically weighted importance sampling

In this section, we provide a brief review for the theory of dynamically weighted importance sampling [14]. Let $g_t(\theta, w)$ denote the joint density of (θ, w), an individual state of $(\boldsymbol{\theta}_t, \boldsymbol{w}_t)$, and let $\psi(\theta)$ denote the target distribution we are working with. Dynamically weighted importance sampling differs from conventional importance sampling [17] in that for any given θ, the weight w is a random variable instead of a constant defined as the ratio of the true and trial densities at θ.

Definition 2.1 *The distribution $g_t(\theta, w)$ defined on $\Theta \times (0, \infty)$ is called correctly weighted with* *respect to $\psi(\theta)$ if the following conditions hold,*

$$\int w g_t(\theta, w) dw = c_{t\theta}\psi(\theta), \tag{3}$$

$$\frac{\int_{\mathcal{A}} c_{t\theta}\psi(\theta)d\theta}{\int_{\Theta} c_{t\theta}\psi(\theta)d\theta} = \int_{\mathcal{A}} \psi(\theta)d\theta, \tag{4}$$

where $\mathcal{A}$ is any Borel set, $\mathcal{A} \subseteq \Theta$.

Definition 2.2 *If $g_t(\theta, w)$ is correctly weighted with respect to $\psi(\theta)$, and samples $(\theta_{t,i}, w_{t,i})$ are simulated from $g_t(\theta, w)$ for $i = 1, 2, \ldots, n_t$, then $(\boldsymbol{\theta}_t, \boldsymbol{w}_t) = (\theta_{t,1}, w_{t,1}; \cdots; \theta_{t,n_t}, w_{t,n_t})$ is called a correctly weighted population with respect to $\psi(\theta)$.*

Let $(\boldsymbol{\theta}_t, \boldsymbol{w}_t)$ be a correctly weighted population with respect to $\psi(\theta)$, and let $\theta'_1, \ldots, \theta'_m$ be distinct states in $\boldsymbol{\theta}_t$. Generate a random variable/vector ϑ such that

$$P\{\vartheta = \theta'_i\} = \frac{\sum_{j=1}^{n_t} w_j I(\theta_j = \theta'_i)}{\sum_{j=1}^{n_t} w_j}, \quad i = 1, 2, \ldots, m, \tag{5}$$

where $I(\cdot)$ is an indicator function. Then ϑ is approximately distributed as $\psi(\cdot)$ if n_t is large. This can be expressed in the following theorem:

Theorem 2.1 *As the population size $n_t \to \infty$, the random variable ϑ generated in (5) converges in distribution to a random variable θ which is distributed with the pdf $\psi(\theta)$.*

Let $(\boldsymbol{\theta}_1, \boldsymbol{w}_1)$, ..., $(\boldsymbol{\theta}_N, \boldsymbol{w}_N)$ be a series of correctly weighted populations generated by a DWIS algorithm with respect to $\psi(\theta)$. Then the quantity $\mu = E_\psi \rho(\theta)$, provided its existence, can be estimated by

$$\widehat{\mu} = \frac{\sum_{t=1}^{N} \sum_{i=1}^{n_t} w_{t,i}\rho(\theta_{t,i})}{\sum_{t=1}^{N} \sum_{i=1}^{n_t} w_{t,i}}. \tag{6}$$

Let $U_t = \sum_{i=1}^{n_t} w_{t,i}\rho(\theta_{t,i})$, $S_t = \sum_{i=1}^{n_t} w_{t,i}$, $S = ES_t$, and $V_t = U_t - \mu S_t$. If the variance of U_t and V_t are both finite, then the standard error of $\widehat{\mu}$ can be calculated using the ratio estimate as in finite population sampling [18]. As shown by Liang[14], $\widehat{\mu}$ is consistent and asymptotically normally distributed; that is,

$$\sqrt{N}(\widehat{\mu} - \mu) \to N(0, \sigma^2), \tag{7}$$

where σ^2 is defined as $\text{Var}(V_t)/S^2$.

Definition 2.3 *A transition rule for a population $(\boldsymbol{\theta}, \boldsymbol{w})$ is said to be* **invariant with respect to the dynamic importance weights** *(***IDIW***) if the joint density of $(\boldsymbol{\theta}, \boldsymbol{w})$ remains correctly weighted whenever the initial joint density is correctly weighted.*

3. Monte Carlo dynamically weighted importance sampling

In this section, we describe a Monte Carlo dynamic weighting (MCDW) sampler and a population control scheme for the MCDWIS, and show that they both are IDIW. Let $(\boldsymbol{\theta}_t, \boldsymbol{w}_t)$ denote the current population, let (θ, w) denote an individual state of $(\boldsymbol{\theta}_t, \boldsymbol{w}_t)$, and let (θ', w') denote the individual state transmitted from (θ, w) in one transition step.

3.1. A general Monte Carlo dynamic weighting sampler

A general Monte Carlo dynamic weighting scheme can be described as follows.

1. Draw θ^* from some proposal distribution $T(\theta^*|\theta)$.
2. Simulate auxiliary samples $\boldsymbol{y}_1, \ldots, \boldsymbol{y}_m$ from $f(\boldsymbol{y}|\theta^*)$ using a MCMC algorithm, say, the MH algorithm. Estimate the normalizing constant ratio $R_t(\theta, \theta^*) = Z(\theta)/Z(\theta^*)$ by
$$\widehat{R}_t(\theta, \theta^*) = \frac{1}{m} \sum_{i=1}^{m} \frac{p(\boldsymbol{y}_i, \theta)}{p(\boldsymbol{y}_i, \theta^*)}, \tag{8}$$
which is also known as the importance sampling (IS) estimator of $R_t(\theta, \theta^*)$.
3. Calculate the Monte Carlo dynamic weighting ratio
$$r_d = r_d(\theta, \theta^*, w) = w\widehat{R}_t(\theta, \theta^*) \frac{p(\boldsymbol{x}, \theta^*)}{p(\boldsymbol{x}, \theta)} \frac{T(\theta|\theta^*)}{T(\theta^*|\theta)}.$$
4. Choose $\beta_t = \beta_t(\boldsymbol{\theta}_t, \boldsymbol{w}_t) \geq 0$ and draw $U \sim unif(0,1)$. Update (θ, w) as (θ', w')
$$(\theta', w') = \begin{cases} (\theta^*, r_d/a), & \text{if } U \leq a, \\ (\theta, w/(1-a)), & \text{otherwise,} \end{cases}$$
where $a = r_d/(r_d + \beta_t)$; β_t is a function of $(\boldsymbol{\theta}_t, \boldsymbol{w}_t)$, but remains a constant for each individual state of the same population.

The sampler is termed "Monte Carlo" in the sense that $R_t(\theta, \theta^*)$ is substituted by its Monte Carlo estimator in calculation of r_d. If $R_t(\theta, \theta^*)$ is available analytically, then the sampler is reduced to the dynamic weighting sampler [14]. For this sampler, we have two remarks as follows.

Remark 1. As pointed out in [19], $\widehat{R}_t(\theta, \theta^*)$ is an unbiased and consistent estimator of $R_t(\theta, \theta^*)$. Following from the central limit theorem,
$$\sqrt{m}\left(\widehat{R}_t(\theta, \theta^*) - R_t(\theta, \theta^*)\right) \to N\left(0, \sigma_t^2\right), \tag{9}$$
where σ_t^2 can be expressed as
$$\sigma_t^2 = \mathrm{Var}\left(\frac{p(\boldsymbol{y}_1, \theta)}{p(\boldsymbol{y}_1, \theta^*)}\right) + 2\sum_{i=2}^{\infty} \mathrm{Cov}\left(\frac{p(\boldsymbol{y}_1, \theta)}{p(\boldsymbol{y}_1, \theta^*)}, \frac{p(\boldsymbol{y}_i, \theta)}{p(\boldsymbol{y}_i, \theta^*)}\right).$$
Alternatively, we can write $\widehat{R}_t(\theta, \theta^*) = R_t(\theta, \theta^*)(1 + \epsilon_t)$, where $\epsilon_t \sim N(0, \sigma_t^2/[mR_t^2(\theta, \theta^*)])$.

Remark 2. The parameter β_t can be chosen as a function of the population $(\boldsymbol{\theta}_t, \boldsymbol{w}_t)$. For simplicity, this article concentrates only on the cases where $\beta_t = 0$ or 1. If $\beta_t \equiv 0$, then the sampler is essentially a random walk induced by $T(\cdot, \cdot)$ on the space Θ. If $\beta_t \equiv 1$, then the sampler is essentially the R-type move [15], which has the capability of moving across energy barriers by self-adjusting the weight values.

The following theorem shows that the Monte Carlo dynamic weighting sampler is IDIW when the initial condition
$$\int_{\Theta} c_{t_0,\theta} T(\theta|\theta') d\theta = c_{t_0,\theta'} \tag{10}$$
holds for some population $(\boldsymbol{\theta}_{t_0}, \boldsymbol{w}_{t_0})$.

Journal of Physics: Conference Series **197** (2009) 012004 doi:10.1088/1742-6596/197/1/012004

Theorem 3.1 *The Monte Carlo dynamic weighting sampler is IDIW; that is, if the joint distribution $g_t(\theta, w)$ for $(\boldsymbol{\theta}_t, \boldsymbol{w}_t)$ is correctly weighted with respect to $\pi(\theta|\boldsymbol{x})$ and (10) holds, after one Monte Carlo dynamic weighting step, the new joint density $g_{t+1}(\theta', w')$ for $(\boldsymbol{\theta}_{t+1}, \boldsymbol{w}_{t+1})$ is also correctly weighted with respect to $\pi(\theta|\boldsymbol{x})$, and (10) still holds.*

PROOF: For the case $\beta_t > 0$,

$$\int_0^\infty w' g_{t+1}(\theta', w')dw' = \int_\Theta \int_0^\infty \int_{-\infty}^\infty [r_d(\theta, \theta', w) + \beta_t] g_t(\theta, w) T(\theta'|\theta)\varphi(\epsilon_t) \frac{r_d(\theta, \theta', w)}{r_d(\theta, \theta', w) + \beta_t} d\epsilon_t dw d\theta$$
$$+ \int_\Theta \int_0^\infty \int_{-\infty}^\infty \frac{w[r_d(\theta', \vartheta, w) + \beta_t]}{\beta_t} g_t(\theta', w) T(\vartheta|\theta')\varphi(\epsilon_t) \frac{\beta_t}{r_d(\theta', \vartheta, w) + \beta_t} d\epsilon_t dw d\vartheta$$
$$= (I) + (II),$$

where $\varphi(\cdot)$ denotes the density of ϵ_t. For (I), we have

$$\begin{aligned}
(I) &= \int_\Theta \int_0^\infty \int_{-\infty}^\infty wR(\theta, \theta')(1 + \epsilon_t) \frac{p(\boldsymbol{x}, \theta')\pi(\theta')}{p(\boldsymbol{x}, \theta)\pi(\theta)} \frac{T(\theta|\theta')}{T(\theta'|\theta)} g_t(\theta, w) T(\theta'|\theta)\varphi(\epsilon_t) d\epsilon_t dw d\theta \\
&= \int_\Theta \int_0^\infty wR(\theta, \theta') \frac{p(\boldsymbol{x}, \theta')\pi(\theta')}{p(\boldsymbol{x}, \theta)\pi(\theta)} T(\theta|\theta') g_t(\theta, w) dw d\theta \\
&= \int_\Theta \frac{\pi(\theta'|\boldsymbol{x})}{\pi(\theta|\boldsymbol{x})} T(\theta|\theta') \left(\int_0^\infty w g_t(\theta, w) dw \right) d\theta \\
&= \pi(\theta'|\boldsymbol{x}) \int_\Theta c_{t,\theta} T(\theta|\theta') d\theta \\
&= c_{t,\theta'} \pi(\theta'|\boldsymbol{x}).
\end{aligned}$$

For (II), we have

$$\begin{aligned}
(II) &= \int_\Theta \int_0^\infty \int_{-\infty}^\infty w g_t(\theta', w) T(\vartheta|\theta')\varphi(\epsilon_t) d\epsilon_t dw d\vartheta \\
&= \int_\Theta \int_0^\infty w g_t(\theta', w) T(\vartheta|\theta') dw d\vartheta \\
&= \int_\Theta c_{t,\theta'} \pi(\theta'|\boldsymbol{x}) T(\vartheta|\theta') d\vartheta \\
&= c_{t,\theta'} \pi(\theta'|\boldsymbol{x}).
\end{aligned}$$

In summary of (I) and (II), we have

$$\int_0^\infty w' g_{t+1}(\theta', w') dw' = 2c_{t,\theta'} \pi(\theta'|\boldsymbol{x}).$$

For the case $\beta_t = 0$, only the term (I) remains. Thus,

$$\int_0^\infty w' g_{t+1}(\theta', w') dw' = c_{t,\theta'} \pi(\theta'|\boldsymbol{x}).$$

By defining $c_{t+1,\theta'} = 2c_{t,\theta'}$ for the case $\beta_t > 0$ and $c_{t+1,\theta'} = c_{t,\theta'}$ for the case $\beta_t = 0$, it is easy to see that the condition (10) still holds for the new population. Hence, $g_{t+1}(\theta', w')$ is still correctly weighted with respect to $\pi(\theta|\boldsymbol{x})$. □

Journal of Physics: Conference Series **197** (2009) 012004 doi:10.1088/1742-6596/197/1/012004

In practice, the initial condition (10) can be easily satisfied. For example, choose $g_0(\theta, w) = g(\theta)$ and set $w = \widehat{R}p(\boldsymbol{x}, \theta)\pi(\theta)/g(\theta)$, where $\widehat{R}$ denotes an unbiased estimator of $1/Z(\theta)$, then $\int\int w g_0(\theta, w)\varphi(\epsilon)d\epsilon dw = \pi(\theta|\boldsymbol{x})$; that is, $c_{t,\theta} \equiv 1$. The initial condition is satisfied because $\int T(\theta|\theta')d\theta = 1$ always holds.

To avoid an extremely large weight caused by a nearly zero divisor, similarly to [14, 16], we make the following assumption for the target distribution: For any pair of $(\theta, \theta^*) \in \Theta \times \Theta$ and any sample $\boldsymbol{y} \in \mathcal{X}$, there exists a constant r_1 such that

$$r_1 \leq \frac{p(\boldsymbol{y}, \theta)}{p(\boldsymbol{y}, \theta^*)} \leq \frac{1}{r_1}. \tag{11}$$

This condition can be easily satisfied by restricting both Θ and $\mathcal{X}$ to be compact. This is natural for many spatial problems, such as the autologistic model studied in Section 4. In addition, for the proposal distribution, we assume that there exists a constant r_2 such that for any pair $(\theta, \theta^*) \in \Theta \times \Theta$,

$$r_2 \leq \frac{T(\theta|\theta^*)}{T(\theta^*|\theta)} \leq \frac{1}{r_2}. \tag{12}$$

This condition can be easily satisfied by using a symmetric proposal, which implies that $T(\theta|\theta^*)/T(\theta^*|\theta) = 1$, or using a bounded proposal, e.g., $\theta^* \sim Unif(\theta_l, \theta_r)$ with θ_l and θ_r being fixed constants. Following from (8), (11) and (12), it is easy to see that there exists a constant r_0 such that for any pair $(\theta, \theta^*) \in \Theta \times \Theta$,

$$r_0 \leq \widehat{R}_t(\theta, \theta^*)\frac{p(\boldsymbol{x}, \theta^*)}{p(\boldsymbol{x}, \theta)}\frac{T(\theta|\theta^*)}{T(\theta^*|\theta)} \leq \frac{1}{r_0}. \tag{13}$$

3.2. A population control scheme

Given the current population $(\boldsymbol{\theta}_t, \boldsymbol{w}_t)$, to adjust the dynamic weights and the population size to suitable ranges, we propose the following adaptive pruned-enriched population control scheme (APEPCS), which is similar to the one used in [14].

Let $(\theta_{t,i}, w_{t,i})$ be the ith individual state of the population, let n_t and n_t' denote the current and new population sizes, let $W_{low,t}$ and $W_{up,t}$ denote the lower and upper weight control bounds, let $n_{\min}$ and $n_{\max}$ denote the minimum and maximum population size allowed by the user, and let n_{low} and n_{up} denote the lower and upper reference bound of the population size.

1. (Initialization) Initialize the parameters $W_{low,t}$ and $W_{up,t}$ by

$$W_{low,t} = \sum_{i=1}^{n_t} w_{t,i}/n_{up}, \quad W_{up,t} = \sum_{i=1}^{n_t} w_{t,i}/n_{low}.$$

   Set $n_t' = 0$ and $\lambda > 1$. Do steps 2–4 for $i = 1, 2, \cdots, n_t$.
2. (Pruned) If $w_{t,i} < W_{low,t}$, prune the state with probability $q = 1 - w_{t,i}/W_{low,t}$. If it is pruned, drop $(\theta_{t,i}, w_{t,i})$ from $(\boldsymbol{\theta}_t, \boldsymbol{w}_t)$; otherwise, update $(\theta_{t,i}, w_{t,i})$ as $(\theta_{t,i}, W_{low,t})$ and set $n_t' = n_t' + 1$.
3. (Enriched) If $w_{t,i} > W_{up,t}$, set $d = [w_{t,i}/W_{up,t} + 1]$, $w_{t,i}' = w_{t,i}/d$, replace $(\theta_{t,i}, w_{t,i})$ by d identical states $(\theta_{t,i}, w_{t,i}')$, and set $n_t' = n_t' + d$, where $[z]$ denotes the integer part of z.
4. (Unchanged) If $W_{low,t} \leq w_{t,i} \leq W_{up,t}$, keep $(\theta_{t,i}, w_{t,i})$ unchanged, and set $n_t' = n_t' + 1$.
5. (Checking) If $n_t' > n_{\max}$, set $W_{low,t} \leftarrow \lambda W_{low,t}$, $W_{up,t} \leftarrow \lambda W_{up,t}$ and $n_t' = 0$, do step 2–4 again for $i = 1, 2, \cdots, n_t$. If $n_t' < n_{\min}$, set $W_{low,t} \leftarrow W_{low,t}/\lambda$, $W_{up,t} \leftarrow W_{up,t}/\lambda$ and $n_t' = 0$, do step 2–4 again for $i = 1, 2, \cdots, n_t$. Otherwise, stop.

In this scheme, λ is required to set to a number greater than 1. In all examples of this article, we set $\lambda = 2$. In addition, n_{low}, n_{up}, $n_{\min}$ and $n_{\max}$ are required to satisfy the constraint $n_{\min} < n_{low} < n_{up} < n_{\max}$. With the APEPCS, the population size is strictly controlled to the range $[n_{min}, n_{max}]$, and the weights are adjusted to the range $[W_{low,t}, W_{up,t}]$. Therefore, the APEPCS avoids the possible overflow or extinction of a population in simulations.

The APEPCS is closely related to the resampling technique that is often used in the context of importance sampling [20]. It can also be viewed as a generalization of the pruned-enriched Rosenbluth method [21] and the rejection controlled sequential importance sampler [22]. Refer to [14] for more discussions on the relationship between APEPCS and the latter two methods.

The following theorem shows that the APEPCS is IDIW, whose proof can be found in [14]. To make the paper self-contained, we still provide a proof below.

Theorem 3.2 *The APEPCS is IDIW; that is, if the joint distribution $g_t(\theta, w)$ for $(\boldsymbol{\theta}_t, \boldsymbol{w}_t)$ is correctly weighted with respect to $\pi(\theta|\boldsymbol{x})$, then after one run of the scheme, the new joint distribution $g_{t+1}(\theta', w')$ for $(\boldsymbol{\theta}_{t+1}, \boldsymbol{w}_{t+1})$ is also correctly with respect to $\pi(\theta|\boldsymbol{x})$.*

PROOF: In the population control step, only the weights w's are possibly modified, whereas the θ's are not changed. Thus,

$$\int_0^\infty w' g_{t+1}(\theta', w') dw' = \int_0^{W_{low,t}} 0 g_t(\theta', w)\left(1 - \frac{w}{W_{low,t}}\right) dw + \int_0^{W_{low,t}} W_{low,t} g_t(\theta', w) \frac{w}{W_{low,t}} dw$$
$$+ \int_{W_{low,t}}^{W_{up,t}} w g_t(\theta', w) dw + \int_{W_{up,t}}^{\infty} \left(\sum_{i=1}^{d} \frac{w}{d}\right) g_t(\theta', w) dw$$
$$= \int_0^\infty w g_t(\theta', w) dw = c_{t,\theta'} \psi(\theta').$$

Hence, $g_{t+1}(x', w')$ is still correctly weighted with respect to $\psi(\cdot)$ by setting $c_{t+1,\theta'} = c_{t,\theta'}$. □

3.3. A Monte Carlo dynamically weighted importance sampler

Since both the MCDW move and the population control scheme are IDIW, they can be used together in any fraction, while leaving the IDIW property unchanged. Based on this observation, we compose the following sampler, which alters between the MCDW step and the population control step. Let W_c denote a dynamic weighting move switching parameter, which switches the value of β_t between 0 and 1 depending on the value of $W_{up,t}$.

- (Move type setting) If $W_{up,t} \le W_c$, then set $\beta_t = 1$. Otherwise, set $\beta_t = 0$.
- (MCDW) Apply the Monte Carlo dynamic weighting move to the population $(\boldsymbol{\theta}_t, \boldsymbol{w}_t)$. The new population is denoted by $(\boldsymbol{\theta}'_{t+1}, \boldsymbol{w}'_{t+1})$.
- (Population Control) Apply APEPCS to $(\boldsymbol{\theta}'_{t+1}, \boldsymbol{w}'_{t+1})$. The new population is denoted by $(\boldsymbol{\theta}_{t+1}, \boldsymbol{w}_{t+1})$.

Let $(\boldsymbol{\theta}_1, \boldsymbol{w}_1), \ldots, (\boldsymbol{\theta}_N, \boldsymbol{w}_N)$ denote a series of populations generated by MCDWIS. Then, according to (6), the quantity $\mu = E_\pi \rho(\theta)$ can be estimated by

$$\widehat{\mu} = \frac{\sum_{t=N_0+1}^{N} \sum_{i=1}^{n_t} w_{t,i} \rho(\theta_{t,i})}{\sum_{t=N_0}^{N} \sum_{i=1}^{n_t} w_{t,i}}, \tag{14}$$

where N_0 denotes the number of burn-in iterations. Following the generalized theory of importance sampling, $\widehat{\mu}$ is consistent and asymptotically normally distributed.

3.4. Weight behavior analysis

To analyze the weight behavior of the MCDWIS, we first introduce the following lemma.

Lemma 3.1 *Let $f(\boldsymbol{x}|\theta) = p(\boldsymbol{x},\theta)/Z(\theta)$ denote the likelihood function of $\boldsymbol{x}$, let $\pi(\theta)$ denote the prior distribution of θ, and let $T(\cdot|\cdot)$ denote a proposal distribution of θ. Define $p(\theta,\theta'|\boldsymbol{x}) = p(\boldsymbol{x},\theta)\pi(\theta)T(\theta'|\theta)$, and $r(\theta,\theta') = \widehat{R}(\theta,\theta')p(\theta',\theta|\boldsymbol{x})/p(\theta,\theta'|\boldsymbol{x})$ to be a Monte Carlo MH ratio, where $\widehat{R}(\theta,\theta')$ denotes an unbiased estimator of $Z(\theta)/Z(\theta')$. Then*

$$e_0 = E\log r(\theta,\theta') \leq 0,$$

where the expectation is taken with respect to the joint density $\varphi(\widehat{R}) \times p(\theta,\theta'|\boldsymbol{x})/Z(\theta)$.

Proof: By Jensen's inequality,

$$e_0 = E\log\left[\widehat{R}(\theta,\theta')\frac{p(\theta',\theta|\boldsymbol{x})}{p(\theta,\theta'|\boldsymbol{x})}\right] \leq \log E\left[\widehat{R}(\theta,\theta')\frac{p(\theta',\theta|\boldsymbol{x})}{p(\theta,\theta'|\boldsymbol{x})}\right] = 0,$$

where the equality holds when $p(\theta',\theta|\boldsymbol{x}) = p(\theta,\theta'|\boldsymbol{x})$, and $\varphi(\cdot)$ is a Dirac measure with $\varphi(\widehat{R}=R) = 1$ and 0 otherwise. □

For simplicity, let (θ_t, w_t) denote an individual state of the population $(\boldsymbol{\theta}_t, \boldsymbol{w}_t)$. When $\beta_{t-1} = 0$, the weights in the MCDW moves evolve as

$$\log w_t = \log w_{t-1} + \log r(\theta_{t-1},\theta_t),$$

which results in

$$\log w_t = \log w_0 + \sum_{s=1}^{t}\log r(\theta_{s-1},\theta_s). \tag{15}$$

Following from Lemma 1 and the ergodicity theorem (under stationarity),

$$\frac{1}{t}\sum_{s=1}^{t}\log r(\theta_{s-1},\theta_s) \to e_0 < 0, \quad a.s. \tag{16}$$

as $t \to \infty$. Hence, w_t will go to 0 almost surely as t becomes large.

When $\beta_{t-1} = 1$, the expectation of w_t, conditioned on θ_{t-1}, θ_t and w_{t-1}, can be calculated as

$$\begin{aligned} E[w_t|\theta_{t-1},\theta_t,w_{t-1}] &= (r_d+1)\frac{r_d}{r_d+1} + w_{t-1}(r_d+1)\frac{1}{r_d+1} = r_d + w_{t-1} \\ &= w_{t-1}[1 + r(\theta_{t-1},\theta_t)]. \end{aligned} \tag{17}$$

Since $r(\theta_{t-1},\theta_t) \geq 0$, the weight process is driven by an inflation drift. Hence, w_t will go to infinity almost surely as t becomes large.

To prevent the weight process from going to 0 and ∞, the MCDWIS alters the use of $\beta_t = 0$ and $\beta_t = 1$. When $W_{up,t} > W_c$, β_t is set to 0, so the weight process of MCDWIS can be bounded above by

$$\log w_{t+s} = \log w_t + \sum_{j=1}^{s}\log r(\theta_{t+j-1},\theta_{t+j}) - \sum_{j=1}^{s}\log(d_{t+j}),$$

assuming that $\beta_t = \beta_{t+1} = \cdots = \beta_{t+s} = 0$, where d_{t+j} is a positive integer as defined in the APEPCS. When $W_{up,t+s} < W_c$, β_{t+s} is set to 1, so the weight can be boosted to a large value. In summary, the weight process of MCDWIS can be characterized by the following process:

$$Z_t = \begin{cases} Z_{t-1} + \log r(\theta_{t-1}, \theta_t) - \log(d_t), & \text{if } Z_{t-1} > 0 , \\ 0, & \text{if } Z_{t-1} < 0 , \end{cases} \tag{18}$$

where there exists a constant C such that $|Z_t - Z_{t-1}| \leq C$ almost surely. The existence of C follows directly from condition (13). Let $T_0 = 0$, $T_i = \min\{t : t > T_{i=1}, Z_t = 0\}$, and $L_i = T_i - T_{i-1}$ for $i \geq 1$. From (16) and the fact that $d_t \geq 1$, it is easy to see that L_i is almost surely finite; that is, there exists an integer M such that $P(L_i < M) = 1$. This implies that for any fixed $\eta > 0$,

$$E \exp(\eta Z_t) \leq \exp(\eta MC), \quad a.s.$$

This leads to the following theorem:

Theorem 3.3 *Under the assumptions (11) and (12), the MCDWIS almost surely has finite moments of any order.*

Finally, we would like to mention that the ratio $\kappa = W_{up,t}/W_{low,t}$ determines the moving ability of the MCDWIS. In [14], κ is called the freedom parameter of DWIS. In the APEPCS, κ is also equal to the ratio n_{up}/n_{low}. As suggested by Liang [14], when choosing the value of κ, the efficiency of the resulting sampler should be taken into account. A large κ will ease the difficulty of escaping from local energy minima, but the variability of the weights will also increase accordingly. The efficiency of the MCDWIS may be reduced by an excessively large κ. In this paper, we set $\kappa = 2.5$ in all simulations by setting $n_{low} = 200$ and $n_{up} = 500$.

4. Bayesian analysis for spatial autologistic models

The autologistic model [1] has been widely used for spatial data analysis [23–25]. Let $\boldsymbol{x} = \{x_i : i \in D\}$ denote the observed binary data, where x_i is called a spin and D is the set of indices of the spins. Let $|D|$ denote the total number of spins in D, and let $n(i)$ denote the set of neighbors of spin i. The likelihood function of the model is

$$f(\boldsymbol{x}|\theta) = \frac{1}{Z(\theta)} \exp\left\{ \theta_a \sum_{i \in D} x_i + \frac{\theta_b}{2} \sum_{i \in D} x_i \Big(\sum_{j \in n(i)} x_j \Big) \right\}, \quad (\theta_a, \theta_b) \in \Theta, \tag{19}$$

where $\theta = (\theta_a, \theta_b)$, the parameter θ_a determines the overall proportion of $x_i = +1$, the parameter θ_b determines the intensity of interaction between x_i and its neighbors, and $Z(\theta)$ is the intractable normalizing constant defined by

$$Z(\theta) = \sum_{\text{for all possible } \boldsymbol{x}} \exp\left\{ \theta_a \sum_{j \in D} x_j + \frac{\theta_b}{2} \sum_{i \in D} x_i \Big(\sum_{j \in n(i)} x_j \Big) \right\}.$$

An exact evaluation of $Z(\theta)$ is impossible even for a moderate system.

To conduct a Bayesian analysis for the model, we assume a uniform prior on

$$(\theta_a, \theta_b) \in \Theta = [-1, 1] \times [0, 1]$$

for all examples studied in this section. Then the MCDWIS can be applied to simulate from the posterior distribution $\pi(\theta|\boldsymbol{x})$. Since, for this model, $\mathcal{X}$ is finite and Θ is compact, the condition (11) is satisfied. As long as the proposal distribution satisfies the condition (12), the weights will have finite moments of order. In all simulations of this model, we used a Gaussian random walk proposal. Thus, $T(\theta|\theta^*)/T(\theta^*|\theta) = 1$, and condition (12) is satisfied.

4.1. U.S. cancer mortality data

United States cancer mortality maps have been compiled for investigating possible association of cancer with unusual demographic, environmental, industrial characteristics, or employment patterns [26]. Figure 2(a) shows the mortality map of liver and gallbladder (including bile ducts) cancers for white males during the decade 1950-1959, which indicates some apparent geographic clustering. Refer to [25] for more descriptions of the data. As in [25], we modeled the data by a spatial autologistic model. The total number of spins is $|D| = 2293$. A free boundary condition is assumed for the model, under which the boundary points have less neighboring points than the interior points. The assumption is natural to this example, as the lattice has an irregular shape.

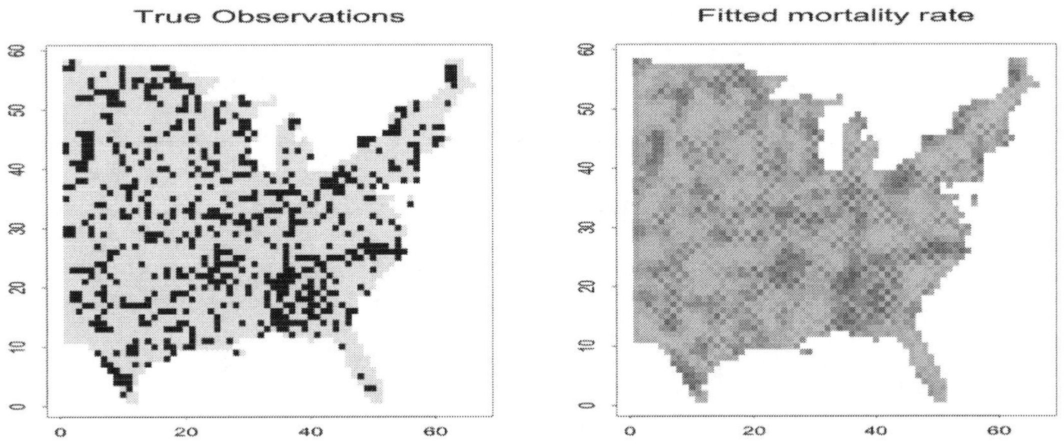

Figure 2. US cancer mortality data. (a) The mortality map of liver and gallbladder cancers (including bile ducts) for white males during the decade 1950-1959. Black squares denote counties of high cancer mortality rate, and white squares denote counties of low cancer mortality rate. (b) Fitted cancer mortality rates by the autologistic model with the parameters being replaced by its approximate Bayesian estimates. The cancer mortality rate of each county is represented by the gray level of the corresponding square.

The MCDWIS was applied to this example with the setting as described below. In each run, the MCDWIS was initialized with $n_0 = 250$ random samples of θ generated in a short run of the double MH algorithm [10], with each sample being assigned an equal weight of 1. Refer to Appendix for the details of the double MH algorithm. To generate the initial samples, the double MH algorithm was run for 3500 iterations, where the first 1000 iterations were discarded for the burn-in process, and then 250 samples were collected equally-spacedly from the remaining 2500 iterations and were used as the initial samples. In each iteration, the auxiliary variable is generated with a single MH update. As aforementioned, the double MH algorithm is very fast, but only can sample from the posterior distribution $\pi(\theta|\boldsymbol{x})$ approximately, even when the run becomes very long.

After initialization, the MCDWIS iterates between the MCDW and population control steps. In the MCDW step, the normalizing constant ratio $R_t(\theta, \theta^*)$ was estimated using 50 auxiliary samples, which were generated from 50 cycles of Gibbs updates starting with $\boldsymbol{x}$; and $T(\cdot|\cdot)$ was a Gaussian random walk proposal $N_2(\theta, s^2 I_2)$, where s is the step size, and I_2 is the 2×2 identity matrix. In this article, we set $s = 0.03$ for all examples. In the population control step, we set $\log(W_c) = 5$, $n_{low} = 200$, $n_{up} = 500$, $n_{\min} = 100$ and $n_{\max} = 1000$. The MCDWIS was run 10 times independently, and each run consisted of 100 iterations. The CPU time cost by each run was about 5.8m on a 3.0GHz personal computer (all computations reported in this paper were done in the same computer). Figure 3 shows the time plots of the population size, β_t and $W_{up,t}$ produced in a MCDWIS run. It indicates that after a certain number of burn-in iterations, the population size and the magnitude of the weights can evolve stably with iterations. In each run, the first 20 iterations were discarded for the burn-in process, and the samples generated from the remaining iterations were used for inference. Averaging over the estimates obtained from the ten runs, we got the following estimate: $(\widehat{\theta}_a, \widehat{\theta}_b) = (-0.3016, 0.1229)$ with the standard error $(1.9 \times 10^{-3}, 8.2 \times 10^{-4})$.

For comparison, the exchange algorithm [12] was also applied to this example. Refer to the Appendix for the details of the algorithm. As aforementioned, the exchange algorithm is an auxiliary variable MCMC algorithm, which requires a perfect sampler for generating auxiliary variables, but can sample correctly from the posterior distribution when the number of iterations is large. Hence, following the standard theory of Markov chain Monte Carlo [27], the estimates produced by the exchange algorithm in long runs will converge to their true values, and can be used as a test standard for assessing whether the MCDWIS works for this example. As in [12], we adopted the summary state algorithm [28] as the perfect sampler for the exchange algorithm. The exchange algorithm was also run 10 times. Each run consisted of 55000 iterations, where first 5000 iterations were discarded for the burn-in process, and the samples generated from the remaining iterations were used for inference. The CPU time cost by each run was about 6.5m. The overall acceptance rate of the perfect auxiliary variables was 0.2. Averaging over the estimates obtained from the ten runs, we got the estimate $(\widehat{\theta}_a, \widehat{\theta}_b) = (-0.3018, 0.1227)$ with the standard error $(2.7 \times 10^{-4}, 1.4 \times 10^{-4})$.

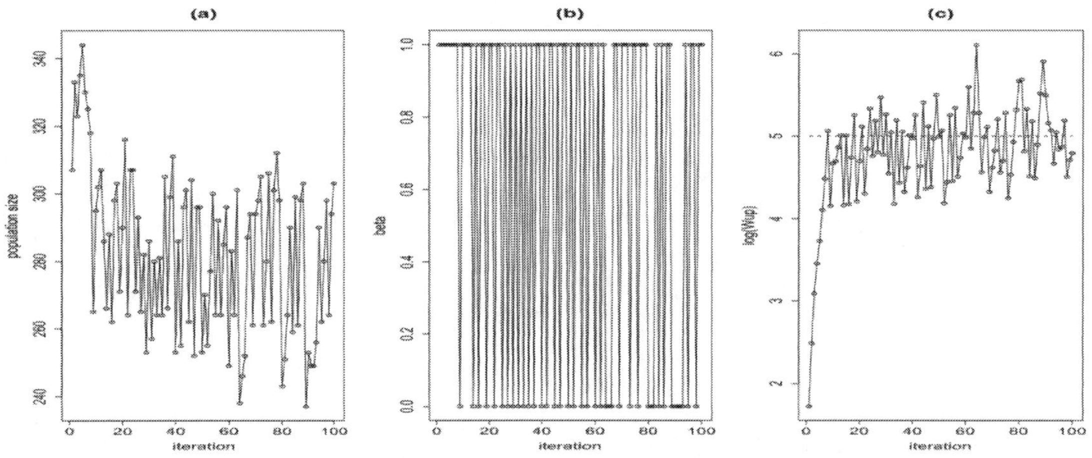

Figure 3. Simulation results of the MCDWIS for the U.S. Cancer Mortality example: (a) time plot of population size; (b) time plot of β_t; and (c) time plot of $\log(W_{up,t})$. The dotted line in plot (c) shows the value of $\log(W_c)$.

It is easy to see that the MCDWIS and the exchange algorithm produced almost identical estimates for this example. These estimates are also almost identical to the estimate $(-0.3008, 0.1231)$ obtained using contour Monte Carlo [9], and the estimate $(-0.2999, 0.1234)$ obtained using stochastic approximation Monte Carlo [8]. We note that both the contour Monte Carlo and stochastic approximation Monte Carlo algorithms try to first approximate the unknown normalizing constant function, and then estimate the parameters based on the approximated normalizing constant function. As reported by the authors, both the algorithms take hours of CPU time to approximate the normalizing constant function. This data has also been analyzed in [25] using the Monte Carlo maximum likelihood algorithm [7], resulting in the estimate $(-0.304, 0.117)$ which is slightly far from the estimate $(-0.3018, 0.1227)$ resultant from the exchange algorithm.

In summary, for this example the MCDWIS produced almost identical estimates of θ with the exchange algorithm, although whose estimates have slightly larger variation. This is reasonable, as the MCDWIS is an importance sampling algorithm. It is known that an importance sampling algorithm is usually less efficient than a MCMC algorithm for the problems for which the latter works. The advantage of the MCDWIS can be seen in the next subsection, where for some cases the exchange algorithm does not work while the MCDWIS still works well.

4.2. Simulation studies

To assess general accuracy of the estimates produced by the MCDWIS, we simulated 50 independent samples for the U.S. cancer mortality data under each setting of (θ_a, θ_b) given in Table 1. Since the lattice is irregular, the free boundary condition was again assumed in the simulations. We then re-estimated the parameters using the MCDWIS. The sampler was run as for the previous example. The computational results were summarized in Table 1.

For comparison, the exchange algorithm [12] was also applied to this example. It was run as for the previous example, except that each run was shortened to 10500 iterations. In each run, the first 500 iterations were discarded for the burn-in process, and the samples generated in the remaining iterations were used for inference. The results were summarized in Table 1. For a thorough comparison, we also included in Table 1 the maximum pseudo-likelihood estimators (MPLE) of θ, which were obtained in [10] for the same datasets. The MPLE were obtained by maximizing the pseudo-likelihood function

$$\tilde{L}(\theta|\boldsymbol{x}) = \prod_{i\in D}\left[\frac{e^{\theta_a x_i+\theta_b\sum_{j\in n(i)} x_i x_j}}{e^{\theta_a+\theta_b\sum_{j\in n(i)} x_j}+e^{-\theta_a-\theta_b\sum_{j\in n(i)}\sum x_j}}\right]$$

using the downhill simplex method [29]. The advantage of this optimization method is that it does not require gradient information of the objective function, and can thus be easily applied to the constraint optimization problems.

Table 1 indicates that the MCDWIS can produce almost the same accurate results as does the exchange algorithm, and more accurate results than does the MPLE especially when both θ_a and θ_b are large. It is remarkable that the CPU time cost by the MCDWIS is independent of the values of θ. Whilst as θ_b increases, the CPU time cost by the exchange algorithm increases exponentially. Childs et al. [28] studied the behavior of the perfect sampler for the Ising model, a simplified autologistic model with θ_a being restricted to 0. For the Ising model, they fitted an exponential law for the convergence time, and reported that the perfect sampler may diverge at a value of θ_b lower than the critical value (≈ 0.44). Their finding is consistent with our results. It takes an extremely long CPU time for the perfect sampler to generate a sample under the settings $(0, 0.4)$ and $(0.5, 0.5)$. We note that due to the effect of θ_a, it usually takes a longer CPU time for the perfect sampler to generate a sampler under the setting $(0, \theta_b)$ than under the setting (θ_a, θ_b); and that when both θ_a and θ_b are large, the accuracy of the estimates tend to be reduced by their correlation.

5. Conclusion

In this paper, we have proposed the MCDWIS as a general MCMC algorithm for simulating from distributions with intractable normalizing constants. The algorithm is illustrated with spatial autologistic models. Unlike other auxiliary variable MCMC algorithms, the new

Table 1. Computational results for the simulated U.S. cancer mortality data. The numbers in the parentheses denote the standard error of the estimates. Notes: —, not available; [a] the CPU time (in minutes) cost by a single run of the algorithm'; [b] the data were simulated using the perfect sampler; [c] the data were simulated using the Gibbs sampler, starting with a random configuration and then iterating for 100000 Gibbs cycles.

	MCDWIS			Exchange algorithm			MPLE	
(θ_a, θ_b)	$\widehat{\theta}_a$	$\widehat{\theta}_b$	CPU[a]	$\widehat{\theta}_a$	$\widehat{\theta}_b$	CPU[a]	$\widehat{\theta}_a$	$\widehat{\theta}_b$
(0,0.1)[b]	−0.0041	0.1004	5.8	−0.0038	0.1004	1.2	−0.0035	0.1016
	(0.0025)	(0.0019)		(0.0024)	(0.0018)		(0.0024)	(0.0019)
(0,0.2)[b]	−0.0022	0.2020	5.8	−0.0023	0.2007	2.8	−0.0024	0.2025
	(0.0021)	(0.0019)		(0.0020)	(0.0019)		(0.0022)	(0.0022)
(0,0.3)[b]	−0.0013	0.2980	5.8	−0.0011	0.2975	7.9	−0.0019	0.2981
	(0.0013)	(0.0017)		(0.0014)	(0.0017)		(0.0016)	(0.0022)
(0,0.4)[b]	−0.0004	0.3963	5.8	0.0001	0.3979	145.8	0.0020	0.4013
	(0.0012)	(0.0020)		(0.0005)	(0.0012)		(0.0012)	(0.0020)
(0.1,0.1)[b]	0.1029	0.0986	5.8	0.1032	0.0985	1.1	0.1023	0.0999
	(0.0025)	(0.0023)		(0.0025)	(0.0022)		(0.0025)	(0.0023)
(0.3,0.3)[b]	0.2947	0.3035	5.8	0.3008	0.3008	3.5	0.2904	0.3041
	(0.0105)	(0.0045)		(0.0097)	(0.0043)		(0.0102)	(0.0046)
(0.5,0.5)[c]	0.4948	0.5171	5.8	—	—	—	0.5610	0.4847
	(0.0347)	(0.0122)		—	—		(0.0393)	(0.0123)

algorithm removes the need of perfect sampling, and thus can be applied to a wide range of problems for which perfect sampling is not available or very expensive. Examples include image segmentation [2], social network modeling [30], genetic analysis [4], among others.

The novelty of the MCDWIS is that it allows for the use of Monte Carlo estimates in MCMC simulations, while still leaving the target distribution invariant under the criterion of dynamically weighted importance sampling. In addition to the applications mentioned above, this novelty also leads to potential applications of the algorithm to Bayesian inference for missing data problems, where it often involves simulating from a posterior distribution with intractable integrals.

Acknowledgments
Liang's research was partially supported by grants from the National Science Foundation (DMS-0607755) and the award (KUS-C1-016-04) made by King Abdullah University of Science and Technology (KAUST).

Journal of Physics: Conference Series **197** (2009) 012004 doi:10.1088/1742-6596/197/1/012004

Appendix

A. The Exchange Algorithm [12] Let θ_t denote the sample generated from $\pi(\theta|\boldsymbol{x})$ at iteration t. One iteration of the algorithm consists of the following steps.

(a) Propose $\theta' \sim q(\theta'|\theta_t, \boldsymbol{x})$, where $q(\cdot|\cdot)$ denotes the proposal distribution.

(b) Generate an auxiliary variable $\boldsymbol{y} \sim f(\boldsymbol{y}|\theta')$ using an perfect sampler.

(c) Accept θ' with probability $\min\{1, r(\theta_t, \theta', \boldsymbol{y}|\boldsymbol{x})\}$, where

$$r(\theta_t, \theta', \boldsymbol{y}|\boldsymbol{x}) = \frac{\pi(\theta')f(\boldsymbol{x}|\theta')f(\boldsymbol{y}|\theta_t)q(\theta_t|\theta', \boldsymbol{x})}{\pi(\theta_t)f(\boldsymbol{x}|\theta_t)f(\boldsymbol{y}|\theta')q(\theta'|\theta_t, \boldsymbol{x})}.$$

If it is accepted, set $\theta_{t+1} = \theta'$. Otherwise, set $\theta_{t+1} = \theta_t$.

Since swapping changes are involved in the simulations, it is called the exchanged algorithm.

B. The Double MH Algorithm [10] Let θ_t denote the sample generated at iteration t of the algorithm.

(a) Simulate a new sample θ' from $\pi(\theta)$ using the MH algorithm starting with θ_t.

(b) Generate an auxiliary variable $\boldsymbol{y}$ from $f(\cdot|\theta')$ through m MH updates starting with the observation $\boldsymbol{x}$. The probability of transition from $\boldsymbol{x}$ to $\boldsymbol{y}$ is

$$P_{\theta'}^{(m)}(\boldsymbol{y}|\boldsymbol{x}) = K_{\theta'}(\boldsymbol{x} \to \boldsymbol{x}_1) \cdots K_{\theta'}(\boldsymbol{x}_{m-1} \to \boldsymbol{y}),$$

where $K_{\theta'}(\cdot \to \cdot)$ denotes a MH transition kernel. Accept y with probability $\min\{1, r(\theta_t, \theta', \boldsymbol{y}|\boldsymbol{x})\}$, where

$$r(\theta_t, \theta', \boldsymbol{y}|\boldsymbol{x}) = \frac{f(\boldsymbol{y}|\theta_t)P_{\theta'}^{(m)}(\boldsymbol{x}|\boldsymbol{y})}{f(\boldsymbol{x}|\theta_t)P_{\theta'}^{(m)}(\boldsymbol{y}|\boldsymbol{x})} = \frac{f(\boldsymbol{y}|\theta_t)f(\boldsymbol{x}|\theta')}{f(\boldsymbol{x}|\theta_t)f(\boldsymbol{y}|\theta')}.$$

(c) Set $\theta_{t+1} = \theta'$ if the auxiliary variable is accepted in step (b), and set $\theta_{t+1} = \theta_t$ otherwise.

Since two types of MH updates are performed in step (b), one for generating the auxiliary variable $\boldsymbol{y}$ and one for acceptance of θ', the algorithm is called the double MH algorithm in [10].

References

[1] Besag J E 1974 *J. R. Stat. Soc.* B **36** 192
[2] Hurn M, Husby O and Rue H 2003 *Lecture Notes in Statistics* **173** 87
[3] Green P J and Richardson S 2002 *J. Am. Stat. Assoc.* **97** 1055
[4] Francois O, Ancelet S and Guillot G 2006 *Genetics* **174** 805
[5] Metropolis N, Rosenbluth A W, Rosenbluth M N, Teller A H and Teller E 1953 *J. Chem. Phys.* **21** 1087
[6] Hastings W K 1970 *Biometrika* **57** 97
[7] Geyer C J and Thompson E A 1992 *J. R. Stat. Soc.* B **54** 657
[8] Liang F, Liu C and Carroll R J 2007 *J. Am. Stat. Assoc.* **102** 305
[9] Liang F 2007 *J. Comput. Graph. Stat.* **16** 608
[10] Liang F 2009 *J. Stat. Comput. Sim.* In press
[11] Møller J, Pettitt A N, Reeves R and Berthelsen K K 2006 *Biometrika* **93** 451
[12] Murray I, Ghahramani Z and MacKay D J C 2006 *Proc. 22nd Annual Conf. on Uncertainty in Artificial Intelligence* (AUAI Press) p 359
[13] Propp G and Wilson D B 1996 *Random Struct. Algor.* **9** 223
[14] Liang F 2002 *J. Am. Stat. Assoc.* **97** 807
[15] Wong W H and Liang F 1997 *Proc. Natl. Acad. Sci., USA* **94** 14220
[16] Liu J S, Liang F and Wong W H 2001 *J. Am. Stat. Assoc.* **96** 561
[17] Geweke J 1989 *Econometrica* **57** 1317
[18] Ripley B D 1987 *Stochastic Simulation* (New York: John Wiley)

[19] Chen M H, Shao Q M and Ibrahim J G 2000 *Monte Carlo Methods in Bayesian Computation* (New York: Springer)
[20] Liu J S and Chen R 1998 *J. Am. Stat. Assoc.* **93** 1032
[21] Grassberger P 1997 *Phys. Rev.* E **56** 3682
[22] Liu J S, Chen R and Wong W H 1998 *J. Am. Stat. Assoc.* **93** 1022
[23] Preisler H K 1993 *Appl. Statist.* **42** 501
[24] Wu H and Huffer F W 1997 *Ecological Statistics* **4** 49
[25] Sherman M, Apanasovich T V and Carroll R J 2006 *J. Stat. Comput. Sim.* **76** 167
[26] Riggan W B, Creason J P, Nelson W C, Manton K G, Woodbury M A, Stallard E, Pellom A C and Beaubier J 1987 *U.S. Cancer Mortality Rates and Trends, 1950-1979. Vol. IV: Maps* (Washington D.C.: U.S. Government Printing Office)
[27] Tierney L 1994 *Ann. Stat.* **22** 1701
[28] Childs A M, Patterson R B and MacKay D J C 2001 *Phys. Rev.* E **63** 036113
[29] Press W H, Teukolsky S A, Vetterling W T and Flannery B P 1992 *Numerical Recipes in C (2nd ed)* (Cambridge University Press)
[30] Wasserman S and Faust K 1994 *Social Network Analysis: Methods and Applications* (Cambridge: Cambridge University Press)

Journal of Physics: Conference Series **197** (2009) 012005 doi:10.1088/1742-6596/197/1/012005

Inference of a random potential from random walk realizations: Formalism and application to the one-dimensional Sinai model with a drift

S Cocco [1,3] **and R Monasson** [2,3]

[1] Laboratoire de Physique Statistique de l'ENS, CNRS, UMPC, 24 rue Lhomond, 75005 Paris, France

[2] Laboratoire de Physique Théorique de l'ENS, CNRS, UPMC, 24 rue Lhomond, 75005 Paris, France

[3] The Simons Center for Systems Biology, Institute for Advanced Study, Einstein Drive, Princeton NJ 08540, USA

E-mail: `cocco@lps.ens.fr, monasson@lpt.ens.fr`

Abstract. We consider the Sinai model, in which a random walker moves in a random quenched potential $\mathbf{V}$, and ask the following questions: 1. how can the quenched potential $\mathbf{V}$ be inferred from the observations of one or more realizations of the random motion? 2. how many observations (walks) are required to make a reliable inference, that is, to be able to distinguish between two similar but distinct potentials, $\mathbf{V}_1$ and $\mathbf{V}_2$? We show how question 1 can be easily solved within the Bayesian framework. In addition, we show that the answer to question 2 is, in general, intimately connected to the calculation of the survival probability of a fictitious walker in a potential $\mathbf{W}$ defined from $\mathbf{V}_1$ and $\mathbf{V}_2$, with partial absorption at sites where $\mathbf{V}_1$ and $\mathbf{V}_2$ do not coincide. For the one-dimensional Sinai model, this survival probability can be analytically calculated, in excellent agreement with numerical simulations.

1. Introduction

Random walks in random media have been intensively studied in the past decades in statistical physics, as well as in the theory of probability [1, 2]. Most efforts have been devoted to the characterization of the statistical properties of the walks, such as the velocity, the correlation functions, ... from the ones of the environment, that is, the energy potential. Fewer studies (see however [3]) have tackled the inverse problem: given one (or more) realizations of the random walks, can we guess the values of the potential?

The issue of the inference of a potential from the observation of a random, noisy motion naturally arises in the analysis of experiments on biophysics or biology. A nice illustration is provided by single-molecule experiments, in which biological objects, such as macro-molecules (DNA,RNA, enzymes) can be individually tracked. These objects, of the micrometer or nanometer size, are submitted to the thermal agitation of the solvent, and to intra- or inter-molecular interactions. Examples of interactions are ligand-receptor interactions in experiments dealing with the rupture of non-covalent bonds [4], base-pairing interactions in polynucleotides unzipping, *i.e.* the mechanical separation of DNA/RNA strands [5, 6]. The resulting potentials can be considered as disordered since they depend on the details of the molecular constituents,

Journal of Physics: Conference Series **197** (2009) 012005 doi:10.1088/1742-6596/197/1/012005

and may vary a lot among ligand-receptor systems and DNA/RNA molecules. In addition, in order to modify the strength of the interactions between constituents, external forces can be exerted on the pico-Newton scale. The presence of such an external force tilts the potential, and affect the observed motion, which, in turn, should provide additional information on the interaction potential.

In a recent letter [8], motivated by the idea of infering the sequence of DNA molecule from the signal of the unzipping force in separation experiments [7], we have studied the case of a random walker moving in a quenched random potential $\mathbf{V}$, and asked the following two questions: **1.** how can the quenched potential $\mathbf{V}$ be inferred from the observations of one or more realizations of the random motion? **2.** how many observations (walks) are required to make a reliable inference, that is, to be able to distinguish between two similar but distinct potentials, $\mathbf{V}_1$ and $\mathbf{V}_2$? The present paper is intended as a pedagogical introduction to those questions and our results, as well as a numerical check of our main result, that is, the close and abtract relationship between question 2 and the survival probability of a fictitious random walker in a partially absorbing potential. Definitions are given in Section 2, questions 1 and 2 are studied in Sections 3 and 4 respectively, while the relationship mentioned above is exposed in Section 5.

2. Definitions of the random potentials, and random walks

The model we look at is the discrete space-continuous time Sinai model. A walker is moving on the set of non-negative integers $x = 0, 1, 2, 3, \ldots$, according to the set of values of the potential on those sites, $\mathbf{V} = \{V_x,\ x \geq 0\}$.

2.1. Dynamical rules for the random walk

Given the potential $\mathbf{V}$, the rate (probability per unit of time) for the walker to jump from x onto a neighbouring site $x' = x \pm 1$ reads

$$r_{\mathbf{V}}(x', x) = e^{\gamma(V_x - V_{x'})} \tag{1}$$

which ensures that detailed balance is satisfied at temperature $1/(2\gamma)$. We impose a reflecting boundary condition in $x = 0$, by setting $V_{-1} = +\infty$. The rate transition matrix $\hat{M}_{\mathbf{V}}$ is tridiagonal with non zero elements $M_{\mathbf{V}}(x \pm 1, x) = r_{\mathbf{V}}(x \pm 1, x)$, $M_{\mathbf{V}}(x, x) = -M_{\mathbf{V}}(x+1, x) - M_{\mathbf{V}}(x-1, x)$. An important property of $\hat{M}_{\mathbf{V}}$ is that the sum of entries along each column is equal to zero, which ensures the conservation of the probability for the walker to be somewhere on the postive half-line. A realization of the random walk is the set of positions of the walker at all times smaller than the duration T of the walk, $\mathbf{X} = \{x(t),\ 0 \leq t \leq T\}$. For definiteness, we assume that the walker is initially at the origin: $x(0) = 0$.

The above dynamics is defined in continuous time. It may be convenient in some cases to consider that time is discrete. Observations of the position are then made at times multiple of Δt. The number of observations is $T/\Delta t$. The transition matrix $\hat{Q}_{\mathbf{V},\Delta t}$ is simply obtained from the exponential of the rate transition matrix $M_{\mathbf{V}}$ through

$$\hat{Q}_{\mathbf{V},\Delta t} = \exp\left[\Delta t\ \hat{M}_{\mathbf{V}}\right] . \tag{2}$$

When Δt is very small, $\hat{Q}_{\mathbf{V},\Delta t} \simeq \mathrm{Id} + \Delta t\ \hat{M}_{\mathbf{V}}$, and is almost tridiagonal.

2.2. Random force model

The potential is itself a quenched random variable drawn as follows. On each bond $(x, x+1)$ with $x \geq 0$, we define a force f_x. The value of the potential at site x is defined through

$$V_x = -\sum_{y<x} f_y , \tag{3}$$

with $V_0 \equiv 0$. The forces are i.i.d. stochastic variables taking ± 1 values, and drawn from the law

$$P(f_x = f) = \frac{1+b}{2}\,\delta_{f,+1} + \frac{1-b}{2}\,\delta_{f,-1}\ , \tag{4}$$

where b is called tilt in the following. The tilt is the opposite of the average slope of the potential. Note that (3) defines a one-to-one mapping between the potential $\mathbf{V}$ and the set of forces $\{f_x\}$. Inferring the former is therefore equivalent to inferring the latter.

Hereafter, we focus on the $b > 0$ case, for which the random walk is transient. We show a sample of potential drawn with $b = 0.4$ in Figure 1A, as well as several realizations of random walks of duration $T = 5000$ in this potential, for $\gamma = 0.5$, in Figure 1B. The dynamics is affected by the presence of many deep local minima, in which the walked gets trapped for long times. The control parameter for the dynamics is $\alpha = \frac{T}{2}\log(\frac{1+b}{1-b})$[2]. When $\alpha > 1$, the velocity of the walk is finite, and self-averaging, *i.e.* it depends only on b and T, and not on the sample of the potential. When $\alpha < 1$, the velocity is vanishing. The average location of the walker increases as t^α only, due to the presence of deep traps with long sojourn times τ. As we will see later, while the identification of these two regimes is critical to the description of the random walk motion, it is less important as far as the inference problem is concerned.

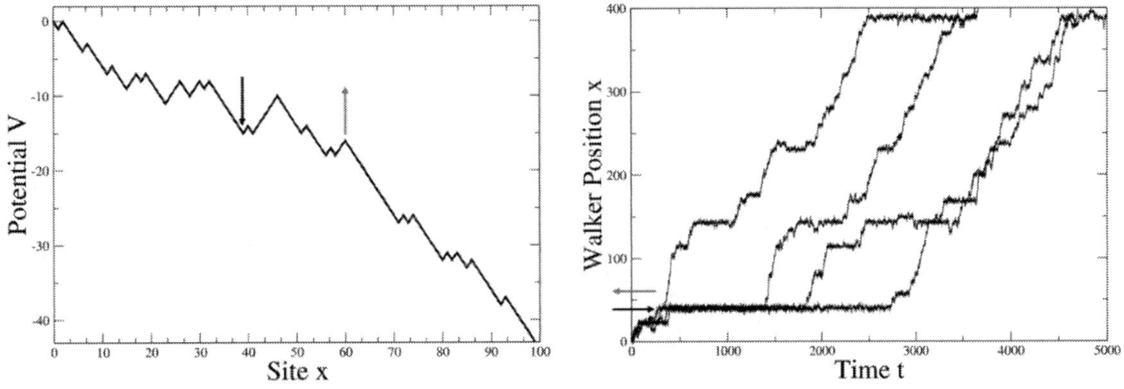

Figure 1. A. example of potential with $b = 0.4$. **B.** four realizations of random walks of duration $T = 5000$, and for $\gamma = 0.5$ in this potential. The up and right arrows point to a bond whose attached force is easy to infer, the down and left arrows point to a bond corresponding to a hardly predictable force, see Figure 2.

3. Question 1: How to infer the potential from one or more realizations in practice?

3.1. On realizations and signals

In this subsection, we focus on the discrete-time dynamics, and eventually send the time step $\Delta t \to 0$ at the end of the calculation. The probability of a time-trace (realization) $\mathbf{X}$ in the potential $\mathbf{V}$ reads

$$P[\mathbf{X}|\mathbf{V}] = \prod_{m=0}^{T/\Delta t - 1} \hat{Q}_{\mathbf{V},\Delta t}(x_{m+1}, x_m)\ . \tag{5}$$

where the transition matrix $\hat{Q}_{\mathbf{V},\Delta t}$ is defined in (2). As the product over times in the above expression can be reordered, a sufficient statistics for $\mathbf{X}$ is the set $\mathbf{S} = \{N_x, u_x^+, u_x^-\}$, hereafter

Journal of Physics: Conference Series **197** (2009) 012005 doi:10.1088/1742-6596/197/1/012005

called signal, where N_x is the number of time-steps m such that $x_{m+1} = x_m = x$, *i.e.* during which the walker remains at site x, and $u_x^{\pm}$ are the numbers of jumps from $x_m = x$ to $x_{m+1} = x \pm 1$. In term of the signal **S**, the probability of **X** can be rewritten as

$$P[\mathbf{X}|\mathbf{V}] = \prod_{x\geq 0} \hat{Q}_{\mathbf{V},\Delta t}(x,x)^{N_x}\ \hat{Q}_{\mathbf{V},\Delta t}(x-1,x)^{u_x^-}\ \hat{Q}_{\mathbf{V},\Delta t}(x+1,x)^{u_x^+}\ . \tag{6}$$

with the obvious definition $u_0^- = 0$. Note that the above expression could be extended to the case of a walker not constrained to nearest-neighbour jumps. The signal S would then be a matrix, whose elements $S(x',x)$ would count the number of jumps (in the realization **X**) from site x to site x' during a time-step.

Assume now that after time T (the duration of one walk), the walker is put again at the origin, and the random process is resumed. This way, more realizations of the random walk, $\mathbf{X}^1, \mathbf{X}^2, \ldots, \mathbf{X}^R$, can be obtained in the same potential. As the random walks are independent of each other, their probabilities factorize, and the corresponding signals simply add up.

3.2. Bayesian expression for the most likely forces

Let us now take the $\Delta t \to 0$ limit. We expect $u_x^{\pm}$ to remain finite, and N_x to scale as $t_x/\Delta t$, where t_x is the local time at site x. The signal attached to a time-trace is now the set $\mathbf{S} = \{t_x, u_x^+, u_x^-\}$, and, from (6), we have

$$P[\mathbf{X}|\mathbf{V}] = \prod_{x\geq 0} \exp\left(-t_x\left(e^{\gamma f_x} + e^{-\gamma f_{x-1}}\right) - \gamma f_{x-1}\, u_x^- + \gamma f_x\, u_x^+\right) = C\ \exp\left(\gamma \sum_{x\geq 0} f_x\, h_x\right) \tag{7}$$

where C is a signal-independent factor, and the local field acting on force f_x is

$$h_x = \frac{\sinh\gamma}{\gamma}\,(t_{x+1} - t_x) + u_x^+ - u_x^-\ . \tag{8}$$

Finding the set of forces f_x^* with maximal *a priori* probabilities is straightforward: for each x, f_x^* is given by the sign of h_x. In other words, the forces are uncoupled in (7). Hence, the random force model is the simplest model one can think of from the inference point of view. More complicated models, with short-range couplings between neighboring forces, can be solved using transfer matrix techniques [7].

From a strictly Bayesian point of view, the *a posteriori* probability of the potential should include the prior over the potentials. Here, the prior is given by the distribution of forces (4). It is easy to see that this extra and multiplicative contribution to the probability amounts to an additive contribution, equal to α, to the local fields h_x acting on the forces f_x.

We show in Figure 2 the probabilities of inferring the wrong force f_x on bond $(x, x+1)$, $P_{error}(x)$, as a function of x for the potential **V** of Figure 1A. Those probabilities were calculated from 10,000 samples of $R = 1$ or 4 realizations of the random walk of equal duration T. For each sample, we extract the local times, t_x, and numbers of jumps, $u_x^{\pm}$, and compare the forces $f_x^* = \text{sign}(h_x)$ (8) to their true values. We see that $P_{error}(x)$ widely fluctuates from bond to bond: it is small close to local minima of **V**, where the walker spends much time, and large in the vicinity of maxima or on the descending flanks of the potential.

As the random walk is transient, the walker visits each site only a finite number of times. The information about the values of the potential in the vicinity of a site can therefore only be increased by collecting more random walk realizations. Figure 2 shows that the probability of error strongly decreases when the number of realizations on which the inference is based grows from $R = 1$ to $R = 4$. A systematic numerical study shows that, for a given bond $(x, x+1)$, $P_{error}(x)$ asymptotically decays as $e^{-R/R_c(x)}$, where $R_c(x)$ depends on x and on the whole potential **V**. The following section is devoted to the calculation of $R_c(x)$, the number of realizations necessary for an accurate inference of the force on the $(x, x+1)$ bond .

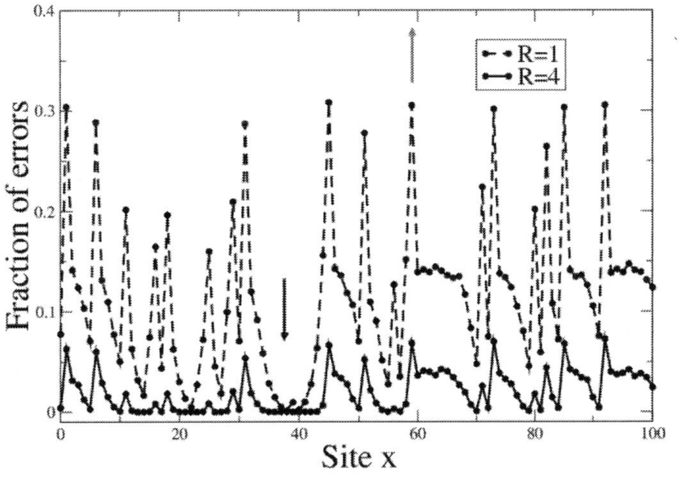

Figure 2. Fraction of errors, $P_{error}(x)$, for the forces over the first 100 bonds inferred from $R = 1$ and $R = 4$ realizations. The potential is the one shown in Figure 1. Parameters are $b = 0.4$, $\gamma = 0.5$, $T = 5000$. The fractions are calculated from the average over 10,000 samples of R random walks. The arrows locate the same bonds as in Figure 1, respectively, $(x, x+1) = (38, 39)$ and $(59, 60)$.

4. Question 2: how many realizations are necessary for a reliable inference?

4.1. A (much) simpler problem: 2-coin testing

To start with, let us consider the following and simpler problem. Suppose one has two biased coins $i = 1, 2$; a draw of coin i will give head with probability p_i, and tail with probability $1 - p_i$. The values p_1 and p_2 are known to the player. Then one picks up one of the two coins (with equal probability one half), and launch it R times. Can we infer which coin was picked up from the observation of the sequence of the R draws? How big should be R to ensure that the probability of guessing the wrong coin is smaller than, say, ϵ? These questions, called hypothesis testing, are very fundamental ones in statistics, see [9] for instance.

Let H be the number of heads in the sequence of R draws. As the draws are independent from each other, H is binomially distributed, see Figure 3A. When R is large, the probability of H can be approximated through

$$P(H = R\,h|i) = \exp\left(- R\ D(h|p_i) + o(R)\right)\ , \tag{9}$$

where i is the label of the 'true' coin, and

$$D(h|p_i) = h\ \ln(h/p_i) + (1-h)\ \ln\left((1-h)/(1-p_i)\right) \tag{10}$$

is the Kullback-Leibler divergence between h and p_i. $D(h|p_i)$ is a convex and positive function of h over the $[0;1]$ range, vanishing in $h = p_i$ only (Figure 3B). In the Bayesian inference framework [9], the *a posteriori* probability for the coin label i given the value of H is

$$P(i|H) = \frac{P(H|i)}{P(H|1) + P(H|2)}\ . \tag{11}$$

Consequently, the most likely coin i is the one maximizing $P(H|i)$. For large R, one infers $i = 1$ if $H < h^* R$, $i = 2$ if $H > h^* R$, where h^* is the fraction of heads at which the two Kullback-Leibler divergences cross: $D(h^*|p_1) = D(h^*|p_2)$ (Figure 3B).

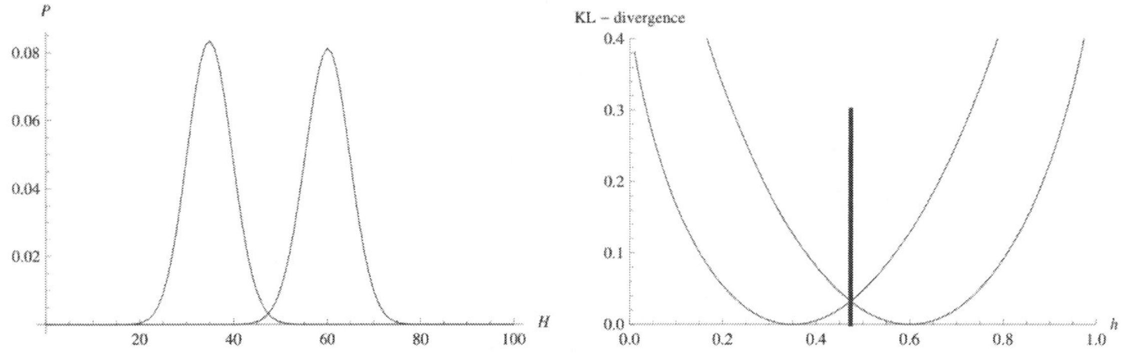

Figure 3. The two-coin testing problem. **A.** Probability of the number of head draws, H, after $R = 100$ draws for coins $i = 1, 2$, with $p_1 = 0.35 < p_2 = 0.6$. **B.** Kullback-Leibler divergences associated to the two coins, $D(h = H/R|p_i)$, see(10). The vertical line shows the location of the crossing point, $h^* \simeq 0.4739$. The inverse of the height of the crossing point is R_c (13).

What confidence can one have about the above inference? The probability of wrongly inferring, say, coin $i = 2$, while the true coin is $i = 1$ is given by

$$P_{error} = \sum_{H > h^* R} P(H|1) = \exp\left(- R/R_c(p_1, p_2) + o(R)\right) . \qquad (12)$$

with

$$R_c(p_1, p_2) = \frac{1}{D(h^*|p_1)} = \frac{1}{D(h^*|p_2)} . \qquad (13)$$

The above expression clearly shows that the inference error is a symmetric function of the coin probabilities when R is large. The height of the crossing point of the two Kullback-Leibler divergences is called Chernoff information in information theory [9]. Its inverse, R_c, can be interpreted as the order of magnitude of the number of draws needed to have a reliable inference. More precisely, the probability of error will be smaller than ϵ if the number of draws exceeds

$$R > \log(1/\epsilon) \times R_c(p_1, p_2) . \qquad (14)$$

4.2. Back to the inverse Sinai problem: tiling of the signal space
Despite its simplicity, the 2-coin testing problem introduces the main idea needed to treat the more complicated case of the inverse Sinai problem. This key idea is the following: errors in the inference are due to very atypical observations. When the number of observations, R, is large, the main contribution to the probability of error comes from observations equally likely to be generated by any of the two candidate distributions *i.e.* the two coins in the previous section, or the two potentials $\mathbf{V}_1$, $\mathbf{V}_2$ for the Sinai problem. Those critical observations (and their probability) can be estimated based on extremal arguments, called large-deviation theory in probability literature [10].

Table 1 gives the correspondence between the quantities defined and studied in the 2-coin testing problem, and their counterparts in the problem of the inference of a potential from some realizations of a random walk.

As explained in Section 3.1, the information on a potential which can be extracted from the observation of the motion of the walker is contained in the signal $\mathbf{S}$. Signals corresponding to independent random walks in the same potential sum up. We thus expect that $\mathbf{S}$ will be an

Journal of Physics: Conference Series **197** (2009) 012005 doi:10.1088/1742-6596/197/1/012005

Table 1. Correspondence between 2-coin testing and the potential inference problem.

	2-coin testing problem	potential inference problem
quantity to infer	coin i	potential $\mathbf{V}$
observation	sequence of draws	random walk $\mathbf{X}$
sufficient statistics	number H of heads	signal $\mathbf{S} = \{N_x, u_x^{\pm}; x \geq 0\}$
asymptotic probability for the sufficient statistics	$\exp\left(-R\, D(\frac{H}{R}; p) + o(R)\right)$	$\exp\left(-R\, \omega_{\mathbf{V}}(\frac{\mathbf{S}}{R}) + o(R)\right)$
boundary between candidates	fraction h^*	critical signal $\mathbf{s}^*$

extensive function of the number R of realizations. In the following, we will denote the intensive ratio $\mathbf{S}/R$ by $\mathbf{s}$.

We now think of a signal $\mathbf{S}$ as a vector of components $\{\mathbf{S}_{x,0} = N_x, \mathbf{S}_{x,+} = u_x^+, \mathbf{S}_{x,-} = u_x^-; x \geq 0\}$ in a (infinite-dimensional) space. In the same way, we define the vector $\mathbf{v}$ attached to each potential $\mathbf{V}$, with components $\{\mathbf{v}_{x,0} = \ln \hat{Q}_{\mathbf{V}_1,\Delta t}(x,x), \mathbf{v}_{x,+} = \ln \hat{Q}_{\mathbf{V}_1,\Delta t}(x+1,x), \mathbf{v}_{x,-} = \ln \hat{Q}_{\mathbf{V}_1,\Delta t}(x-1,x); x \geq 0\}$ attached to each potential $\mathbf{V}$. From (6), the likelihood of the signal given the potential is simply the exponential of the dot product between the signal and potential vectors,

$$P[\mathbf{X}|\mathbf{V}] = \exp\left(\sum_{x\geq 0} \mathbf{S}_x\, \mathbf{v}_x\right) . \qquad (15)$$

It is therefore natural to partition the space of signals into Voronoi cells $C_{\mathbf{v}}$. $C_{\mathbf{v}}$ is the set of signals having a larger dot product with $\mathbf{v}$ than with any other $\mathbf{v}'(\neq \mathbf{v})$, see Figure 4. It is also the set of signals from which the potential $\mathbf{V}$ will be inferred.

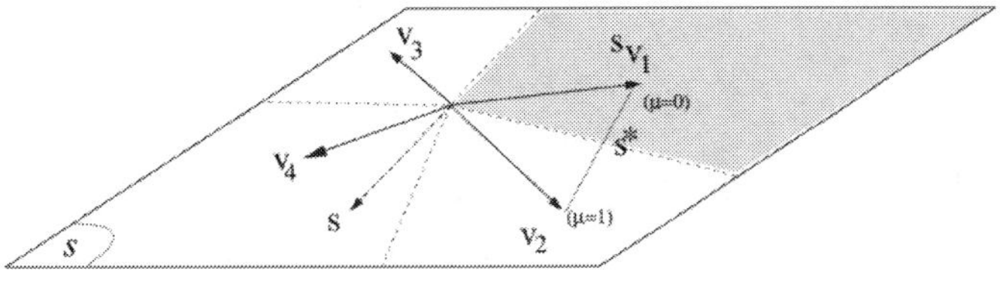

Figure 4. The space of signals, $\mathbf{s}$, tiled with Voronoi cells attached to every possible potentials. The cell corresponding to potential $\mathbf{V}_i$ is the set of signals $\mathbf{s}$ having a larger dot product with $\mathbf{v}_i$ than with any other potential vectors. The line joining the potential vectors $\mathbf{v}_1$ and $\mathbf{v}_2$, parametrized by μ, is shown. Its intersection with the boundary between the cells is the critical signal $\mathbf{s}^*$.

Let $\mathbf{V}_1$ be the 'true' potential in which the walker moves. We know that, in the large R limit, the signals $\mathbf{S}$ most likely to lead to the erroneous inference of $\mathbf{V}_2$ are located on the boundary between cells $C_{\mathbf{v}_1}$ and $C_{\mathbf{v}_2}$. The probability of erroneously predicting $\mathbf{V}_2$ will therefore be given, when R is large, by the probability of signals extracted from potential $\mathbf{V}_1$, and constrained to have equal dot products with both potential vectors,

$$P_{error} = \sum_{\mathbf{S}} P_R(\mathbf{S}|\mathbf{V}_1)\ \delta_{\mathbf{S}\cdot(\mathbf{v}_1-\mathbf{v}_2),0}\ . \tag{16}$$

It is important to distinguish the probability $P(\mathbf{X}|\mathbf{V}_1)$ of a realization $\mathbf{X}$, given by (15), from the probability $P_R(\mathbf{S}|\mathbf{V}_1)$ of the signal $\mathbf{S}$ itself. The latter is equal to the former multiplied by the number of random walks sharing the same signal. We denote this multiplicity by $\mathcal{M}_R(\mathbf{S})$, where we make explicit the dependence on the number R of realizations.

When R gets large, the multiplicity $\mathcal{M}_R(\mathbf{S})$ is expected to scale as $\exp(R\,\sigma(\mathbf{s})+o(R))$, where the growth function σ depends on the duration T of the walk. Hence the probability of a signal given the 'true' potential is

$$P_R(\mathbf{S}|\mathbf{V}_1) = \exp\left\{R\left[\sigma(\mathbf{s}) + \mathbf{s}\cdot\mathbf{v}_1\right] + o(R)\right\}\ . \tag{17}$$

Usinf a Lagrange multiplier to enforce the constraint in the sum on the right-hand side of (16), we conclude that

$$P_{error} = \min_{\mu}\sum_{\mathbf{s}} \exp\left\{R\left[\sigma(\mathbf{s}) + \mathbf{s}\cdot((1-\mu)\,\mathbf{v}_1 + \mu\,\mathbf{v}_2)\right] + o(R)\right\}\ . \tag{18}$$

The role of the optimization over μ is precisely to locate the critical signal $\mathbf{s}^*$ on the boundary between the two cells attached to the two potentials, see Figure 4.

Interestingly, there is no need to calculate the growth function σ to exploit the above formula. An immediate comparison of (18) and (17) shows that P_{error} is, apart the minimization over μ, the sum over all possible signals of the probability of signals attached to a fictitious 'potential', whose vector is $(1-\mu)\,\mathbf{v}_1 + \mu\,\mathbf{v}_2$. This potential is fictitious in that it is generally not (for $\mu \neq 0, 1$) one of the potentials attached to a sequence of binary forces. The nature of the fictitious potential and of the random walk taking place therein, is unveiled in the next section.

5. Relationship with a fictitious and partially absorbing random walk

5.1. Interpretation of P_{error} as a probability of survival

The second member of (18) defines an effective transition matrix, whose elements are given by, to the first order in Δt,

$$\begin{aligned}\hat{Q}_{\mathbf{V}_1,\mathbf{V}_2,\mu,\Delta t}(x',x) &= \hat{Q}_{\mathbf{V_1},\Delta t}(x',x)^{1-\mu}\,\hat{Q}_{\mathbf{V_2},\Delta t}(x',x)^{\mu} \\ &= \begin{cases} \Delta t\; r_{\mathbf{V}_1}(x',x)^{1-\mu}\, r_{\mathbf{V}_2}(x',x)^{\mu} & \text{if} \quad x' = x \pm 1 \\ 1 - \Delta t \displaystyle\sum_{y=\pm 1} ((1-\mu)\, r_{\mathbf{V}_1}(x+y,x) + \mu\, r_{\mathbf{V}_2}(x+y,x)) & \text{if} \quad x' = x \end{cases}\end{aligned} \tag{19}$$

Hence, in the $\Delta t \to 0$ limit, we may interpret $\hat{Q}_{\mathbf{V}_1,\mathbf{V}_2,\mu,\Delta t}$ as a transition matrix describing a one-dimensional random walk on the set of non-negative integers. The associated potential, $\mathbf{W}$, can be read from the off-diagonal transition rates. For instance

$$r_{\mathbf{W}}(x+1,x) = \left(e^{\gamma f_{1,x}}\right)^{1-\mu}\left(e^{\gamma f_{2,x}}\right)^{\mu} = e^{\gamma\left((1-\mu) f_{1,x} + \mu f_{2,x}\right)}\ , \tag{20}$$

where $f_{i,x}$ is the force acting on bond $(x, x+1)$ in the potential $\mathbf{V}_i$. A similar result holds for the transition from x to $x-1$. Hence $\mathbf{W} = (1-\mu)\mathbf{V}_1 + \mu\mathbf{V}_2$, as could be guessed from the comparison of (17) and (18).

However, $\hat{Q}_{\mathbf{V}_1,\mathbf{V}_2,\mu,\Delta t}$ is not equal to $\hat{Q}_{\mathbf{W},\Delta t}$, since the latter conserves the probability, while the former does not. Indeed, according to (19),

$$\sum_{x'} \hat{Q}_{\mathbf{V}_1,\mathbf{V}_2,\mu,\Delta t}(x',x) = 1 - \Delta t\; d_{\mathbf{V}_1,\mathbf{V}_2,\mu}(x) + O(\Delta t^2)\ , \tag{21}$$

where

$$d_{\mathbf{V}_1,\mathbf{V}_2,\mu}(x) = \sum_{y=\pm 1} \Big[(1-\mu)\, r_{\mathbf{V}_1}(x+y,x) + \mu\, r_{\mathbf{V}_2}(x+y,x) - r_{\mathbf{W}}(x+y,x) \Big] \ . \tag{22}$$

When $0 < \mu < 1$, by convexity of the exponential function, $d_{\mathbf{V}_1,\mathbf{V}_2,\mu}$ is a positive quantity, which vanishes only if $\mathbf{V}_1$ and $\mathbf{V}_2$ coincide around site x, that is, if the forces $f_{i,x-1}$ and $f_{i,x}$ on, respectively, bonds $(x-1,x)$ and $(x,x+1)$ take the same values in both potentials. Hence $d_{\mathbf{V}_1,\mathbf{V}_2,\mu}$ can be interpreted as the rate of absorption at site x for the random walk process associated to $\hat{Q}$. In other words, it is the probability per unit of time for the walker to die at site x. We conclude that the probability of error (18) is

$$P_{error} = \pi^{R+o(R)} \ , \tag{23}$$

where π is the probability that the walker, initially at site $x = 0$, survives till time T. We find back the exponential decay of the probability of error with R, with a decay constant $R_c = -1/\ln \pi$.

We emphasize that the more different are the potentials $\mathbf{V}_1$ and $\mathbf{V}_2$ in the neighbourhood of x, the higher is the absorption rate $d_{\mathbf{V}_1,\mathbf{V}_2,\mu}(x)$, and the more likely is the walker to die in x. Hence the probability of survival, $\pi(0,T)$, is large, *i.e.* close to 1, when the potentials $\mathbf{V}_1$ and $\mathbf{V}_2$ are difficult to tell from one another, and small when they are easily distinguishable based on the observation of one random walk realization.

5.2. Calculation of the survival probability for the one-dimensional Sinai model

The results derived in the above section are easily generalizable to the case of random walk in spaces of dimension larger than one. Formula (23) then gives a practical way to estimate R_c by means of numerical simulations. In the case of the one-dimensional Sinai model, the survival probability can be analytically calculated as we now show.

Let $\pi(x,t)$ be the probability that a random walker, initially in x, has survived until time t; we are eventually interested in calculating $\pi = \pi(0,T)$ (23). To obtain the equation fulfilled by $\pi(x,t)$, we write that, in order to be alive until time $t + \Delta t$ where Δt is an infinitesimal time interval, the walker can first move during the first Δt period, and then has to survive during a period of duration equal to t,

$$\pi(x, t+\Delta t) = \sum_{x'} \pi(x',t)\ \hat{Q}_{\mathbf{V}_1,\mathbf{V}_2,\mu,\Delta t}(x',x) \ , \tag{24}$$

In the $\Delta t \to 0$ limit, we obtain the following backward Fokker-Planck equation,

$$\frac{d\pi}{dt}(x,t) = \sum_{x'(\neq x)} r_{\mathbf{W}}(x',x)\ (\pi(x',t) - \pi(x,t)) - d_{\mathbf{V}_1,\mathbf{V}_2,\mu}(x)\ \pi(x,t) \ , \tag{25}$$

with the initial condition $\pi(x,0) = 1$.

Consider the case of where the differences between the two sequences of forces are localized on a subset of bonds arount site z, *e.g.* $f_{1,x} = f_{2,x}$ for all x but $x = z$, and $f_{1,z} = -f_{2,z}$. The absorption rates vanish everywhere, except in the vicinity of z. Let t_z denotes the mean first-passage time at site z in one of the two potentials. If T is much smaller than t_z, the random walker is alive with high probability since the neighborhood of z has not been visited. On the contrary, when T is much larger than t_z (or, more precisely, the average last-passage time in z), the random walk is very likely not to visit z anymore, and the walker will wander to infinity without risking of being absorbed.

To calculate the probability of survival in this large T limit, $\pi = \pi(0,\infty)$, we look for a stationary solution of (25), denoted by $\pi_s(x)$ for a given μ. Defining $q^{\pm}(x) = r_{\mathbf{W}}(x \pm 1, x)/(r_{\mathbf{W}}(x+1,x) + r_{\mathbf{W}}(x-1,x) + d_{\mathbf{V}_1,\mathbf{V}_2,\mu}(x))$, the ratio $\rho(x) = \pi_s(x)/\pi_s(x+1)$ satisfies the recurrence equation

$$\rho(x) = \frac{q^+(x)}{1 - q^-(x)\,\rho(x-1)} \,, \tag{26}$$

for $x \geq 1$. We iterate this recursive equation, with the boundary condition $\rho(0) = q^+(0)$, to get $\pi(0) = \prod_{x\geq 0} \rho(x)$. The probability of survival can then be minimized over μ to obtain π. Figure 5 shows the application of this procedure to the potential of Figure 1A.

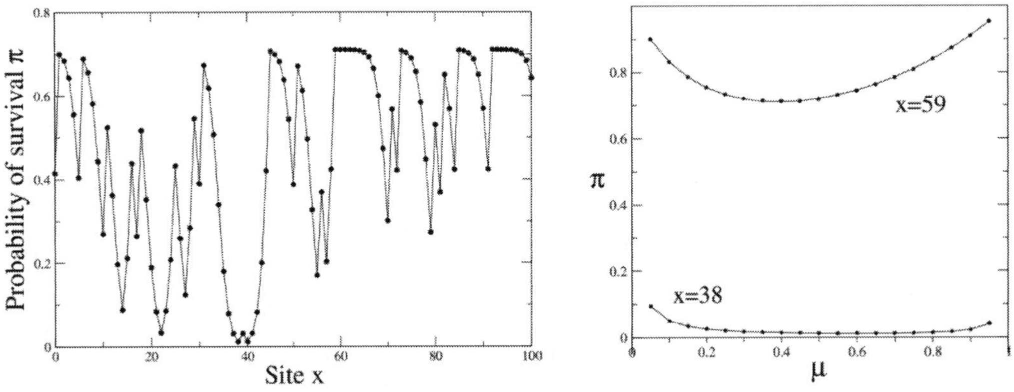

Figure 5. A. Probability of survival π corresponding to the comparison of the potential of Figure 1 and the same potential with the force on bond $(x, x+1)$ inverted. The probability of error from R realizations decays as π^R for large R, compare to Figure 2. **B.** Probability of survival π as a function of μ for sites $x = 38$ and $x = 59$. Lines are theoretical predictions. Dots show results from numerical simulations averaged over 100,000 realizations of the absorbing random walk; error bars are smaller than the dot size. The survival probability corresponds to the minimum of the curves over μ.

6. Conclusion

In this paper, we have shown how the observation of a random walker in a potential can be used to infer the value of this potential. We have also investigated the issue of the reliability of the inferred potential, and, more precisely, of how many observations are necessary to distinguish between two candidate potentials. From a technical point of view, R_c is obtained from the probability of survival of a random walker in a fictitious and partially absorbing potential. The interest is two-fold. First, it allowed us to avoid the calculation of the entropy σ of signals. Secondly, it gives a practical way to estimate R_c for any model, and not only for the one-dimensional Sinai model, where many quantities can be calculated exactly.

We have here focused on the case of transient walks. The issue of inferring a random potential from recurrent walks, *i.e.* when the tilt b vanishes, is a very interesting one from a mathematical point of view, and was recently studied by Andreoletti [3]. In this case, the motion is very slow: very few sites are visited (of the order of $(\log T)^2$ after a time T), but the forces on the corresponding bonds can be accurately inferred. From a practical point of view, it may be more profitable to work at a finite tilt, so that the walk acquires a finite velocity. The number of visited bonds grows linearly with T and, though the information collected by a single random walk realization is incomplete, the strategy is on the overall faster as far as the inference is concerned [8].

Acknowledgement: We thank P. Andreoletti and R. Diel for useful discussions. This work was partially funded by the ANR 06-JCJC-051 contract.

References

[1] Hughes B D 1996 *Random Walks and Random Environments* (Oxford University Press)
[2] Bouchaud J P and Georges A 1990 *Phys. Rep.* **195** 127
[3] Andreoletti P 2006 *Preprint* arxiv:math.PR/0612208
[4] Evans E and Ritchie K 1997 *Biophys. J.* **72** 1541
[5] Essevaz-Roulet B, Bockelmann U and Heslot F 1997 *Proc. Natl. Acad. Sci. USA* **94** 11935
[6] Danilowitcz C *et al* 2003 *Proc. Natl. Acad. Sci. USA* **100** 1694
[7] Baldazzi V *et al* 2006 *Phys. Rev. Lett.* **96** 128102; 2007 *Phys. Rev.* E **75** 011904
[8] Cocco S and Monasson R 2008 *Europhys. Lett.* **81** 20002
[9] Cover T M and Thomas J A 1991 *Elements of Information Theory* (Wiley)
[10] Dembo A and Zeitouni O 1998 *Large Deviations Techniques and Applications* (Springer-Verlag)
[11] Chernoff H 1952 *Ann. Math. Stat.* **23** 493

International Workshop on Statistical-Mechanical Informatics 2009 (IW-SMI 2009) IOP Publishing
Journal of Physics: Conference Series **197** (2009) 012006 doi:10.1088/1742-6596/197/1/012006

Generating functional analysis of complex formation and dissociation in large protein interaction networks

A C C Coolen and S Rabello

Department of Mathematics, King's College London, The Strand, London WC2R 2LS, U.K. and Randall Division of Cell and Molecular Biophysics, King's College London, New Hunt's House, London SE1 1UL, UK

E-mail: ton.coolen@kcl.ac.uk

Abstract. We analyze large systems of interacting proteins, using techniques from the non-equilibrium statistical mechanics of disordered many-particle systems. Apart from protein production and removal, the most relevant microscopic processes in the proteome are complex formation and dissociation, and the microscopic degrees of freedom are the evolving concentrations of unbound proteins (in multiple post-translational states) and of protein complexes. Here we only include dimer-complexes, for mathematical simplicity, and we draw the network that describes which proteins are reaction partners from an ensemble of random graphs with an arbitrary degree distribution. We show how generating functional analysis methods can be used successfully to derive closed equations for dynamical order parameters, representing an exact macroscopic description of the complex formation and dissociation dynamics in the infinite system limit. We end this paper with a discussion of the possible routes towards solving the nontrivial order parameter equations, either exactly (in specific limits) or approximately.

1. Introduction

It is safe to say that biomedicine is undergoing a profound and irreversible change, brought about by what some coined a 'data tsunami'. A spectacular increase of both the quality and quantity of data on cellular signalling, molecular structure and gene expressions, including *in vivo* data, have prompted biomedicine to transform rapidly into a much more quantitative as opposed to descriptive science, and theory now lags behind experiment to an almost embarrassing extent. The biological systems thought to be responsible for generating these data tend to be complex and nonlinear, involving many variables and nested processes acting on widely separated timescales. Also, biology lacks many of the simplifying features of many large physical systems, such as detailed balance, the identical nature of the degrees of freedom, or translation invariance. In biological systems, the interacting objects and their mutual forces are as a rule non-identical; in biology this would be called 'inhomogeneity', where in physics one would speak of 'disorder'.

One particular biological system of great relevance to medicine is the proteome, the collection of about 2.10^4 cellular proteins and their molecular complexes, which constitute the main work force of the cell, responsible for executing most of the tasks that allow it to function, and for inter- and intra-cellular communication. The vast increase of proteomic data allow us to refine our understanding of its pathways and reactions, but we are running into new barriers: 'The most significant challenges that the mechanistic modelling might be facing are the lack of quantitative kinetic data and the combinatorial increase in the number of emerging distinct species and states of the protein network being simulated' [1]. Our aim in this paper is to contribute to our quantitative understanding of the proteome, by exploring the potential of quantitative macroscopic regularities emerging in very large as opposed to small proteomic systems. To do so we construct dynamical single-compartment models of large systems of interacting proteins, and analyze these in the limit of an infinite number of reaction partners, adapting techniques from the non-equilibrium statistical mechanics of disordered physical many-particle systems.

2. Model definitions

We imagine a cell in which N different proteins can be expressed, labeled by Roman indices $i \in \{1,\ldots,N\}$. Each protein can be in at most q post-translational states, labeled by Greek indices $\alpha \in \{1,\ldots,q\}$. The concentration in the cell of (i,α), the α-th post-translational state of protein i, will be written as x_i^α. There are in principle $\frac{1}{2}N(N+1)$ possible complexes that these proteins could form by pairwise binding, $\frac{1}{2}N(N-1)$ hetero-dimers and N homo-dimers. We denote the complex where i binds to j as $(i{\asymp}j)$, and the concentration in the cell of the complex $(i{\asymp}j)$ as x_{ij}. For mathematical convenience we can always take $x_{ij} = x_{ji}$.

2.1. Elementary processes and dynamical equations

We aim to construct and analyze single-compartment type dynamical equations for the concentrations $\{x_i^\alpha\}$ of unbound proteins and $\{x_{ij}\}$ of their binary complexes, that capture the following elementary proteomic processes:

elementary process:	*symbolic notation:*	*process rate:*
binary complex formation:	$(i,\alpha)+(j,\beta)\to(i{\asymp}j)$	$k_{ij}^{\alpha\beta+}x_i^\alpha x_j^\beta$
binary complex dissociation:	$(i{\asymp}j)\to(i,\alpha)+(j,\beta)$	$k_{ij}^{\alpha\beta-}x_{ij}$
protein degradation/removal:	$(i,\alpha)\to\emptyset$	$\gamma_i^\alpha x_i^\alpha$
protein synthesis:	$\emptyset\to(i,\alpha)$	θ_i^α

All rate parameters $\{k_{ij}^{\alpha\beta\pm},\gamma_i^\alpha,\theta_i^\alpha\}$ are by definition non-negative. Of these parameters only $\{\gamma_i^\alpha,\theta_i^\alpha\}$ are allowed to depend on time, in order to incorporate possible changes in membrane properties and gene expression levels; the complex formation/dissociation parameters $\{k_{ij}^{\alpha\beta\pm}\}$, in contrast, depend only on the structural characteristics of the reaction partners, via the indices (i,α) and (j,β). Consistency demands that $k_{ij}^{\alpha\beta\pm} = k_{ji}^{\beta\alpha\pm}$. Not all pairs (i,j) can in practice form a stable complex $(i{\asymp}j)$, so we need (fixed) variables $c_{ij} \in \{0,1\}$ to define which are potential binding partners: if $(i{\asymp}j)$ is possible we put $c_{ij}=c_{ji}=1$, otherwise $c_{ij}=c_{ji}=0$. The non-directed graph $\mathbf{c} = \{c_{ij}\}$ thus defines the cell's protein interaction network.

Combining the four elementary processes above with the structural information in $\mathbf{c}$ regarding binding partners, we are then led to the following set of Michaelis-Menten reaction equations:

$$\frac{\mathrm{d}}{\mathrm{d}t}x_i^\alpha = \sum_j c_{ij}\sum_\beta[k_{ij}^{\alpha\beta-}x_{ij}-k_{ij}^{\alpha\beta+}x_i^\alpha x_j^\beta] + \theta_i^\alpha - \gamma_i^\alpha x_i^\alpha \tag{1}$$

$$\frac{\mathrm{d}}{\mathrm{d}t}x_{ij} = c_{ij}\sum_{\alpha\beta}[k_{ij}^{\alpha\beta+}x_i^\alpha x_j^\beta - k_{ij}^{\alpha\beta-}x_{ij}] \tag{2}$$

If $c_{ij} = 0$ one will indeed have $x_{ij} = 0$ at all times, as one should. Equations (1,2) satisfy the relevant mass conservation constraints: the total amount of any protein i (whether bound in dimers or in unbound form) changes only due to production/degradation imbalance, viz.

$$\frac{\mathrm{d}}{\mathrm{d}t}\Big[\sum_\alpha x_i^\alpha + \sum_j x_{ij}\Big] = \sum_\alpha(\theta_i^\alpha - \gamma_i^\alpha x_i^\alpha) \tag{3}$$

The presence of concentrations x_{ij} of complexes, with their two protein indices, could complicate a direct statistical mechanical analysis of equations (1,2). However, the x_{ij} are seen to obey linear equations, which we can simply solve:

$$x_{ij}(t) \;=\; c_{ij}\sum_{\rho\lambda} k_{ij}^{\rho\lambda+}\int_{-\infty}^{t}\mathrm{d}s\;\mathrm{e}^{-k_{ij}^{-}(t-s)}x_i^{\rho}(s)x_j^{\lambda}(s) \tag{4}$$

with the short-hand $k_{ij}^{-}=\sum_{\alpha\beta}k_{ij}^{\alpha\beta-}$. If we substitute (4) into equations (1) we obtain

$$\frac{\mathrm{d}}{\mathrm{d}t}x_i^{\alpha}(t) \;=\; \sum_j c_{ij}\int\!\mathrm{d}s\;\sum_{\rho\lambda}W_{\alpha;\rho\lambda}(t-s|\mathbf{k}_{ij})x_i^{\rho}(s)x_j^{\lambda}(s)+\theta_i^{\alpha}(t)-\gamma_i^{\alpha}(t)x_i^{\alpha}(t) \tag{5}$$

with a partially retarded effective free protein interaction kernel

$$W_{\alpha;\rho\lambda}(\tau|\mathbf{k})=k^{\rho\lambda+}\Big[\sum_{\beta}k^{\alpha\beta-}\theta[\tau]\mathrm{e}^{-k^{-}\tau}-\delta_{\alpha\rho}\delta(\tau)\Big] \tag{6}$$

which obeys $\sum_{\alpha}\int\!\mathrm{d}\tau\;W_{\alpha;\rho\lambda}(\tau|\mathbf{k})=0$, for all (ρ,λ).

The above definitions involve several simplifications, the main ones being that the only complexes are dimers, and that we disregard all spatial variations of protein and complex concentrations. However, we would argue that our definitions incorporate the main events in the proteome, while still allowing (as we will see) for an exact generating functional analysis in the limit $N\to\infty$ (under suitable conditions on the statistical properties of the model parameters).

2.2. Statistics of microscopic parameters - interaction network and reaction rates

It is not possible at present to solve equations such as (5) analytically for arbitrary choices of the microscopic model parameters $\{c_{ij},k_{ij}^{\alpha\beta\pm},\theta_i^{\alpha},\gamma_i^{\alpha}\}$. Neither can we benefit from built-in parameter regularities or symmetries (in contrast to e.g. physical systems on lattices). This leaves the route of disordered systems theory, where one exploits the property that large many-particle systems often exhibit universal behaviour of macroscopic observables, whose values will in the infinite system size limit depend only on the *statistics* of the microscopic parameters rather than their precise realization. It follows that these values can then be calculated by performing suitable averages over all microscopic parameter realizations with the correct statistics.

Thus we take the parameters $\{c_{ij},k_{ij}^{\alpha\beta\pm}\}$ to be generated randomly from some appropriate distribution, which should ideally incorporate as much of our available biological information as possible. We will in this paper draw the protein interaction network $\mathbf{c}$ at random from an ensemble of random graphs with prescribed degrees $(k_1,\ldots,k_N)$, that are in turn drawn randomly from an as yet unspecified degree distribution $p(k)$:

$$P(\mathbf{c}) \;=\; \frac{1}{Z}\prod_i \delta_{k_i,\sum_{j\neq i}c_{ij}}\cdot\prod_i\Big[c_0\delta_{c_{ii},1}+(1-c_0)\delta_{c_{ii},0}\Big] \tag{7}$$

The self-connections c_{ii} are included to allow for homo-dimers (with $c_0\in[0,1]$ representing the fraction of proteins that can form homo-dimers), $\langle k\rangle=\sum_k p(k)k>0$ gives the average number of hetero-dimer binding partners per protein, and Z is a normalization constant. In (7) all graphs that exhibit the enforced degree sequence $(k_1,\ldots,k_N)$ are generated equally likely.

We similarly draw for each index pair (i,j) with $i\leq j$ the complex formation/dissociation rates $\mathbf{k}_{ij}=\{k_{ij}^{\alpha\beta\pm}\}$ randomly and independently from a joint distribution $P(\mathbf{k})$ (we can use the same symbol P here as in (7), as the arguments will prevent ambiguity). Averages over $P(\mathbf{k})$ will be written as $\int\!\mathrm{d}\mathbf{k}\;P(\mathbf{k})f(\mathbf{k})=\langle f(\mathbf{k})\rangle_{\mathbf{k}}$. Although the rates for reactions involving one protein pair (i,j) with $i \leq j$ are now statistically independent from those involving any other pair (k,ℓ) with $k \leq \ell$, we do not assume statistical independence at the level of post-translational states or off-rates versus on-rates, i.e. we do *not* assume that $P(\mathbf{k}) = \prod_{\alpha\beta}[P(k^{\alpha\beta+})P(k^{\alpha\beta-})]$. There are only two internal symmetries we have to insist on. First, when referring to homo-dimers we must have $k^{\alpha\beta\pm} = k^{\beta\alpha\pm}$, for reasons of consistency. Second, in the case of hetero-dimers, if we define $S\mathbf{k}$ via $(S\mathbf{k})^{\alpha\beta\pm} = k^{\beta\alpha\pm}$ then we will require that $P(S\mathbf{k}) = P(\mathbf{k})$ for all $\mathbf{k}$, for technical reasons that will become clear later. This implies that the *probability* of assigning specific on/off rates to any reaction $(i,\alpha) + (j,\beta) \leftrightarrow i{\asymp}j$ is the same as the probability to find these rates for the reaction $(i,\beta) + (j,\alpha) \leftrightarrow i{\asymp}j$, but not that the actual rates are themselves identical. We note that, although the number q of *potential* post-translational states labeled by $\alpha = 1\ldots q$ is the same for all proteins, the *actual* number of post-translations states can be made to vary from one protein to another, by appropriate choices for the rate statistics $P(\mathbf{k})$.

The production and decay rates $(\theta_i^\alpha, \gamma_i^\alpha)$ need not be drawn randomly for our methods to apply (as these have only a single protein index i), although one could do so for mathematical convenience. Here we will allow the rates $(\theta_i^\alpha, \gamma_i^\alpha)$ to be time-dependent, to incorporate the effects of proteome perturbations or receptor triggering. If each protein is always assembled in one unique conformation, we must choose the production rates such that for each i only one index α can have $\theta_i^\alpha \neq 0$ (representing the protein's native state).

3. Generating functional analysis

In the spirit of [2] we define a generating functional that allows us to calculate the time-dependent statistics of unbound protein concentrations; from these follow, via (4), also those of all protein complexes. This functional $Z[\boldsymbol{\psi}]$ is a straightforward generalization to time dependent stochastic variables of the conventional moment generating function of random variables.

3.1. The disorder-averaged generating functional

We first discretize time according to $t_i = i\Delta$, with an elementary time step Δ that will be sent to zero, and use δ-functions to enforce at each time the validity of the equations (5):

$$\begin{aligned} Z[\boldsymbol{\psi}] &= \int\Big[\prod_{i\alpha t}\mathrm{d}x_i^\alpha(t)\;\mathrm{e}^{\mathrm{i}\Delta\psi_i^\alpha(t)x_i^\alpha(t)}\Big] \\ &\quad\times\prod_{i\alpha t}\delta\Big[x_i^\alpha(t{+}\Delta)-x_i^\alpha(t)-\Delta\Big(\sum_j c_{ij}\int\!\mathrm{d}s\sum_{\rho\lambda}W_{\alpha;\rho\lambda}(t{-}s|\mathbf{k}_{ij})x_i^\rho(s)x_j^\lambda(s)+\theta_i^\alpha(t)-\gamma_i^\alpha(t)x_i^\alpha(t)\Big)\Big] \\ &= \int\prod_{i\alpha t}\Big[\frac{\mathrm{d}x_i^\alpha(t)\mathrm{d}\hat{x}_i^\alpha(t)}{2\pi}\;\mathrm{e}^{\mathrm{i}\Delta\psi_i^\alpha(t)x_i^\alpha(t)+\mathrm{i}\hat{x}_i^\alpha(t)\left(x_i^\alpha(t+\Delta)-x_i^\alpha(t)-\Delta[\theta_i^\alpha(t)-\gamma_i^\alpha(t)x_i^\alpha(t)]\right)}\Big].\mathrm{e}^{\Xi[\mathbf{c},\{\mathbf{k}\}]} \end{aligned} \tag{8}$$

in which only the following object depends on the network $\mathbf{c}$ and the process rates $\{\mathbf{k}\}$:

$$\begin{aligned} \Xi[\mathbf{c},\{\mathbf{k}\}] &= -\mathrm{i}\Delta\sum_{ij}c_{ij}\sum_{\alpha\rho\lambda}\sum_t\hat{x}_i^\alpha(t)\int\!\mathrm{d}s\;W_{\alpha;\rho\lambda}(t{-}s|\mathbf{k}_{ij})x_i^\rho(s)x_j^\lambda(s) \\ &= -\mathrm{i}\sum_i c_{ii}\Xi_{ii}(\mathbf{k}_{ii})-\mathrm{i}\sum_{i<j}c_{ij}\Xi_{ij}(\mathbf{k}_{ij}) \end{aligned} \tag{9}$$

with

$$\Xi_{ii}(\mathbf{k}) = \sum_{\alpha\rho\lambda}\int\!\mathrm{d}t\mathrm{d}\tau\;W_{\alpha;\rho\lambda}(\tau|\mathbf{k})\hat{x}_i^\alpha(t)x_i^\rho(t{-}\tau)x_i^\lambda(t{-}\tau) \tag{10}$$

$$\begin{aligned} \Xi_{ij}(\mathbf{k}) &= \int\!\mathrm{d}t\mathrm{d}\tau\sum_{\rho\lambda}k^{\rho\lambda+}x_i^\rho(t{-}\tau)x_j^\lambda(t{-}\tau) \\ &\quad\times\Big\{\theta[\tau]\mathrm{e}^{-k^-\tau}\sum_{\alpha\beta}k^{\alpha\beta-}[\hat{x}_i^\alpha(t){+}\hat{x}_j^\beta(t)]-\delta(\tau)[\hat{x}_i^\rho(t){+}\hat{x}_j^\lambda(t)]\Big\} \end{aligned} \tag{11}$$

To arrive at (9,10,11) we have used $k_{ij}^{\alpha\beta\pm}=k_{ji}^{\beta\alpha\pm}$. Note that $Z[\mathbf{0}]=1$, by construction.

We next average the generating functional over the disorder, i.e. over the randomly generated interaction networks (7) and reaction rates $\{\mathbf{k}_{ij}\}$. The result will be written as $\overline{Z[\boldsymbol{\psi}]}$. The idea behind definition (8) is that calculating $\overline{Z[\boldsymbol{\psi}]}$ gives us access to averages of observables that evolve according to the equations (5) without actually solving these equations, e.g.

$$\overline{x_i^\alpha(t)} = -\lim_{\boldsymbol{\psi}\to\mathbf{0}}\lim_{\Delta\to 0}\frac{\mathrm{i}}{\Delta}\frac{\partial}{\partial\psi_i^\alpha(t)}\overline{Z[\boldsymbol{\psi}]} \tag{12}$$

$$\overline{x_i^\alpha(t)x_j^\beta(t')} = -\lim_{\boldsymbol{\psi}\to\mathbf{0}}\lim_{\Delta\to 0}\frac{1}{\Delta^2}\frac{\partial^2}{\partial\psi_i^\alpha(t)\partial\psi_j^\beta(t')}\overline{Z[\boldsymbol{\psi}]} \tag{13}$$

The disorder occurs only in $\Xi[\mathbf{c},\{\mathbf{k}\}]$, so we have to calculate $\overline{\exp\Xi[\mathbf{c},\{\mathbf{k}\}]}$. For the present reaction rates, drawn independently for each (i,j) with $i\leq j$, we may use the fact that in (7) the homo-dimer entries c_{ii} are statistically independent of those representing hetero-dimers:

$$\overline{\mathrm{e}^{\Xi[\mathbf{c},\{\mathbf{k}\}]}} = \overline{\mathrm{e}^{-\mathrm{i}\sum_i c_{ii}\Xi_{ii}(\mathbf{k}_{ii})}}\,.\,\overline{\mathrm{e}^{-\mathrm{i}\sum_{i<j}c_{ij}\Xi_{ij}(\mathbf{k}_{ij})}} \tag{14}$$

We tackle the two factors separately, using the fact that the relevant matrices have entries $\Xi_{ij}(\mathbf{k}_{ij})=\mathcal{O}(N^0)$. The first factor gives a simple expression which factorizes over the proteins:

$$\overline{\mathrm{e}^{-\mathrm{i}\sum_i c_{ii}\Xi_{ii}(\mathbf{k}_{ii})}} = \prod_i\left[1-c_0+c_0\langle\mathrm{e}^{-\mathrm{i}\Xi_{ii}(\mathbf{k})}\rangle_{\mathbf{k}}\right] \tag{15}$$

The second factor contains the main complications. To average over the random interaction networks we first use the fact that a mathematically equivalent way to write (7) is

$$P(\mathbf{c}) = \frac{1}{\mathcal{Z}}\prod_i\delta_{k_i,\sum_{j\neq i}c_{ij}}\cdot\prod_i\left[c_0\delta_{c_{ii},1}+(1-c_0)\delta_{c_{ii},0}\right]\cdot\prod_{i<j}\left[\frac{\langle k\rangle}{N}\delta_{c_{ij},1}+\left(1-\frac{\langle k\rangle}{N}\right)\delta_{c_{ij},0}\right] \tag{16}$$

The reason is that the extra Poissonnian factor can be written in terms of the average degree $\langle k\rangle=N^{-1}\sum_i k_i$ (and hence absorbed in the normalization constant $\mathcal{Z}$), via the identity

$$\prod_{i<j}\left[\frac{\langle k\rangle}{N}\delta_{c_{ij},1}+\left(1-\frac{\langle k\rangle}{N}\right)\delta_{c_{ij},0}\right] = \left(1-\frac{\langle k\rangle}{N}\right)^{\frac{1}{2}N(N-1)}\mathrm{e}^{\frac{1}{2}N\langle k\rangle\log[\langle k\rangle/(N-\langle k\rangle)]} \tag{17}$$

(since $k_i=\sum_{j\neq i}c_{ij}$ for all i, by virtue of the constrained network degrees). We use integral representations to implement the degree constraints, i.e.

$$\prod_i\delta_{k_i,\sum_{j\neq i}c_{ij}} = \prod_i\int_{-\pi}^{\pi}\frac{\mathrm{d}\omega_i}{2\pi}\,\mathrm{e}^{\mathrm{i}\omega_i(k_i-\sum_{j\neq i}c_{ij})} = \int_{-\pi}^{\pi}\prod_i\left[\frac{\mathrm{d}\omega_i}{2\pi}\mathrm{e}^{\mathrm{i}\omega_i k_i}\right]\mathrm{e}^{-\mathrm{i}\sum_{i<j}c_{ij}(\omega_i+\omega_j)} \tag{18}$$

In combination these ingredients allow us to write

$$\begin{aligned}\overline{\mathrm{e}^{-\mathrm{i}\sum_{i<j}c_{ij}\Xi_{ij}(\mathbf{k}_{ij})}} &= \frac{1}{\mathcal{Z}}\int_{-\pi}^{\pi}\prod_i\left[\frac{\mathrm{d}\omega_i}{2\pi}\mathrm{e}^{\mathrm{i}\omega_i k_i}\right]\prod_{i<j}\Big\langle\sum_{c_{ij}}\left[\frac{\langle k\rangle}{N}\mathrm{e}^{-\mathrm{i}[\omega_i+\omega_j+\Xi_{ij}(\mathbf{k})]}\delta_{c_{ij},1}+\left(1-\frac{\langle k\rangle}{N}\right)\delta_{c_{ij},0}\right]\Big\rangle_{\mathbf{k}}\\ &= \frac{1}{\mathcal{Z}}\int_{-\pi}^{\pi}\prod_i\left[\frac{\mathrm{d}\omega_i}{2\pi}\mathrm{e}^{\mathrm{i}\omega_i k_i}\right]\prod_{i<j}\left[1+\frac{\langle k\rangle}{N}\left(\mathrm{e}^{-\mathrm{i}(\omega_i+\omega_j)}\langle\mathrm{e}^{-\mathrm{i}\Xi_{ij}(\mathbf{k})}\rangle_{\mathbf{k}}-1\right)\right]\\ &= \frac{1}{\mathcal{Z}}\int_{-\pi}^{\pi}\prod_i\left[\frac{\mathrm{d}\omega_i}{2\pi}\mathrm{e}^{\mathrm{i}\omega_i k_i}\right]\mathrm{e}^{\frac{\langle k\rangle}{N}\sum_{i<j}[\mathrm{e}^{-\mathrm{i}(\omega_i+\omega_j)}\langle\mathrm{e}^{-\mathrm{i}\Xi_{ij}(\mathbf{k})}\rangle_{\mathbf{k}}-1]+\mathcal{O}(N^0)}\end{aligned} \tag{19}$$

Journal of Physics: Conference Series **197** (2009) 012006 doi:10.1088/1742-6596/197/1/012006

We may now use $\Xi_{ij}(\mathbf{k}) = \Xi_{ji}(S\mathbf{k})$ (with the previously introduced post-translational state swap operator S defined via $(S\mathbf{k})^{\alpha\beta\pm} = k^{\beta\alpha\pm}$) and the convention that $P(S\mathbf{k}) = P(\mathbf{k})$, which allows us to write $\langle e^{-i\Xi_{ij}(\mathbf{k})}\rangle_{\mathbf{k}} = \langle e^{-i\Xi_{ji}(S\mathbf{k})}\rangle_{\mathbf{k}} = \langle e^{-i\Xi_{ji}(\mathbf{k})}\rangle_{\mathbf{k}}$, and symmetrize the summation $\sum_{i<j}$:

$$\overline{e^{-i\sum_{i<j} c_{ij}\Xi_{ij}(\mathbf{k}_{ij})}} = \frac{e^{-\frac{1}{2}\langle k\rangle N}}{\mathcal{Z}} \int_{-\pi}^{\pi}\prod_i \Big[\frac{d\omega_i}{2\pi} e^{i\omega_i k_i}\Big] e^{\frac{\langle k\rangle}{2N}\sum_{ij} e^{-i(\omega_i+\omega_j)}\langle e^{-i\Xi_{ij}(\mathbf{k})}\rangle_{\mathbf{k}} + \mathcal{O}(N^0)} \tag{20}$$

Obtaining full factorization over protein variables in $\overline{Z[\boldsymbol{\psi}]}$, in leading order in N, requires finding a way to disentangle the different protein variables that appear in the quantity

$$\begin{aligned} &\frac{1}{N^2}\sum_{ij} e^{-i(\omega_i+\omega_j)}\langle e^{-i\Xi_{ij}(\mathbf{k})}\rangle_{\mathbf{k}} = \\ &\frac{1}{N^2}\sum_{ij} e^{-i(\omega_i+\omega_j)}\Big\langle e^{-i\int dt d\tau \sum_{\rho\lambda} k^{\rho\lambda+} x_i^\rho(t-\tau)x_j^\lambda(t-\tau)\left\{\theta[\tau]e^{-k^-\tau}\sum_{\alpha\beta} k^{\alpha\beta-}[\hat{x}_i^\alpha(t)+\hat{x}_j^\beta(t)]-\delta(\tau)[\hat{x}_i^\rho(t)+\hat{x}_j^\lambda(t)]\right\}}\Big\rangle_{\mathbf{k}} \end{aligned} \tag{21}$$

3.2. Introduction of dynamical order parameters

To achieve factorization we introduce an appropriate dynamical order parameter and isolate the joint distribution of unbound protein concentration 'paths' $\{x_i^\alpha\} = \{x_i^\alpha(t)\}$ (with $t \in \mathbb{R}$), for each of their post-translational variants, and their conjugate concentration paths:

$$P[\{x,\hat{x}\}|\{\mathbf{x},\hat{\mathbf{x}}\},\boldsymbol{\omega}] = \frac{1}{N}\sum_i \prod_\alpha \delta[\{x_\alpha\}-\{x_i^\alpha\}]\delta[\{\hat{x}_\alpha\}-\{\hat{x}_i^\alpha\}]e^{-i\omega_i} \tag{22}$$

With this macroscopic object we can write, upon defining the short-hand $\{dx\} = \prod_{t\alpha} dx_\alpha(t)$ and sending $\Delta \to 0$ wherever feasible,

$$\begin{aligned} \frac{1}{N^2}\sum_{ij} e^{-i(\omega_i+\omega_j)}\langle e^{-i\Xi_{ij}(\mathbf{k})}\rangle_{\mathbf{k}} &= \int\{dx d\hat{x} dx' d\hat{x}'\} P[\{x,\hat{x}\}|\{\mathbf{x},\hat{\mathbf{x}}\},\boldsymbol{\omega}]P[\{x',\hat{x}'\}|\{\mathbf{x},\hat{\mathbf{x}}\},\boldsymbol{\omega}] \\ &\times\Big\langle e^{-i\int dt d\tau \sum_{\rho\lambda} k^{\rho\lambda+} x_\rho(t-\tau)x'_\lambda(t-\tau)\left\{\theta[\tau]e^{-k^-\tau}\sum_{\alpha\beta} k^{\alpha\beta-}[\hat{x}_\alpha(t)+\hat{x}'_\beta(t)]-\delta(\tau)[\hat{x}_\rho(t)+\hat{x}'_\lambda(t)]\right\}}\Big\rangle_{\mathbf{k}} \end{aligned} \tag{23}$$

To transform $\overline{Z[\boldsymbol{\psi}]}$ into an expression that can be evaluated for $N \to \infty$ by steepest descent, we have to relocate the kernels $P[\{x,\hat{x}\}|\ldots]$, via the insertion (for each joint path $\{x,\hat{x}\}$) of

$$\begin{aligned} 1 &= \int dP[\{x,\hat{x}\}]\delta\Big[P[\{x,\hat{x}\}] - P[\{x,\hat{x}\}|\{\mathbf{x},\hat{\mathbf{x}}\},\boldsymbol{\omega}]\Big] \\ &= \int \frac{dP[\{x,\hat{x}\}]d\hat{P}[\{x,\hat{x}\}]}{2\pi/N} e^{iN\hat{P}[\{x,\hat{x}\}][P[\{x,\hat{x}\}]-P[\{x,\hat{x}\}|\{\mathbf{x},\hat{\mathbf{x}}\},\boldsymbol{\omega}]]} \end{aligned} \tag{24}$$

After appropriate re-scaling of the conjugate kernels $\hat{P}[\{x,\hat{x}\}]$ (involving factors N and Δ) in order to obtain well-defined functional integrals with the correct N-scaling, we then arrive at

$$\begin{aligned} e^{\frac{\langle k\rangle}{2N}\sum_{ij} e^{-i(\omega_i+\omega_j)}\langle e^{-i\Xi_{ij}(\mathbf{k})}\rangle_{\mathbf{k}}} &= \int\{dP d\hat{P}\}\, e^{iN\int\{dx d\hat{x}\}\hat{P}[\{x,\hat{x}\}]P[\{x,\hat{x}\}]} \prod_i e^{-i\hat{P}[\{x_i,\hat{x}_i\}]e^{-i\omega_i}} \\ &\times \exp\Big\{\frac{1}{2}\langle k\rangle N\int\{dx d\hat{x} dx' d\hat{x}'\}P[\{x,\hat{x}\}]P[\{x',\hat{x}'\}] \\ &\times\Big\langle e^{-i\int dt d\tau \sum_{\rho\lambda} k^{\rho\lambda+} x_\rho(t-\tau)x'_\lambda(t-\tau)\left\{\theta[\tau]e^{-k^-\tau}\sum_{\alpha\beta} k^{\alpha\beta-}[\hat{x}_\alpha(t)+\hat{x}'_\beta(t)]-\delta(\tau)[\hat{x}_\rho(t)+\hat{x}'_\lambda(t)]\right\}}\Big\rangle_{\mathbf{k}}\Big\} \end{aligned} \tag{25}$$

Only the last factor of the first line contains the protein variables $\{x_i, \hat{x}_i, \omega_i\}$, and in a factorized form. This allows us to combine our intermediate results (8,9,10,11,15,20,25) into

$$
\begin{aligned}
\overline{Z[\boldsymbol{\psi}]} \;=\; & \frac{1}{\mathcal{Z}}\int\{\mathrm{d}P\mathrm{d}\hat{P}\}\exp\Big\{\mathcal{O}(N^0)+\;\mathrm{i}N\int\{\mathrm{d}x\mathrm{d}\hat{x}\}\hat{P}[\{x,\hat{x}\}]P[\{x,\hat{x}\}]-\frac{1}{2}N\langle k\rangle \\
& +\frac{1}{2}N\langle k\rangle\int\{\mathrm{d}x\mathrm{d}\hat{x}\mathrm{d}x'\mathrm{d}\hat{x}'\}P[\{x,\hat{x}\}]P[\{x',\hat{x}'\}] \\
& \quad\times\Big\langle \mathrm{e}^{-\mathrm{i}\int\mathrm{d}t\mathrm{d}\tau\sum_{\rho\lambda}k^{\rho\lambda+}x_\rho(t-\tau)x'_\lambda(t-\tau)\left\{\theta[\tau]\mathrm{e}^{-k^{-}\tau}\sum_{\alpha\beta}k^{\alpha\beta-}[\hat{x}_\alpha(t)+\hat{x}'_\beta(t)]-\delta(\tau)[\hat{x}_\rho(t)+\hat{x}'_\lambda(t)]\right\}}\Big\rangle_{\mathbf{k}}\Big\} \\
& \times\prod_i\Big\{\int_{-\pi}^{\pi}\frac{\mathrm{d}\omega}{2\pi}\mathrm{e}^{\mathrm{i}\omega k_i}\int\{\mathrm{d}x\mathrm{d}\hat{x}\}\;\mathrm{e}^{\mathrm{i}\int\mathrm{d}t\sum_\alpha\psi_i^\alpha(t)x_\alpha(t)+\mathrm{i}\int\mathrm{d}t\sum_\alpha\hat{x}_\alpha(t)\left[\frac{\mathrm{d}}{\mathrm{d}t}x_\alpha(t)-\theta_i^\alpha(t)+\gamma_i^\alpha(t)x_\alpha(t)\right]} \\
& \qquad\times\mathrm{e}^{-\mathrm{i}\hat{P}[\{x,\hat{x}\}]\mathrm{e}^{-\mathrm{i}\omega}}\Big[1-c_0+c_0\langle\mathrm{e}^{-\mathrm{i}\sum_{\alpha\rho\lambda}\int\mathrm{d}t\mathrm{d}\tau\;W_{\alpha;\rho\lambda}(\tau|\mathbf{k})\hat{x}_\alpha(t)x_\rho(t-\tau)x_\lambda(t-\tau)}\rangle_{\mathbf{k}}\Big]\Big\}
\end{aligned}
\tag{26}
$$

Now we can for $N\to\infty$ carry out the functional integrations over $\{P,\hat{P}\}$ by steepest descent. Upon defining $\hat{P}[\{x,\hat{x}\}]=\mathrm{i}Q[\{x,\hat{x}\}]$ the outcome takes the form

$$
\lim_{N\to\infty}\frac{1}{N}\log\overline{Z[\psi]} \;=\; \mathrm{extr}_{\{P,Q\}}\Big\{\Psi[\{P,Q\}]+\Phi[\{P\}]+\Omega[\{Q\}]+\mathrm{const}\Big\} \tag{27}
$$

where the constant follows from the identity $\log\overline{Z[\mathbf{0}]}=0$, and with

$$
\Psi[\{P,Q\}] \;=\; -\int\{\mathrm{d}x\mathrm{d}\hat{x}\}P[\{x,\hat{x}\}]Q[\{x,\hat{x}\}]-\frac{1}{2}\langle k\rangle \tag{28}
$$

$$
\begin{aligned}
\Phi[\{P\}] \;=\; & \frac{1}{2}\langle k\rangle\int\{\mathrm{d}x\mathrm{d}\hat{x}\mathrm{d}x'\mathrm{d}\hat{x}'\}P[\{x,\hat{x}\}]P[\{x',\hat{x}'\}]\times \\
& \langle\mathrm{e}^{-\mathrm{i}\int\mathrm{d}t\mathrm{d}\tau\sum_{\rho\lambda}k^{\rho\lambda+}x_\rho(t-\tau)x'_\lambda(t-\tau)\left\{\theta[\tau]\mathrm{e}^{-k^{-}\tau}\sum_{\alpha\beta}k^{\alpha\beta-}[\hat{x}_\alpha(t)+\hat{x}'_\beta(t)]-\delta(\tau)[\hat{x}_\rho(t)+\hat{x}'_\lambda(t)]\right\}}\rangle_{\mathbf{k}}
\end{aligned}
\tag{29}
$$

$$
\begin{aligned}
\Omega[\{Q\}] \;=\; & \Big\langle\log\int_{-\pi}^{\pi}\frac{\mathrm{d}\omega}{2\pi}\mathrm{e}^{\mathrm{i}\omega k}\int\{\mathrm{d}x\mathrm{d}\hat{x}\}\;\mathrm{e}^{\mathrm{i}\int\mathrm{d}t\sum_\alpha\left[\psi_\alpha(t)x_\alpha(t)+\hat{x}_\alpha(t)\left(\frac{\mathrm{d}}{\mathrm{d}t}x_\alpha(t)-\theta_\alpha(t)+\gamma_\alpha(t)x_\alpha(t)\right)\right]} \\
& \qquad\times\mathrm{e}^{Q[\{x,\hat{x}\}]\mathrm{e}^{-\mathrm{i}\omega}}\Big\langle\mathrm{e}^{-\mathrm{i}s\sum_{\alpha\rho\lambda}\int\mathrm{d}t\mathrm{d}\tau\;W_{\alpha;\rho\lambda}(\tau|\mathbf{k})\hat{x}_\alpha(t)x_\rho(t-\tau)x_\lambda(t-\tau)}\Big\rangle_{\mathbf{k},s}\Big\rangle_{\{\psi,\theta,\gamma,k\}}
\end{aligned}
\tag{30}
$$

with the short-hands $\langle f[\{\psi_\alpha,\theta_\alpha,\gamma_\alpha\},k]\rangle_{\{\psi,\theta,\gamma,k\}}=\lim_{N\to\infty}N^{-1}\sum_i f[\{\psi_i^\alpha,\theta_i^\alpha,\gamma_i^\alpha\},k_i]$, and with $\langle f(s)\rangle_s=c_0f(1)+(1-c_0)f(0)$. The random variable s controls whether ($s=1$) or not ($s=0$) a given protein species can form homo-dimers. The functional saddle-point equations that define the extremum in (27), and are fully closed by definition, are the following:

$$
\frac{\delta}{\delta P[\{x,\hat{x}\}]}\Psi[\{P,Q\}]+\frac{\delta}{\delta P[\{x,\hat{x}\}]}\Phi[\{P\}]=0 \tag{31}
$$

$$
\frac{\delta}{\delta Q[\{x,\hat{x}\}]}\Psi[\{P,Q\}]+\frac{\delta}{\delta Q[\{x,\hat{x}\}]}\Omega[\{Q\}]=0 \tag{32}
$$

From now on we analyze equations (31,32) only for $\boldsymbol{\psi}=\mathbf{0}$, the generating fields $\boldsymbol{\psi}$ will be used in due course only to identify the physical meaning of our order parameters.

3.3. Saddle point equations

For $N\to\infty$ the dynamical order parameters $P[\{x,\hat{x}\}]$ and $Q[\{x,\hat{x}\}]$ take well-defined values that depend only on the statistical properties of the microscopic system parameters, and which

Journal of Physics: Conference Series **197** (2009) 012006 doi:10.1088/1742-6596/197/1/012006

follow upon solving the coupled equations (31,32). Working out these fully exact formulae gives:

$$
\begin{aligned}
Q[\{x,\hat{x}\}] &= \langle k\rangle \int \{\mathrm{d}x'\mathrm{d}\hat{x}'\} P[\{x',\hat{x}'\}] \times \qquad (33)\\
&\quad \langle \mathrm{e}^{-\mathrm{i}\int \mathrm{d}t\mathrm{d}\tau \sum_{\rho\lambda} k^{\rho\lambda +} x_\rho(t-\tau)x'_\lambda(t-\tau)\left\{\theta[\tau]\mathrm{e}^{-k^- \tau}\sum_{\alpha\beta} k^{\alpha\beta -}[\hat{x}_\alpha(t)+\hat{x}'_\beta(t)]-\delta(\tau)[\hat{x}_\rho(t)+\hat{x}'_\lambda(t)]\right\}}\rangle_{\mathbf{k}}\\
P[\{x,\hat{x}\}] &= \Big\langle \frac{\int_{-\pi}^{\pi}\frac{\mathrm{d}\omega}{2\pi}\mathrm{e}^{\mathrm{i}\omega(k-1)+Q[\{x,\hat{x}\}]\mathrm{e}^{-\mathrm{i}\omega}} M[\{x,\hat{x},\theta,\gamma\}]}{\int\{\mathrm{d}x'\mathrm{d}\hat{x}'\}\int_{-\pi}^{\pi}\frac{\mathrm{d}\omega}{2\pi}\mathrm{e}^{\mathrm{i}\omega k+Q[\{x',\hat{x}'\}]\mathrm{e}^{-\mathrm{i}\omega}} M[\{x',\hat{x}',\theta,\gamma\}]}\Big\rangle_{\{\theta,\gamma,k\}} \qquad (34)
\end{aligned}
$$

in which

$$
M[\{x,\hat{x},\theta,\gamma\}] = \Big\langle \mathrm{e}^{\mathrm{i}\int \mathrm{d}t\sum_\alpha \hat{x}_\alpha(t)\left[\frac{\mathrm{d}}{\mathrm{d}t}x_\alpha(t)-\theta_\alpha(t)+\gamma_\alpha(t)x_\alpha(t)-s\sum_{\rho\lambda}\int \mathrm{d}\tau\; W_{\alpha;\rho\lambda}(\tau|\mathbf{k})x_\rho(t-\tau)x_\lambda(t-\tau)\right]}\Big\rangle_{\mathbf{k},s} \qquad (35)
$$

The identity $(2\pi)^{-1}\int_{-\pi}^{\pi}\mathrm{d}\omega\;\exp[\mathrm{i}\omega k+z\mathrm{e}^{-\mathrm{i}\omega}]=z^k/k!$ allows us to simplify (34) to

$$
P[\{x,\hat{x}\}] = \Big\langle \frac{kQ^{k-1}[\{x,\hat{x}\}]M[\{x,\hat{x},\theta,\gamma\}]}{\int\{\mathrm{d}x'\mathrm{d}\hat{x}'\}Q^k[\{x',\hat{x}'\}]M[\{x',\hat{x}',\theta,\gamma\}]}\Big\rangle_{\{\theta,\gamma,k\}} \qquad (36)
$$

It seems appropriate at this stage to switch from $P[\{x,\hat{x}\}]$ to its partial Fourier transform, which will lead to equations involving only real-valued order parameters that can be interpreted (as will be shown) as conditional path probabilities in the sense suggested by our adopted notation:

$$
W[\{x\}|\{y\}] = \int\{\mathrm{d}\hat{x}\}P[\{x,\hat{x}\}]\mathrm{e}^{-\mathrm{i}\int \mathrm{d}t\sum_\alpha y_\alpha(t)\hat{x}_\alpha(t)} \qquad (37)
$$

Adopting this new definition converts our closed equations (33,36) into

$$
\begin{aligned}
W[\{x\}|\{y\}] &= \Big\langle \frac{k\int\{\mathrm{d}\hat{x}\}Q^{k-1}[\{x,\hat{x}\}]M[\{x,\hat{x},\theta+y,\gamma\}]}{\int\{\mathrm{d}x'\mathrm{d}\hat{x}'\}Q^k[\{x',\hat{x}'\}]M[\{x',\hat{x}',\theta,\gamma\}]}\Big\rangle_{\{\theta,\gamma,k\}} \qquad (38)\\
Q[\{x,\hat{x}\}] &= \langle k\rangle\int\{\mathrm{d}x'\mathrm{d}y\}W[\{x'\}|\{y\}]\Big\langle \mathrm{e}^{-\mathrm{i}\int \mathrm{d}t\sum_\alpha \hat{x}_\alpha(t)\int \mathrm{d}\tau\sum_{\rho\lambda}W_{\alpha;\rho\lambda}(\tau|\mathbf{k})x_\rho(t-\tau)x'_\lambda(t-\tau)}\\
&\quad\times\prod_{\alpha t}\delta\Big[y_\alpha(t)-\int \mathrm{d}\tau\sum_{\rho\lambda}W_{\alpha;\rho\lambda}(\tau|S\mathbf{k})x'_\rho(t-\tau)x_\lambda(t-\tau)\Big]\Big\rangle_{\mathbf{k}} \qquad (39)
\end{aligned}
$$

with the previously introduced short-hand $(S\mathbf{k})^{\alpha\beta\pm}=k^{\beta\alpha\pm}$, and where we have used the following identity which follows trivially from definition (35),

$$
M[\{x,\hat{x},\theta,\gamma\}]\mathrm{e}^{-\mathrm{i}\int \mathrm{d}t\sum_\alpha y_\alpha(t)\hat{x}_\alpha(t)} = M[\{x,\hat{x},\theta+y,\gamma\}] \qquad (40)
$$

To work out equations (38,39) further we need to evaluate the following integrals, for $k>0$:

$$
\begin{aligned}
&\int\{\mathrm{d}\hat{x}\}Q^k[\{x,\hat{x}\}]M[\{x,\hat{x},\theta,\gamma\}]\\
&= C\langle k\rangle^k\int\prod_{\ell\leq k}\Big[\{\mathrm{d}x_\ell\mathrm{d}y_\ell\}W[\{x_\ell\}|\{y_\ell\}]\Big\langle\prod_{\ell\alpha t}\delta\Big[y_{\ell\alpha}(t)-\int \mathrm{d}\tau\sum_{\rho\lambda}W_{\alpha;\rho\lambda}(\tau|S\mathbf{k}_\ell)x_{\ell\rho}(t-\tau)x_\lambda(t-\tau)\Big]\\
&\quad\times\lim_{\Delta\to 0}\prod_{\alpha t}\delta\Big[x_\alpha(t+\Delta)-x_\alpha(t)-\Delta[\theta_\alpha(t)-\gamma_\alpha(t)x_\alpha(t)]\\
&\quad-\Delta\sum_{\rho\lambda}\int \mathrm{d}\tau\; x_\rho(t-\tau)\Big(sW_{\alpha;\rho\lambda}(\tau|\mathbf{k})x_\lambda(t-\tau)+\sum_\ell W_{\alpha;\rho\lambda}(\tau|\mathbf{k}_\ell)x_{\ell\lambda}(t-\tau)\Big)\Big]\Big\rangle_{\mathbf{k},s;\mathbf{k}_1\ldots\mathbf{k}_k} \qquad (41)
\end{aligned}
$$

Here C is a constant, which will drop out of our equations. Further simplification of equations (38,39) requires knowing the physical interpretation of the order parameter $W[\{x\}|\{y\}]$.

3.4. Physical meaning of order parameters

We can infer the meaning of $W[\{x\}|\{y\}]$, starting from (22,37) and using the manipulations applied to (8), which effectively boils down to making in (26) the substitution

$$e^{i\int dt \sum_{i\alpha} \psi_i^\alpha(t) x_i^\alpha(t)} \rightarrow \frac{1}{N}\sum_j e^{-i\omega_j} \prod_\alpha \left[\delta[\{x_\alpha\}-\{x_j^\alpha\}]\, e^{-i\int dt\, y_\alpha(t)\hat{x}_j^\alpha(t)}\right] \tag{42}$$

Carrying out this substitution shows that insertion of the factor $\exp[-i\omega_j]$ into (26) is equivalent to replacing $k_j \rightarrow k_j - 1$, and that insertion of the factor $\exp[-i\int dt\; y_\alpha(t)\hat{x}_j^\alpha(t)]$ is equivalent to replacing $\theta_j^\alpha(t) \rightarrow \theta_j^\alpha(t) + y_\alpha(t)$. We may therefore conclude that

$$W[\{x\}|\{y\}] \;=\; \frac{1}{N}\sum_j \overline{\langle\delta[\{x\} - \{x_j\}]\rangle}|_{k_j\rightarrow k_j-1,\;\; \theta_j^\alpha(t)\rightarrow\theta_j^\alpha(t)+y_\alpha(t)\;\forall\alpha} \tag{43}$$

where the brackets $\langle\ldots\rangle$ denote evaluation of the argument for the microscopic process (5). $W[\{x\}|\{y\}]$ is apparently a generalized response function. It gives the probability that if we pick at random a protein species j, remove one randomly selected binding partner from it and increase instead its production rates $\{\theta_j^\alpha\}$ by (time-dependent) amounts $\{y_\alpha\}$, we will observe for the post-translational states of that protein the concentration evolution $\{x_\alpha\}$. Similarly,

$$D[\{x\}|\{y\}] \;=\; \frac{1}{N}\sum_j \overline{\langle\delta[\{x\} - \{x_j\}]\rangle}|_{\theta_j^\alpha(t)\rightarrow\theta_j^\alpha(t)+y_\alpha(t)\;\forall\alpha} \tag{44}$$

(which quantifies the response to perturbation of the actual original system, without the node removals). Since the calculation of (44) differs from that of (43) only in the removal of the factors $\exp[-i\omega_i]$ from definition (22), tracing the differences between the two calculations is easy. One finds that it boils down to replacing $\exp[i\omega(k-1)]$ by $\exp[i\omega k]$ in (34), and hence replacing $kQ^{k-1}[\{x,\hat{x}\}]$ by $Q^k[\{x,\hat{x}\}]$ in (38), which leads us directly to

$$D[\{x\}|\{y\}] \;=\; \sum_{k\geq 0} p(k) \Big\langle \frac{\int\{d\hat{x}\}Q^k[\{x,\hat{x}\}]M[\{x,\hat{x},\theta{+}y,\gamma\}]}{\int\{dx'd\hat{x}'\}Q^k[\{x',\hat{x}'\}]M[\{x',\hat{x}',\theta,\gamma\}]}\Big\rangle_{\{\theta,\gamma\}} \tag{45}$$

Both $W[\{x\}|\{y\}]$ and $D[\{x\}|\{y\}]$ are conditional measures in the paths $\{x\}$, so they must obey $\int\{dx\}\; W[\{x\}|\{y\}] = \int\{dx\}\; D[\{x\}|\{y\}] = 1$ for all $\{y\}$. After solving $W[\{x\}|\{y\}]$ from (38,39), one obtains $D[\{x\}|\{y\}]$ via (45), and extracts $D[\{x\}|\{0\}]$ as the measure for the concentrations of unbound proteins in the unperturbed system. It follows from (43), together with the causality of our equations (5) and the fact that we have discretized our microscopic laws according to the Itô convention $dx(t) = x(t+dt) - x(t)$, that each $x_\alpha(t)$ in $W[\{x\}|\{y\}]$ can depend only on those $y_\beta(t')$ that have $t' < t$ (to be precise: that have $t' \leq t - \Delta$ before the $\Delta \rightarrow 0$ limit).

3.5. Implications of causality

Returning to (41), we see that causality allows us to integrate (41) over $\{x\}$. We first discretize time and integrate over the $x_\alpha(t)$ with $t = t_{\max}$, followed by integration over the $x'_{\ell\alpha}(t_{\max})$ and the $y_{\ell\alpha}(t_{\max})$ (in that order). The net result is the same expression we started out with, but with $t_{\max}$ replaced by $t_{\max} - \Delta$. Repetition of the argument then leads to the simple result

$$\int\{dxd\hat{x}\}\; Q^k[\{x,\hat{x}\}]M[\{x,\hat{x},\theta,\gamma\}] \;=\; C\langle k\rangle^k \tag{46}$$

Integration of (38) over $\{x\}$ reproduces the correct normalization $\int\{dx\}W[\{x\}|\{y\}] = 1$. We insert relation (46) into (38), which thereby simplifies to

$$W[\{x\}|\{y\}] \;=\; \sum_{k>0} \frac{p(k)k}{\langle k\rangle} \int\{d\hat{x}\}Q^{k-1}[\{x,\hat{x}\}]\Big\langle M[\{x,\hat{x},\theta{+}y,\gamma\}]\Big\rangle_{\{\theta,\gamma|k\}} \tag{47}$$

Here $\langle\ldots\rangle_{\{\theta,\gamma|k\}}$ denotes averaging over the statistics of production and decay rates of those proteins that have k binding partners. Finally we insert (41) to eliminate the auxiliary kernel $Q[\{x,\hat{x}\}]$, and use $C_k = k!C_0$. This brings us after some simple re-arrangements to a transparent closed equation that involves the physical kernel $W[\{x\}|\{y\}]$ only. This equation, exact in the limit $N\to\infty$, is the final result of the generating functional analysis:

$$
\begin{aligned}
W[\{x\}|\{y\}] &= \sum_{k\geq 1}\frac{p(k)k}{\langle k\rangle} \\
&\times\Big\langle\Big\langle\int\prod_{\ell<k}\Big\{\{\mathrm{d}x_\ell \mathrm{d}y_\ell\}W[\{x_\ell\}|\{y_\ell\}]\prod_{\alpha t}\delta\Big[y_{\ell\alpha}(t)-\int\!\mathrm{d}\tau\sum_{\rho\lambda}W_{\alpha;\rho\lambda}(\tau|S\mathbf{k}_\ell)x_{\ell\rho}(t-\tau)x_\lambda(t-\tau)\Big]\Big\} \\
&\times\prod_{\alpha t}\delta\Big\{\mathrm{d}x_\alpha(t)-\mathrm{d}t\Big[\theta_\alpha(t)+y_\alpha(t)-\gamma_\alpha(t)x_\alpha(t) \\
&+\sum_{\rho\lambda}\int\!\mathrm{d}\tau\ x_\rho(t-\tau)\Big(sW_{\alpha;\rho\lambda}(\tau|\mathbf{k})x_\lambda(t-\tau)+\sum_{\ell<k}W_{\alpha;\rho\lambda}(\tau|\mathbf{k}_\ell)x_{\ell\lambda}(t-\tau)\Big)\Big]\Big\}\Big\rangle_{\mathbf{k},s;\mathbf{k}_1\ldots\mathbf{k}_{k-1}}\Big\rangle_{\{\theta,\gamma|k\}}
\end{aligned}
\tag{48}
$$

From the solution of (48) then follows $D[\{x\}|\{y\}]$, defined in (44), via equation (45), giving

$$
\begin{aligned}
D[\{x\}|\{y\}] &= \sum_{k\geq 0}p(k) \\
&\times\Big\langle\Big\langle\int\prod_{\ell\leq k}\Big\{\{\mathrm{d}x_\ell \mathrm{d}y_\ell\}W[\{x_\ell\}|\{y_\ell\}]\prod_{\alpha t}\delta\Big[y_{\ell\alpha}(t)-\int\!\mathrm{d}\tau\sum_{\rho\lambda}W_{\alpha;\rho\lambda}(\tau|S\mathbf{k}_\ell)x_{\ell\rho}(t-\tau)x_\lambda(t-\tau)\Big]\Big\} \\
&\times\prod_{\alpha t}\delta\Big\{\mathrm{d}x_\alpha(t)-\mathrm{d}t\Big[\theta_\alpha(t)+y_\alpha(t)-\gamma_\alpha(t)x_\alpha(t) \\
&+\sum_{\rho\lambda}\int\!\mathrm{d}\tau\ x_\rho(t-\tau)\Big(sW_{\alpha;\rho\lambda}(\tau|\mathbf{k})x_\lambda(t-\tau)+\sum_{\ell\leq k}W_{\alpha;\rho\lambda}(\tau|\mathbf{k}_\ell)x_{\ell\lambda}(t-\tau)\Big)\Big]\Big\}\Big\rangle_{\mathbf{k},s;\mathbf{k}_1\ldots\mathbf{k}_{k}}\Big\rangle_{\{\theta,\gamma|k\}}
\end{aligned}
\tag{49}
$$

Although different at the level of mathematical details, the structure of the above equations is very similar to what was found in non-equilibrium statistical mechanical studies of spin models on finitely connected random graphs, such as [3, 4, 5, 6]. In the special case of protein interaction networks with Poissonian degree distributions, i.e. $p(k)=\mathrm{e}^{-c}c^k/k!$, where $p(k+1)(k+1)/\langle k\rangle = p(k)$ for all k, a simple transformation $k\to k+1$ shows that the kernels $P[\{x\}|\{y\}]$ and $D[\{x\}|\{y\}]$ will be identical. In all other cases this will not be true.

3.6. Cavity interpretation of the order parameter equations

One can understand our results (48,49) at a more intuitive level, starting from equations (5). Any protein species i interacts with the proteome only via its direct partners, the species in the set $\partial_i=\{\ell|\ c_{i\ell}=1\}$. Given the paths $\{x_\ell\}$ taken by the concentrations of the species $\ell\in\partial_i$, the path $\{x_i\}$ taken by i could in principle be calculated (modulo non-invertibility issues) by solving a linear equation, giving an expression of the form $\{x_i\}=\mathcal{F}_i[\{x_\ell\},\ell\in\partial_i]$. This is illustrated in figure 1[A], where site i is shown in black for an example with $k_i=4$. Next we imagine changing the structure of the graph locally, by removing the information flow from i to ∂_i, see figure 1[B] where the protein species in ∂_i are drawn as ◎. We compensate for this intervention, however, by adjusting the production rates of all $j\in\partial_i$ according to $\theta_j^\alpha(t)\to\theta_j^\alpha(t)+y_j^\alpha(\{x_i,x_j\},t)$ such that we do not change any of the concentration paths in the system, which requires

$$
y_j^\alpha(\{x_i,x_j\},t) = \int\!\mathrm{d}s\sum_{\rho\lambda}W_{\alpha;\rho\lambda}(t-s|\mathbf{k}_{i\ell})x_i^\lambda(s)x_j^\rho(s) \tag{50}
$$

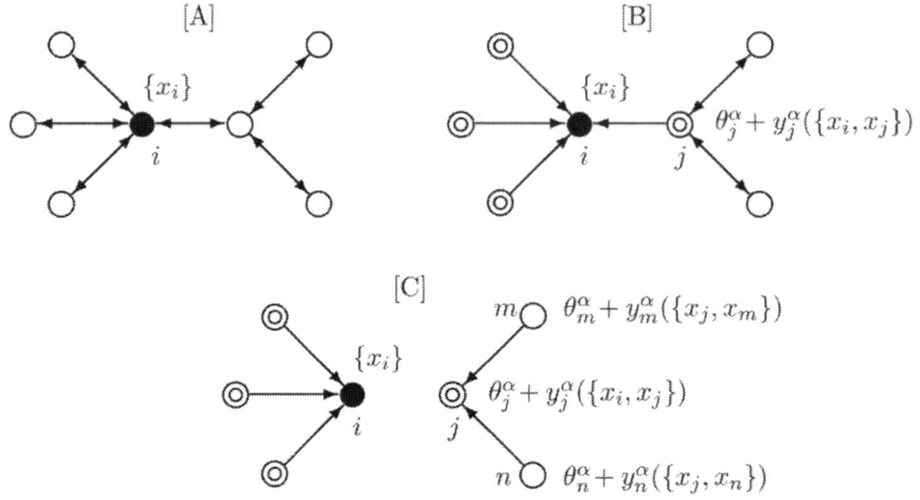

Figure 1. Interpretation of our order parameter equations. Top left [A]: section of the original interaction network, showing a site i which has four interaction partners (these define the set ∂_i). Top right [B]: a locally modified network where information flow from $i \to \partial_i$ is prohibited (so node i is effectively removed), but the production rates of *all* $j \in \partial_i$ (i.e. of all sites marked as ◎) are adjusted in compensation, so that none of the concentrations in the system change, following (50). Bottom [C]: result of a further modification, where also the information flow from $j \to \partial_j$ is prohibited (so also node j is effectively removed), but again compensated for by appropriate adjustment of the production rates of the sites in $\partial_j = \{m, n\}$.

Hence, with the conventions $\partial_i = \{j_{i1}, \ldots, j_{ik_i}\}$ and $D[\{x\}] = D[\{x\}|\{0\}]$ we may write

$$\begin{aligned}
D[\{x\}] &= \lim_{N\to\infty} \frac{1}{N} \sum_i \overline{\delta[\{x\} - \mathcal{F}_i[\{x_{j_{i1}}, \ldots, x_{j_{ik_i}}\}]]} \\
&= \lim_{N\to\infty} \frac{1}{N} \sum_i \overline{\int \prod_{\ell=1}^{k_i} \Big\{\{\mathrm{d}x_\ell\}\delta[\{x_\ell\} - \{x_{j_{i\ell}}\}]\Big\}\, \delta[\{x\} - \mathcal{F}_i[\{x_1, \ldots, x_{k_i}\}]]} \\
&= \lim_{N\to\infty} \frac{1}{N} \sum_i \overline{\int \prod_{\ell=1}^{k_i} \Big[\{\mathrm{d}x_\ell\}\delta[\{x_\ell\} - \{x_{j_{i\ell}}\}]_{i \text{ removed},\ \{\theta_{j_{i\ell}} + y_{j_{i\ell}}(\{x, x_\ell\})\}}\Big]\, \delta[\{x\} - \mathcal{F}_i[\{x_1, \ldots, x_{k_i}\}]]} \\
&= \lim_{N\to\infty} \frac{1}{N} \sum_i \int \prod_{\ell=1}^{k_i} \Big[\{\mathrm{d}x_\ell \mathrm{d}y_\ell\}\Big] \qquad (51) \\
&\quad \times \overline{\prod_{\ell=1}^{k_i} \Big[\delta[\{x_\ell\} - \{x_{j_{i\ell}}\}]_{i \text{ removed},\ \{\theta_{j_{i\ell}} + y_\ell\}} \delta[\{y_\ell\} - \{y_{j_{i\ell}}(\{x, x_\ell\})\}]\Big] \delta[\{x\} - \mathcal{F}_i[\{x_1, \ldots, x_{k_i}\}]]}
\end{aligned}$$

If the original network had no loops, then in the new 'cavity' graph the branches attached to sites $\ell \in \partial_i$ are disconnected, and each will behave as an independent graph, for large N topologically equivalent to the original. It follows that in the cavity graph we can treat the paths $\{x_\ell\}$ with $\ell \in \partial_i$ as statistically independent, and assume they have identical statistical properties, so

$$D[\{x\}] = \lim_{N\to\infty} \frac{1}{N} \sum_i \int \prod_{\ell=1}^{k_i} \Big[\{\mathrm{d}x_\ell \mathrm{d}y_\ell\} \overline{\Big(\frac{1}{N} \sum_j \delta[\{x_\ell\} - \{x_j\}]_{i \text{ removed},\ \{\theta_j + y_\ell\}}\Big)}\Big] \qquad (52)$$

$$
\begin{aligned}
&\overline{\times\delta[\{x\}-\mathcal{F}_i[\{x_1,\ldots,x_{k_i}\}]]\prod_{\ell\leq k_i}\delta[\{y_\ell\}-\{y_{j_{i\ell}}(\{x,x_\ell\})\}]} \\
=\;& \lim_{N\to\infty}\frac{1}{N}\sum_i\int\prod_{\ell=1}^{k_i}\Big[\{\mathrm{d}x_\ell\mathrm{d}y_\ell\}W[\{x_\ell\}|\{y_\ell\}]\Big] \\
&\overline{\times\;\delta[\{x\}-\mathcal{F}_i[\{x_1,\ldots,x_{k_i}\}]]\prod_{\ell\leq k_i}\delta[\{y_\ell\}-\{y_{j_{i\ell}}(\{x,x_\ell\})\}]}
\end{aligned}
\tag{53}
$$

One will now be led to equation (49), upon simply adding $\{y\}$ to the production rates $\{\theta_i\}$ (which brings us from $D[\{x\}]$ to $D[\{x\}|\{y\}]$) and with the last two lines of (49) representing the more explicit representation of our symbolic expression $\delta[\{x\}-\mathcal{F}_i[\{x_1,\ldots,x_{k_i}\}]]$.

What remains is to understand the origin of (48), which requires an expression for the concentration path statistics of nodes in ∂_i, such as j in figure 1[B]. We repeat the process of blocking the information flow away from the node of which we try to calculate the concentration paths, while compensating the production rates of its partners appropriately. Now this means removing j, and adjusting the production rates of the nodes in ∂_j (except for i, which has been removed already); see figure 1[C]. The equation for $W[\{x\}|\{y\}]$ is therefore nearly identical to that of $D[\{x\}|\{y\}]$, with two differences: first, since we are looking strictly at nodes j that were connected to a (now removed) cavity node i, these nodes j no longer have in-degree statistics $p(k)$, and second, only k_j-1 of the original k_j partners of each j contribute to $\{x_j\}$. Since our random graph ensemble has no degree-degree correlations, the modified degree statistics $\tilde{p}(k)$ of the nodes $j\in\partial_i$ that were initially attached to (randomly drawn) 'cavity' sites i follows from

$$
\tilde{p}(k)=\lim_{N\to\infty}\frac{\sum_{ij}c_{ij}\delta_{k_j,k}}{\sum_{ij}c_{ij}}=\lim_{N\to\infty}\frac{k}{N\langle k\rangle}\sum_j\delta_{k_j,k}=\frac{p(k)k}{\langle k\rangle}
\tag{54}
$$

We can thus obtain our equation for $W[\{x\}|\{y\}]$ by making in the right-hand side of (49) the replacements $p(k)\to p(k)k/\langle k\rangle$, $\prod_{\ell\leq k}\to\prod_{\ell<k}$, and $\sum_{\ell\leq k}\to\sum_{\ell<k}$. The result is indeed (48).

It is satisfactory that we have been able to use generating functional analysis (GFA) to obtain (48,49), which is more precise and direct than the above reasoning and did not require the strict absence of loops (in effect, GFA confirms that for $N\to\infty$ any loops generated in (7) have vanishing impact on the process). Second, as soon as we use more complicated ensembles than (7), e.g. those with degree-degree correlations as in [7], or introduce correlations between the reaction rates of distinct protein pairs, the above simple arguments would become prohibitively messy, whereas the GFA route should in principle remain open.

4. Solution of the macroscopic equations

The focus must now turn to solving equation (48) for $W[\{x\}|\{y\}]$, from which $D[\{x\}|\{y\}]$ follows via (49). This is a highly nontrivial problem, on which progress has so far been slow; not just here, but in all GFA studies of processes on finitely connected graphs [3, 5, 6]. The possible routes to be explored come, roughly, in four areas. This paper is only the first step in a research programme, establishing 'proof of principle' that GFA methods can indeed be used to study the proteome, hence in view of space limitations we will here only comment briefly on each area:

4.1. Numerical solution of the macroscopic laws

Due to the exponential increase with time of the number of macroscopic observables concerned, even in studies with discrete time and discrete variables [3, 5, 6], numerical solution of equations such as (48) was possible only for a small number of time steps. In the present problem, where the arguments of $W[\{x\}|\{y\}]$ are continuous paths, numerical solution is not a realistic option.

4.2. Working out solutions in specific simplifying limits

The two main limits where analytical solution is possible are low connectivity and high connectivity (where $\langle k\rangle \to \infty$ must be preceded by a suitable re-scaling $k^{+}_{\alpha\beta} \to k^{+}_{\alpha\beta}/\langle k\rangle$ of all on-rates). If we assume our network has no disconnected nodes, then the lowest overall connectivity is found for $p(k) = \delta_{k1}$, where we find (48,49) reducing to

$$W[\{x\}|\{y\}] = \Big\langle\Big\langle \prod_{\alpha t}\delta\Big\{\mathrm{d}x_\alpha(t) - \mathrm{d}t\Big[\theta_\alpha(t) + y_\alpha(t) - \gamma_\alpha(t)x_\alpha(t) + \sum_{\rho\lambda}\int \mathrm{d}\tau\; x_\rho(t-\tau)\Big(sW_{\alpha;\rho\lambda}(\tau|\mathbf{k})x_\lambda(t-\tau)\Big)\Big]\Big\}\Big\rangle_{\mathbf{k},s}\Big\rangle_{\{\theta,\gamma\}} \tag{55}$$

$$D[\{x\}|\{y\}] = \Big\langle\int\{\mathrm{d}x'\mathrm{d}y'\}W[\{x'\}|\{y'\}]\prod_{\alpha t}\delta\Big[y'_\alpha(t) - \int\mathrm{d}\tau\sum_{\rho\lambda}W_{\alpha;\rho\lambda}(\tau|S\mathbf{k}')x'_\rho(t-\tau)x_\lambda(t-\tau)\Big]$$
$$\times\Big\langle\prod_{\alpha t}\delta\Big\{\mathrm{d}x_\alpha(t) - \mathrm{d}t\Big[\theta_\alpha(t) + y_\alpha(t) - \gamma_\alpha(t)x_\alpha(t) \tag{56}$$
$$+\sum_{\rho\lambda}\int\mathrm{d}\tau\; x_\rho(t-\tau)\Big(sW_{\alpha;\rho\lambda}(\tau|\mathbf{k})x_\lambda(t-\tau) + W_{\alpha;\rho\lambda}(\tau|\mathbf{k}_\ell)x'_\lambda(t-\tau)\Big)\Big]\Big\}\Big\rangle_{\mathbf{k},s;\mathbf{k}'}\Big\rangle_{\{\theta,\gamma\}}$$

These equations are easily interpreted (all cavity sites are now isolated) and converted into a pair of coupled stochastic differential equations, whose statistics represent the diversity of concentration paths in the original N-protein system. For large connectivity we first re-scale the reaction on-rates according to $k^{\alpha\beta^+} = \tilde{k}^{\alpha\beta+}/\langle k\rangle$. One now finds that $D[\{x\}] = D[\{x\}|\{0\}] = W[\{x\}|\{0\}]$ obeys a closed equation as $\langle k\rangle \to \infty$, which using the law of large numbers becomes

$$D[\{x\}] = \Big\langle\prod_{\alpha t}\delta\Big[\mathrm{d}x_\alpha(t) - \mathrm{d}t\Big(\theta_\alpha(t) - \gamma_\alpha(t)x_\alpha(t) + s\sum_{\rho\lambda}\int\mathrm{d}\tau\; W_{\rho\lambda}(\tau|\mathbf{k})x_\rho(t-\tau)x_\lambda(t-\tau)$$
$$+\sum_{\rho\lambda}\int\mathrm{d}\tau\; W_{\rho\lambda}(\tau|S\tilde{\mathbf{k}})x_\lambda(t-\tau)\int\{\mathrm{d}x'\}\; D[\{x'\}]x'_\rho(t-\tau)\Big)\Big]\Big\rangle_{s,\mathbf{k},\tilde{\mathbf{k}},\{\theta,\gamma\}} \tag{57}$$

Again the problem can be converted into a relatively simple stochastic equation. The protein-protein interaction occurs in a 'mean field' way. In fact for $\langle k\rangle \to \infty$ we can extract from (57) a closed equation for the disorder-averaged path $\{X\} = \int\{\mathrm{d}x\}D[\{x\}]\{x\}$, rather than a closed theory in the language of correlation- and response functions (see e.g. [8]) which one would have found upon taking the diverging connectivity limit in models of disordered spin systems or neural networks on finitely connected random graphs.

4.3. Constructing ad-hoc approximations

Various approximations could be considered. First, one could solve (48) iteratively, generating a sequence of measures $W_n[\{x\}|\{y\}]$ by substitution for each n of $W_n[\{x\}|\{y\}]$ in the right-hand side of (48) and defining the left-hand side as $W_{n+1}[\{x\}|\{y\}]$. Upon starting e.g. from the trivial $W_0[\{x\}|\{y\}] = \delta[\{x\}]$, one would find $W_1[\{x\}|\{y\}]$ describing a dimers-only system of proteins, and $W_2[\{x\}|\{y\}]$ describing interacting proteins but without an Onsager reaction term, etc. An alternative would be a linear response approximation, based on truncating the exact expansion

$$W[\{x\}|\{y\}] = W[\{x\}] + \sum_\alpha\int\mathrm{d}t\; y_\alpha(t)\frac{\delta W[\{x\}]}{\delta\theta_\alpha(t)} + \frac{1}{2}\sum_{\alpha\beta}\int\mathrm{d}t\mathrm{d}t'\; y_\alpha(t)y_\beta(t')\frac{\delta^2 W[\{x\}]}{\delta\theta_\alpha(t)\delta\theta_\beta(t')} + \ldots \tag{58}$$

Each such approximation will have specific strengths and weaknesses, which will depend on the statistical characteristics (topology, strength of interactions) of the system being studied.

4.4. Probing further the mathematical structure of the macroscopic laws

An obvious question to be investigated is whether in stationary states there exist exact closed laws for a reduced set of static order parameters (this may require the equivalent of detailed balance, e.g. demanding that $S\mathbf{k} = \mathbf{k}$ for all reaction rates). The natural candidates would be the probability densities $W(\mathbf{x}|\mathbf{y})$ for asymptotic time-averages $\mathbf{x}$ of protein concentrations, given asymptotic time averages $\mathbf{y}$ of production perturbations (as opposed to paths).

One could also try to exploit the origin of (48) as a saddle-point equation for a functional. Upon applying the various transformations of order parameters directly to the function $\Psi[\{P,Q\}] + \Phi[\{P\}] + \Omega[\{Q\}]$ in (27), with an inverse Fourier transform for Q, one obtains

$$\begin{aligned}
\mathcal{L}[\{V,W\}] &= \frac{1}{2} - \int\{\mathrm{d}x\mathrm{d}y\}\, W[\{x\}|\{y\}]V[\{y\}|\{x\}] \\
&\quad + \frac{1}{2}\int\{\mathrm{d}x\mathrm{d}y\mathrm{d}x'\mathrm{d}y'\}\, W[\{x\}|\{y\}]\, \mathcal{M}[\{x,y\};\{x',y'\}]\, W[\{x'\}|\{y'\}] \qquad (59) \\
&\quad + \sum_k \frac{p(k)}{\langle k\rangle}\Big\langle \log \int\{\mathrm{d}x\}\int \prod_{\ell\leq k}\Big[\{\mathrm{d}y_\ell\}\; V[\{y_\ell\}|\{x\}]\Big]\Big\langle \delta\Big[\{x\} - F[\{\theta + \sum_{\ell\leq k} y_\ell\}; s, \mathbf{k}]\Big]\Big\rangle_{\mathbf{k},s}\Big\rangle_{\{\theta,\gamma|k\}}
\end{aligned}$$

with a kernel $\mathcal{M}[\ldots;\ldots]$ that contains all the information on protein-protein interaction,

$$\begin{aligned}
\mathcal{M}[\{x,y\};\{x',y'\}] &= \Big\langle \prod_{\alpha t}\delta\Big[y_\alpha(t) - \int\mathrm{d}\tau\sum_{\rho\lambda} W_{\alpha;\rho\lambda}(\tau|\mathbf{k})x_\rho(t-\tau)x'_\lambda(t-\tau)\Big] \\
&\quad \times \prod_{\alpha t}\delta\Big[y'_\alpha(t) - \int\mathrm{d}\tau\sum_{\rho\lambda} W_{\alpha;\rho\lambda}(\tau|S\mathbf{k})x'_\rho(t-\tau)x_\lambda(t-\tau)\Big]\Big\rangle_{\mathbf{k}} \qquad (60)
\end{aligned}$$

and where $F[\{\theta\}; s, \mathbf{k}]$ denotes the solution of the equation

$$\frac{\mathrm{d}}{\mathrm{d}t}x_\alpha(t) = \theta_\alpha(t) - \gamma_\alpha(t)x_\alpha(t) + s\sum_{\rho\lambda}\int \mathrm{d}\tau\; W_{\alpha;\rho\lambda}(\tau|\mathbf{k})x_\rho(t-\tau)x_\lambda(t-\tau) \qquad (61)$$

One's instinct next would be to attempt a variational formulation: to define on the basis of (59) a function $\mathcal{L}[\{W\}]$ whose minimum would give the true solution. It turns out that this will never be possible here, due to an intriguing property of (59): one can show that for *all* functions that obey causality, the surface $\mathcal{L}[\{V,W\}]$ is no longer dependent upon whether or not the proteins interact. The conclusion must be that all attempts to convert (59) into a variational problem, suitable for generating variational approximations of $W[\{x\}|\{y\}]$, are doomed from the start, since variation within any subset of causal measures will at most give us information on $W[\{x\}|\{y\}]$ that would be true irrespective of whether or not proteins form hetero-dimers.

Ackowledgements

ACCC would like to thank the Engineering and Physical Sciences Research Council (UK) for support in the form of a Springboard Fellowship.

References

[1] Kholodenko B N 2006 *Nat. Rev. Mol. Cell Bio.* **7** 165–76
[2] De Dominicis C 1978 *Phys. Rev.* B **18** 4913–19
[3] Hatchett J P L, Wemmenhove B, Pérez-Castillo I, Nikoletopoulos T, Skantzos N S and Coolen A C C 2004 *J. Phys. A: Math. Gen.* **37** 6201–20
[4] Mozeika A and Coolen A C C 2009 *J. Phys. A: Math. Theor.* **42** 195006
[5] Mimura K and Coolen A C C 2009 *Proc. of 2009 Int. Symp. Information Theory (ISIT)* p 1829
[6] Mimura K and Coolen A C C 2009 *J. Phys. A: Math. Theor.* **42** 415001
[7] Pérez-Vicente C J and Coolen A C C 2008 *J. Phys. A: Math. Theor.* **41** 255003 and 2009 *J. Phys. A: Math. Theor.* **42** 169801
[8] Crisanti A and Sompolinsky H 1987 *Phys. Rev.* A **36** 4922–39

International Workshop on Statistical-Mechanical Informatics 2009 (IW-SMI 2009) IOP Publishing
Journal of Physics: Conference Series **197** (2009) 012007 doi:10.1088/1742-6596/197/1/012007

Differential gene detection incorporating common expression patterns

Shigeyuki Oba[1,2] **and Shin Ishii**[1]

1. Graduate School of Informatics, Kyoto University, Japan.
2. PRESTO, Japan Science and Technology Agency.

E-mail: oba@i.kyoto-u.ac.jp

Abstract. In detection of differentially expressed (DE) genes between different groups of samples based on a high-throughput expression measurement system, we often use a classical statistical testing based on a simple assumption that the expression of a certain DE gene in one group is higher or lower in average than that in the other group. Based on this simple assumption, the theory of optimal discovery procedure (ODP) (Storey, 2005) provided an optimal thresholding function for DE gene detection. However, expression patterns of DE genes over samples may have such a structure that is not exactly consistent with group labels assigned to the samples. Appropriate treatment of such a structure can increase the detection ability. Namely, genes showing similar expression patterns to other biologically meaningful genes can be regarded as statistically more significant than those showing expression patterns independent of other genes, even if differences in mean expression levels are comparable. In this study, we propose a new statistical thresholding function based on a latent variable model incorporating expression patterns together with the ODP theory. The latent variable model assumes hidden common signals behind expression patterns over samples and the ODP theory is extended to involve the latent variables. When applied to several gene expression data matrices which include cluster structures or 'cancer outlier' structures, the newly-proposed thresholding functions showed prominently better detection performance of DE genes than the original ODP thresholding function did. We also demonstrate how the proposed methods behave through analyses of real breast cancer and lymphoma datasets.

1. Introduction

Detection of differentially expressed (DE) genes between different groups of samples is one of the most important issues in gene expression data analyses based on microarray technology. Various statistical thresholding functions have been proposed to evaluate the significance of DE genes, where the thresholding function is a scalar function of the observed data of each gene such that a significant gene is detected if the output of the thresholding function is larger than a threshold. Typical examples are the well-known absolute t-statistic and several modifications of it, such as a statistic called statistical analysis of microarrays (SAM) [1].

Several fundamental ways to design good thresholding functions have been proposed for reducing both type-I (false positive) and type-II (false negative) errors in multiple simultaneous hypothesis testing (MSHT). The empirical Bayes (EB) method provided a non-parametric way to construct a thresholding function called the local false discovery rate for the cases when parametric likelihood functions for null and alternative hypotheses are not directly available [2]. The optimal discovery procedure (ODP) minimized expected number of type-II errors at any pre-fixed number of expected type-I errors provided that all parameters of parametric likelihood functions are exactly known for all pairs of null and alternative hypotheses [3, 4]. Recently, we showed that the EB and ODP are equivalent if the EB is conducted based on a specially designed statistic that corresponds to the ODP [5]. Thus, detection power of the ODP is the theoretical upper limit if the likelihood functions are known. In usual, however, the parameter values in the likelihood functions are unknown, and the estimated thresholding functions based on the estimated parameters may not the optimal. An empirical comparison study showed that the EB and ODP frameworks exhibited near optimal performance in practice [6]. There is another attempt, called Bayesian ODP [7], which also sought better empirical estimation of the theoretically-optimal ODP thresholding function by means of Bayesian shrinkage estimation of each parameter.

Have we really achieved the upper limit of the DE gene detection performance? In this study, we take a different approach which enables us to go over the past theoretical limit of DE gene detection power. In general, detection power in MSHT can be improved by sharing information between multiple tests, and past important improvements proposed various kinds of information to be shared. For example, the SAM used unbalanced numbers of positive and negative DE genes for determining asymmetric thresholds for two-sided test [1]. The EB and ODP incorporated empirical distribution of test statistics of all genes for determining the thresholding function [2]. A common idea in these approaches was that genes whose mean expression levels are similar to those of other DE genes are more likely to be DE genes. We, in this study, incorporate a new aspect of similarity, similarity in *gene expression patterns*; we assume that genes whose expression patterns are similar to those of other DE genes are more likely to be DE genes. In order to implement this idea, we assume a generative model incorporating latent variable that corresponds to the hidden expression patterns whose structures over samples (sub-patterns, see Fig. 1(B)) may be common between genes. The ODP framework is applied for this latent variable model. This, however, demands a general but non-trivial extension of the framework of empirical estimation of ODP thresholding functions.

In Section 2, we review the general framework of the ODP theory and an extension of the ODP to handle a latent variable model, which we designate as hidden variables' ODP (HODP), is explained. In Section 3, a latent variable model is proposed to which the HODP framework is applied. Several heuristics that can be used in the HODP framework are also discussed. Demonstrations on artificial and real microarray datasets are exhibited in Section 4. The article is concluded in Section 5.

1.1. Related works

There have been some other studies assuming that expression patterns of DE genes are not necessarily consistent to the patterns of observed labels of samples. Bair and Tibshirani (2004) proposed a semi-supervised method to perform stable prediction of patient prognosis; they selected a certain number of significant genes by a conventional method and then applied an unsupervised clustering analysis of patients by using the selected genes, instead of directly predicting prognosis labels based on the selected genes (supervised classification), because patients' data represented by the selected genes may have a different cluster structure from that associated by prognosis labels [8]. Genes exhibiting significant expressions in a small number of cancerous tissues are called 'cancer outlier' genes and special statistics to detect them have been proposed [9, 10]. Leek et al. (2008) pointed out that hidden structure can introduce large estimation bias and variance into false discovery rate (FDR), and proposed a correction method which considers hidden structures [11]. By using the correction method, they actually reduced the bias and variance in FDR estimation, although it did not lead to improvement in the detection performance. An application of the ODP framework to a latent variable model was proposed by ourselves [12], in which semi-supervised cases were of particular interest. In this previous study, however, any hidden structure behind the observation was not considered, so that the performance was similar to those by the classical thresholding functions based on the ODP theory. In the current study, essential improvement in DE gene detection ability is sought by directly dealing with hidden variables that reflect hidden structures (sub-patterns) over the samples.

2. General framework of the ODP

2.1. ODP thresholding function

The ODP [3] defines the optimality of statistical thresholding functions in MSHT based on an extension of the Neyman-Pearson's (NP) framework [13].

Let an observation $\boldsymbol{y} \in U$ be sampled randomly from either of null density function $f(\boldsymbol{y})$ or alternative density function $g(\boldsymbol{y})$. U is a set of possible observations. According to a usual statistical testing, iff an observation $\boldsymbol{y}$ is in a pre-determined rejection region, $B \subset U$, the null hypothesis is rejected and the alternative hypothesis is accepted. In this case, type-I and type-II error rates of the test are defined as $\int_B f(\boldsymbol{y})d\boldsymbol{y}$ and $1 - \int_B g(\boldsymbol{y})d\boldsymbol{y}$, respectively. The NP lemma [13] guarantees that the likelihood function $g(\boldsymbol{y})/f(\boldsymbol{y})$ is the optimal thresholding function; namely, when a rejection region B^* is determined by applying a certain threshold to the likelihood ratio, it realizes the smallest type-II error rate at any pre-determined type-I error rate (which is equivalent to setting a specific threshold to the statistic).

In MSHT, we conduct a hypothetical test based on null and alternative likelihood functions, $f_i(\boldsymbol{y}_i)$ and $g_i(\boldsymbol{y}_i)$, for each of the M observations $\boldsymbol{y}_i \in U, i = 1, \cdots, M$, and a common rejection region B is applied for all observations $\boldsymbol{y}_i$ simultaneously.

Storey [3] defined the expected number of true positives (ETP) and the expected number of false positives (EFP) as the counterparts of the type-I error rate and the detection power for MSHT, respectively:

$$\mathrm{ETP} \overset{\mathrm{def}}{=} \int_B \sum_i (1 - w_i) g_i(\boldsymbol{y}) d\boldsymbol{y}, \qquad \mathrm{EFP} \overset{\mathrm{def}}{=} \int_B \sum_i w_i f_i(\boldsymbol{y}) d\boldsymbol{y}, \tag{1}$$

where $w_i = 1$ ($w_i = 0$) holds if the null (alternative) hypothesis is true for the i-th test based on the i-th observation $\boldsymbol{y}_i$. In the followings, w_i is called the weight of the i-th hypothesis. Then, a thresholding function

$$S_{\mathrm{ODP}}(\boldsymbol{y}) \overset{\mathrm{def}}{=} \frac{\sum_i (1 - w_i) g_i(\boldsymbol{y})}{\sum_i w_i f_i(\boldsymbol{y})}, \tag{2}$$

called an optimal discovery procedure (ODP), is proved to be the optimal; if we prepare a rejection region, $B^* = \{\boldsymbol{y} \in U \text{ such that } S_{\mathrm{ODP}}(\boldsymbol{y}) > \lambda\}$ for any threshold $\lambda > 0$, any other rejection region $B(\subset U)$ with equal or smaller EFP makes an ETP equal to or larger than B^*. A proof is given simply by applying the NP lemma (see [3] for details).

2.2. Parametric hypotheses and estimated ODP thresholding function

In actual situations, the null and alternative likelihood functions have parametric forms as $f_i(\boldsymbol{y}; \phi_i)$ and $g_i(\boldsymbol{y}; \theta_i)$ with parameters ϕ_i and θ_i, respectively, so that the parameters are estimated from observed data.

In individual testing, the likelihood ratio (LR) with maximum likelihood (ML) parameters plugged in, is often used as a statistical thresholding function:

$$S_{\mathrm{LR}}(\boldsymbol{y}_i) = \hat{g}_{ii}/\hat{f}_{ii},$$

where $\hat{f}_{ii} \stackrel{\text{def}}{=} f(\boldsymbol{y}_i;\hat{\phi}_i)$ and $\hat{g}_{ii} \stackrel{\text{def}}{=} g(\boldsymbol{y}_i;\hat{\theta}_i)$ are null and alternative likelihoods, respectively, and

$$\hat{\phi}_i \stackrel{\text{def}}{=} \arg\max_{\phi_i} f_i(\boldsymbol{y}_i;\phi_i), \qquad \hat{\theta}_i \stackrel{\text{def}}{=} \arg\max_{\theta_i} g_i(\boldsymbol{y}_i;\theta_i) \tag{3}$$

are ML parameters. When we assume Gaussian noise models for the null and alternative densities, the log-likelihood ratio statistic becomes equivalent to the absolute value of the t-statistic.

In the ODP framework, an empirical approximation of the ODP was formerly proposed [4], which we call the estimated ODP (EODP):

$$S_{\text{EODP}}(\boldsymbol{y}_l) = \frac{\sum_i (1-\hat{w}_i)\hat{g}_{il}}{\sum_i \hat{w}_i \hat{f}_{il}}, \tag{4}$$

where $\hat{f}_{il} \stackrel{\text{def}}{=} f(\boldsymbol{y}_l;\hat{\phi}_i)$ and $\hat{g}_{il} \stackrel{\text{def}}{=} g(\boldsymbol{y}_l;\hat{\theta}_i)$. Since we do not know the true weight, w_i in (2), its estimation $\hat{w}_i$ is used here, which can be estimated by a conventional statistical test; for example, the Kruskall-Wallis test was used in [4] because of its computational convenience. Another way to estimate the weight w_i is to employ EODP in an iterative manner, and actually our simulation study has shown that the performance of the simultaneous testing increases slightly over that of a non-iterative implementation (data not shown). When the ODP score employs the true values of $\hat{w}_i$, $\hat{\theta}_i$, and $\hat{\phi}_i$, it is called TODP, $S_{\text{TODP}}(\boldsymbol{y}_l)$, which is not available in practical situations.

An important point of the ODP is that it includes *cross-likelihood (CL) terms*: $\hat{f}_{il}$ and $\hat{g}_{il}$, $i \neq l$; namely, when evaluating the l-th hypothesis based on an observation of the l-th gene, $\boldsymbol{y}_l$, these CL terms employ the parameters that may well represent the null and alternative distributions of another gene i. This is the major difference of the ODP from the likelihood ratio statistic that consists only of *self-likelihood (SL) terms*: $\hat{f}_{ll}$ and $\hat{g}_{ll}$. The CL terms are important because they transfer the information of the i-th test to the l-th test through the likelihood function; namely, *sharing information* for improving the detection power of MSHT is realized by employing the CL terms.

2.3. The HODP framework

Consider the likelihood function of the alternative model includes not only a parameter θ_i but also a hidden variable $\boldsymbol{x}_i$;

$$g(\boldsymbol{y}_i;\theta_i) = \int d\boldsymbol{x}_i \Pr(\boldsymbol{y}_i|\boldsymbol{x}_i,\theta_i,H_1)\Pr(\boldsymbol{x}_i|\theta_i,H_1),$$

where i.i.d. sampling of $y_{i1},\cdots,y_{iN}$ is each associated with the corresponding hidden variable $x_{i1},\cdots,x_{iN}$.

In this case, there are two different ways to define the CL terms. The first way is to employ the self-marginalized cross-likelihood (SMCL) term, which is the likelihood function marginalized with respect to the hidden variable:

$$\hat{g}_{il}^{\text{SMCL}} \stackrel{\text{def}}{=} g(\boldsymbol{y}_l;\hat{\theta}_i). \tag{5}$$

The SMCL term is a function of an observation $\boldsymbol{y}_l$ and the estimated parameter $\hat{\theta}_i$, and, in this case, the EODP (4) is calculated straightforwardly without considering hidden variables explicitly. The second one is the cross-marginalized cross-likelihood (CMCL):

$$\hat{g}_{il}^{\text{CMCL}} \stackrel{\text{def}}{=} \int d\boldsymbol{x}_i \Pr(\boldsymbol{y}_l|\boldsymbol{x}_i,\hat{\theta}_i,H_1)\Pr(\boldsymbol{x}_i|\boldsymbol{y}_i,\hat{\theta}_i,H_1), \tag{6}$$

that is, the l-th observation $\boldsymbol{y}_l$ is evaluated by using the posterior (alternative) distribution of the i-th hidden variable based on the i-th observation $\boldsymbol{y}_i$, $\Pr(\boldsymbol{x}_i|\boldsymbol{y}_i, \hat{\theta}_i, H_1)$, which brings information of the expression pattern of the i-th gene in addition to the mean and variance of the pattern, i.e., $\hat{\theta}_i$ itself. Consequently, the SMCL and CMCL lead to different thresholding functions from each other and the latter has a potential to allow other tests to share information of hidden structures, which is represented as $\Pr(\boldsymbol{x}_i|\boldsymbol{y}_i, \hat{\theta}_i, H_1)$ in (6), while the SMCL makes the thresholding function identical to the original ODP (4). Note also that the difference between the two methods above arises in simultaneous testing, but not in individual testing, because putting $l = i$ makes the CMCL shrink to the SMCL. The difference between CMCL and SMCL was previously pointed out by ourselves in a more specific form in the context of exploring semi-supervised scores of DE genes [12].

3. Data generative model with latent variable for DE gene detection

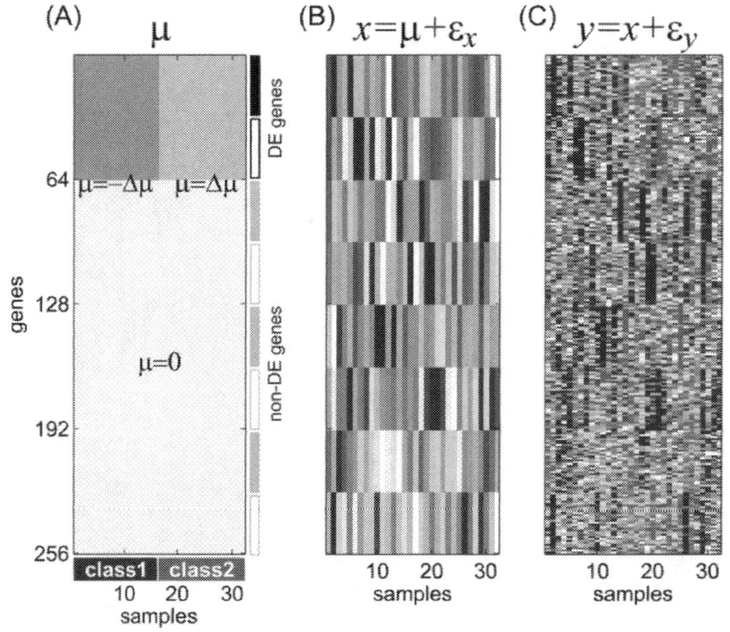

Figure 1. An illustration of the data generation process according to the hierarchical noise model. (A) DE genes have different mean expression levels, $\mu = -\Delta\mu$ and $\mu = \Delta\mu$, according to their sample classes, 1 and 2, while mean expression levels of non-DE genes are consistently zero. (B) In fact, the 256 genes constitute eight clusters, two DE clusters and six non-DE clusters, each of which includes 32 genes. The hidden pattern vectors $\boldsymbol{x}$ of genes are common in each cluster, which have been generated by adding a background pattern ϵ_x to the mean μ. (C) The observation vectors are generated by adding external noise ϵ_y to the hidden pattern $\boldsymbol{x}$.

3.1. Generative model

Let y_{ij} denote an observed expression level of a gene $i = 1, \cdots, M$ in a sample $j = 1, \cdots, N$, and $c_j \in \{1, 2\}$ be the class label of the jth sample. Usually, DE gene detection is based on the generative model:

$$y_{ij} = \mu_i^{(c_j)} + \epsilon_{ij}, \tag{7}$$

and the pairs of hypotheses:

$$H_{0i}: \mu_i^{(1)} = \mu_i^{(2)} (= \mu_i^{(0)}), \qquad H_{1i}: \mu_i^{(1)} \neq \mu_i^{(2)}, \tag{8}$$

where $\mu_i^{(1)}$ and $\mu_i^{(2)}$ are the true mean expressions of the ith gene in the two classes, and ϵ_{ij} is a Gaussian noise with variance σ_i^2, $\epsilon_{ij} \sim N(0, \sigma_i^2)$. The well-known t-statistic, its regularized versions, and standard application of the ODP are based on the above *standard* generative model.

Instead of the standard model, we can assume a *hierarchical* generative model in which the Gaussian noise term ϵ_{ij} consists of two parts,

$$\epsilon_{ij} = \epsilon_{xij} + \epsilon_{yij}, \tag{9}$$

or equivalently, consider a latent variable $x_{ij} = \mu_i^{(c_j)} + \epsilon_{xij}$ so that the observed value is generated by $y_{ij} = x_{ij} + \epsilon_{yij}$. We call ϵ_{yij} and ϵ_{xij} the observation noise and the *background pattern*, and assume that they obey Gaussian distributions with variance σ_y^2 and σ_x^2, respectively. In the hierarchical model, we implicitly assume that the latent variable x_{ij} of a gene i is the same as that of another genes i'.

Figure 1 illustrates a typical expression data matrix generated by the hierarchical generative model. In this figure, there are eight clusters, among which we assume two DE clusters with $\mu_i^{(1)} = -\Delta\mu$ and $\mu_i^{(2)} = \Delta\mu$, and six null clusters with $\mu_i^{(1)} = \mu_i^{(2)} = 0$ (Fig.1(A)). Genes in a cluster share a common hidden pattern which is generated by adding a background pattern [1] $x_{ij} = \mu_i^{(c_j)} + \epsilon_{xij}$ (Fig.1(B)) to the mean $\mu_i^{(c_j)}$, and the observation y_{ij} is generated by further adding observation noise ϵ_{yij}.

This kind of structures should be incorporated in DE gene detection not only because they actually exist but also because appropriate usage of them increases the detection power. In fact, such structures have been observed in various kinds of gene expression datasets, for example, those related to human cancers. In a typical analysis of a cancer-related microarray dataset, each patient (sample) has multiple labels, such as cancer/normal tissues, large/small tumors, favorable/unfavorable prognoses, and some other clinical features. In addition, there must be various characters each of which is shared by some samples and hence represented as a common background pattern of gene expressions over those samples. Thus, even when only one of the labels, patients' prognosis for example, is the target in DE gene detection, there are various other patterns corresponding to other labels underlying the gene expression matrix. If we can appropriately capture them as background patterns, they can be incorporated into MSHT.

Why can the incorporation of background patterns increase the detection ability of DE genes? Usually, many known and unknown factors are involved in the causality network that generates gene expression profile and some of those factors are assumed to be reflected in the gene expression patterns. On the other hand, when we are looking for DE genes whose expression patterns are highly correlated with the target labels, we do NOT always expect those patterns that are exactly the same as the target labels. For example, gene expression patterns in cancer tissues do not tend to show very clear correspondence to their clinical outcome because the clinical outcome depends on treatments for a long period after the measurement of gene expressions. Thus, in this case, true DE genes are not expected to correspond exactly to the label, but to some of several significant factors linking to the outcome. Classically, cluster analysis have been applied to extract such common background patterns or factors of expression profile. In the HODP framework, however, the cluster structure is only implicitly incorporated

[1] Although ϵ_{xij} is a Gaussian noise, it produces a sub-pattern that can be shared by multiple genes. In this sense, it is called the background pattern.

in multiple likelihood functions corresponding to multiple alternative hypotheses; namely, if there are some DE genes with similar expression profiles, CMCL terms put positive bias to each of these genes to be selected. Consequently, a DE gene in a larger cluster of similar expression profile tends to be more significant in the HODP framework, while lone genes tend to be less significant if correspondence to the target labels are identical. This increases detection ability of DE genes in total.

3.2. Maximum likelihood estimation and a control parameter

The similarity in the latent variable patterns is incorporated through the HODP framework by CMCL terms which will be derived in this section.

The likelihood function of the alternative hypothesis based on the above hierarchical model (9) is written as

$$g(\boldsymbol{y}_i, \boldsymbol{x}_i; \theta_i) = \prod_{j=1}^{N} N(y_{ij}; x_{ij}, \sigma_{yi}^2) N(x_{ij}; \mu_i^{(c_j)}, \sigma_{xi}^2), \tag{10}$$

where $\theta_i = (\mu_i^{(1)}, \mu_i^{(2)}, \sigma_{yi}^2, \sigma_{xi}^2)$ are parameters to be estimated from the observations. In the hierarchical noise model (9), two kinds of Gaussian noise are added to an observation y_{ij}, which leads to indeterminacy of the ML estimates for the variance parameters σ_{xi}^2 and σ_{yi}^2. To remove this indeterminacy, a constraint parameter R is introduced as $R = \sigma_{xi}^2/(\sigma_{xi}^2 + \sigma_{yi}^2)$, and called the variance ratio. We also examined another constraint that $\hat{\sigma}_{yi}^2$ is fixed at a certain value for all genes, but found that fixing the R was better than fixing the $\hat{\sigma}_{yi}^2$ in simulated DE gene detection tasks (data not shown).

The ML estimates of the parameters then become

$$\hat{\mu}_i^{(1)} = \operatorname*{mean}_{j;c_j=1} y_{ij}, \qquad \hat{\mu}_i^{(2)} = \operatorname*{mean}_{j;c_j=2} y_{ij}, \qquad \hat{\sigma}_{xi}^2 = R\hat{\sigma}_{0i}^2, \qquad \hat{\sigma}_{yi}^2 = (1-R)\hat{\sigma}_{0i}^2,$$

where $\hat{\sigma}_{0i}^2 \stackrel{\text{def}}{=} \text{mean}_j (y_{ij} - \hat{\mu}_i^{(c_j)})^2$. After the parameters are estimated, the posterior density of the hidden variables is given as

$$\Pr(\boldsymbol{x}_i | \boldsymbol{y}_i, \hat{\theta}_i, H_1) = \prod_{j=1}^{N} N\left(x_{ij};\ \bar{x}_{ij},\ \bar{\sigma}_i^2\right), \tag{11}$$

where $\bar{x}_{ij} \stackrel{\text{def}}{=} (\hat{\sigma}_{xi}^2/\hat{\sigma}_i^2) y_{ij} + (\hat{\sigma}_{yi}^2/\hat{\sigma}_i^2)\hat{\mu}_i^{(c_j)} = R y_{ij} + (1-R)\hat{\mu}_i^{(c_j)}$ and $\bar{\sigma}_i^2 \stackrel{\text{def}}{=} \hat{\sigma}_{xi}^2 \hat{\sigma}_{yi}^2 / \hat{\sigma}_{0i}^2 = R(1-R)\hat{\sigma}_{0i}^2$. It should be noted that the hierarchical noise model (9) does not explicitly define any cluster structure of genes, and no structure is considered in the estimation of the latent variable vector (11). The implicit assumption that the latent variable may be correlated between genes is considered within the ODP framework rather than the ML estimation.

A CMCL term for the alternative model is given by

$$\hat{g}_{il}^{\text{CMCL}} \stackrel{\text{def}}{=} \prod_{j} N\left(y_{lj};\ \bar{x}_{ij},\ \bar{\sigma}_{yi}^2\right), \tag{12}$$

where $\bar{\sigma}_{yi}^2 \stackrel{\text{def}}{=} \hat{\sigma}_{yi}^2 (1 + (\hat{\sigma}_{xi}^2/\hat{\sigma}_{0i}^2)) = (1-R^2)\hat{\sigma}_{0i}^2$.

3.3. Relationship between the ODP and HODP

The EODP for the simple model is obtained as an HODP score with a special setting of $R = 0$. In addition, with another extreme of $R = 1$, the HODP becomes completely insensitive to the class membership label c_j, which leads to an unsupervised thresholding function that evaluates the significance of genes without considering the labels.

The marginalization of (10) with respect to the hidden variables makes the CMCL term identical to the SMCL of the simple noise model (7). Hence, the likelihood ratio score $\hat{g}_{ll}/\hat{f}_{ll}$ based on the hierarchical model (9) becomes exactly the same as the one based on the simple model (7), because the indeterminacy $\sigma_i^2 = \sigma_{yi}^2 + \sigma_{xi}^2$ has no effect on the calculation of the SMCL term.

Consequently, the difference between employing the SMCL and CMCL terms arises only when we conduct simultaneous testing within the ODP framework and the variance ratio parameter is set at $R > 0$. In such a case, the difference between the SMCL and CMCL terms produces an essential difference between the conventional ODP and HODP, even when the ML parameters of CMCL are the same as those of SMCL.

3.4. Determination of the variance ratio R

As shown in the following experiments, the variance ratio R plays an important role, hence should be determined appropriately. To get the improvement in the detection ability, we must avoid extreme values $R = 0$ and $R = 1$. A small value, say $R = 0.1$, makes the HODP's performance similar to the ODP's performance, whereas a large value would not be good because labels are likely ignored in the detection.

Alternatively, an estimation of R is possible by employing Bayesian principal component analysis (BPCA) model [14]. Like singular value decomposition (SVD) [15], PCA employs a linear projection, so that the expression matrix is decomposed into a lower-rank matrix (an essential matrix) and a residual matrix. The BPCA is a Bayesian extension of the PCA so as to make it possible to determine appropriately the rank of the essential matrix. When the BPCA is applied to the expression matrix, the estimated variances, $\hat{\sigma}_E^2$ and $\hat{\sigma}_R^2$, of the elements of the essential and residual matrices correspond to σ_x^2 and σ_y^2 in the above hierarchical model, respectively. Consequently, R can be estimated as $R_{\text{est}} = \hat{\sigma}_E^2/(\hat{\sigma}_E^2 + \hat{\sigma}_R^2)$, which is a fairly good estimation as can be seen in the later simulation results.

3.5. Omitting SL terms in the ODP

The SL terms tend to be larger than the CL terms in average because of two reasons. First, the true hidden variable vector $\boldsymbol{x}_i$ behind an observation $\boldsymbol{y}_i$ may be different from that ($\boldsymbol{x}_l$) underlying another gene l, and, second, even when they happen to be the same, $\boldsymbol{x}_i = \boldsymbol{x}_l$, the estimated latent variable $\hat{\boldsymbol{x}}_i$ is in average closer to $\boldsymbol{y}_i$ than to $\boldsymbol{y}_l$. Because of the second reason, the SL term will be over-estimated in average, in comparison to the CL terms, and hence, the SL terms become dominant in the ODP, which may lose the advantage of incorporating the CL terms. To overcome this problem, we simply omit the SL terms in the EODP thresholding function:

$$S_{\text{EODP}}(\boldsymbol{y}_l) = \frac{\sum_{i \neq l}(1 - \hat{w}_i)\hat{g}_{il}}{\sum_{i \neq l} \hat{w}_i \hat{f}_{il}}.$$

Namely, we remove the SL term $\hat{g}_{ll}$ (or $\hat{f}_{ll}$) which had been added in the numerator (or denominator) from the EODP (4). Consequently, likelihoods of the hypotheses for a gene l are evaluated based on statistics of genes other than l, which encourages information import from other genes. By omitting the SL terms, on the other hand, we do not lose essential information because the CL terms also bring primary information of the gene's significance, i.e., the difference in the mean. Our experimental results show that this modification substantially improves the detection performance of DE genes (data not shown).

3.6. Two options of null hypothesis
There are two possible models for the null hypothesis. The first is the simple noise model itself, and the second is the hierarchical noise model but with the null constraint $\mu^{(1)} = \mu^{(2)}$. When defining HODP thresholding functions, we implemented and compared these two models.

- HODP-A assumes hierarchical noise models for both null and alternative hypotheses. It is calculated with various settings of the intrinsic noise ratio $0 \leq R < 1$, although the BPCA's estimation $R_{\rm est}$ is also available. In this case, the CMCL term for the null model, $\hat{f}_{li}^{\rm CMCL}$, is straightforwardly obtained by (12) with setting $\hat{\mu}_i^{(c)} = 0$ for $c = 1, 2$.
- HODP-B is the same as HODP-A except that it assumes the simple noise model for the null hypothesis.

4. Experimental settings
4.1. Artificial datasets

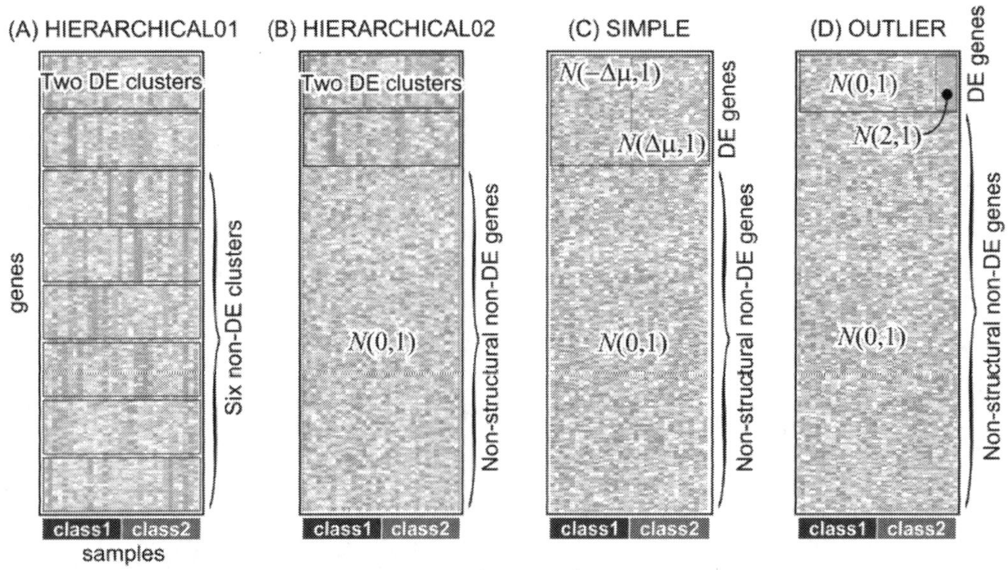

Figure 2. Schematic illustration of four artificial datasets. See the text for details.

To examine the performance of the HODP thresholding functions, we prepared four artificial gene expression data matrices (Figure 2). All of them consisted of expression levels of 256 genes in 32 samples. The 32 samples were composed of 16 'non-tumor' samples (class 1) and 16 'tumor' samples (class 2). The 256 genes included 64 DE genes (or 32 DE genes in the OUTLIER matrix) whose mean expression levels for the tumor and non-tumor samples were different, and the other genes were non-DE genes.

The HIERARCHICAL01 dataset (Fig.1 and Fig.2(A)) was made up of two and six clusters of DE and non-DE genes, respectively. Expression patterns of gene clusters were randomly generated according to the hierarchical noise model (9) with $\sigma_x^2 = \sigma_y^2 = 0.5$, i.e., R was set at 0.5. In a DE gene i, the true mean expression μ_{ij} was set at $-\Delta\mu(= -0.4)$ for non-tumor samples (j s.t. $c_j = 1$), and $\Delta\mu(= 0.4)$ for tumor samples (j s.t. $c_j = 2$), and for non-DE genes, we set

$\mu_{ij} = 0$ for all j (Fig.1(A)). In each cluster, a common hidden pattern $\boldsymbol{x} = (x_1, ..., x_j, ..., x_N)$ was generated by adding a background pattern with variance σ_x^2 to the true mean expression pattern μ_{ij} (Fig.1(B)). Then, observation noise ϵ_{yij}, which is independent from other genes and samples, was generated and added to the hidden pattern x_{ij}. (Fig.1(C)).

The HIERARCHICAL02 dataset (Fig.2(B)) was almost the same as the HIERARCHICAL01 dataset except that the non-DE genes were independently generated from Gaussian distribution of mean zero and variance one, where $\Delta\mu$ for DE genes was set at constant 0.2.

The SIMPLE dataset (Fig.2(C)) was generated from the simple model without any background patterns neither in DE nor non-DE genes. $\Delta\mu$ and σ^2 were set at 0.2 and 1.0, respectively, so that this dataset was similar to HIERARCHICAL02; the difference is that HIERARCHICAL02 includes a background pattern whereas SIMPLE does not.

The OUTLIER dataset (Fig.2(D)) was composed of 32 DE genes (simulating so-called 'cancer outlier' genes) and 224 non-DE genes. This was also non-structural, and generated from Gaussian with the same variance except that, for DE genes, high mean expression level ($\mu_i^{(2)} = 2.0$) was set only for four samples in 16 tumor samples. In this setting, there is no hierarchical noise structure; still however, we can expect good performance of the HODP, because the on/off expression patterns of DE genes are not the same as the class label pattern but common among the DE genes.

4.2. Statistical thresholding functions of gene significance

We compared the following thresholding functions that measure the significance of genes:

- absolute t-score for each gene i, $|\bar{x}_{i1} - \bar{x}_{i2}|/s_{i1}$, where $\bar{x}_{i1}$ and $\bar{x}_{i2}$ are the mean expression levels of sample classes 1 and 2, respectively, and s_{i1} is the intra-class standard deviation.
- Regularized t-score (used in SAM), $|\bar{x}_{i1} - \bar{x}_{i2}|/(s_{i1} + S_0)$, where S_0 is the 10 percentile of all s_{i1}.
- The estimated ODP (EODP), calculated by the Storey's procedure except that the weight of each gene was estimated by iterative application of the ODP (see (4)).
- The true ODP (TODP), calculated based on the true mean and variance in each gene and each class of samples, according to the true data generation process. It gives the theoretical limit of optimality if the latent structures are not considered.

4.3. Microarray datasets on clinical studies

Two published microarray datasets on clinical studies were also used in the performance comparison. The BRCA dataset was from a breast cancer study to examine the difference between BRCA1 ($n = 7$) and BRCA2 ($n = 8$) tumors which include mutations on the BRCA1 and BRCA2 genes, respectively [16]. Expressions of 3226 genes were measured in each tumor sample. The LYMPHOMA dataset consists of expression levels of 7399 probes for 240 cases from diffuse large B-cell lymphoma patients [17]. In the differential analysis, we compared favorable ($n = 102$) and unfavorable ($n = 138$) groups of patients whose outcome was alive and dead at the end of clinical follow-up period, respectively.

5. Results

5.1. Comparisons on artificial datasets

Figures 3(A), (B) and (C) show the performance of the thresholding functions above when applied to the HIERARCHICAL01 dataset. Panels (A) and (B) show the type-I and type-II errors when changing the threshold applied to the thresholding functions. The number of positive calls counts genes whose scores are larger than the threshold, and false discovery proportion (FDP) is the rate of false positives among them. The sensitivity and specificity are proportions

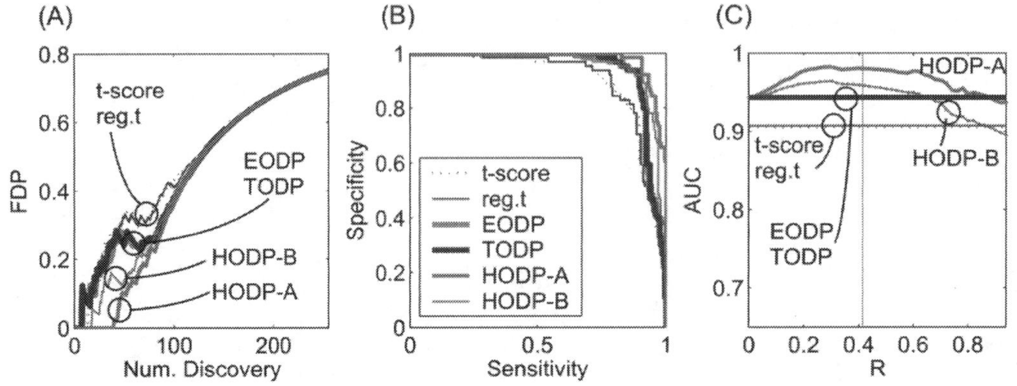

Figure 3. Performance comparison of various thresholding functions on the HIERARCHICAL01 dataset. (A) The true FDP (vertical axis) was calculated for each of the thresholding function output at each number of positive calls (horizontal axis). Each curve denotes a single statistic. For HODP-A and HODP-B, we used an estimated variance ratio $R = R_{\text{est}}$. (B) The ROC curve was drawn for each of the statistics. Line types are consistent with those in panel (A). (C) Area under the ROC curve (AUC) for the statistics with consistent line types with those in panels (A) and (B). Magenta and red curves denote AUCs for HODP-A and HODP-B statistics, respectively, with changing the variance ratio R (horizontal axis) in the HODP models. The red vertical line denotes the estimated value of parameter R, R_{est}. Horizontal lines denote AUCs for other statistics which do not use R. The HODP-A and HODP-B statistics with $R = 0$ are equivalent to the EODP. The HODP statistics outweighed the EODP around the estimated value of R.

of correct calls in truly active and inactive genes, respectively. Note that the errors above are exactly calculated based on true labels of genes, differential or null. Although there are some different aspects between Fig.3(A) and Fig.3(C), it is obvious that the HODPs and the ODPs outweighed the others, and moreover, the HODPs were much better than the ODPs. Note that the HODPs in panels (A) and (B) were calculated by using the estimated parameter R_{est}. In this setting, the EODP exhibited comparative performance to its theoretical limit, the TODP, and the HODP-A outweighed the HODP-B by utilizing the existence of structures in the data. By examining area under the ROC curve (AUC), a measure of overall goodness of statistics considering both type-I and type-II errors, it is shown (Fig.3(C)) that the AUC by HODPs largely depends on the parameter R; the same as that by the ODP at $R = 0$, peaked at around $R = 0.4$ and $R = 0.2$ for the HODP-A and HODP-B, respectively, and degraded with a larger R. However, the estimated parameter R_{est}, denoted by the vertical line, approximately achieved the performance peak in the case of the HODP-A.

Figure 4(A) shows the AUC for the HIERARCHICAL02 dataset, showing the advantage of the HODPs more prominently. On the contrary to the HIERARCHICAL01, the HODP-B was better than the HODP-A in this case, because the assumption of the HODP-B that null genes have no background patterns fits better to this dataset. When applied to this dataset, the performance of the HODP-B monotonically increased as the R increased. When $R = 1.0$, differential property between the two classes defined by the supervised labels, $\mu^{(1)} \neq \mu^{(2)}$, does not make sense in the HODPs (see (11)). Therefore, the performance increase by the HODP-B with large R values is due to that the HODP-B preferred DE genes in an unsupervised manner; namely, DE genes could have obtained higher scores not because their mean expression levels were differential between the two classes, but because those genes shared background patterns

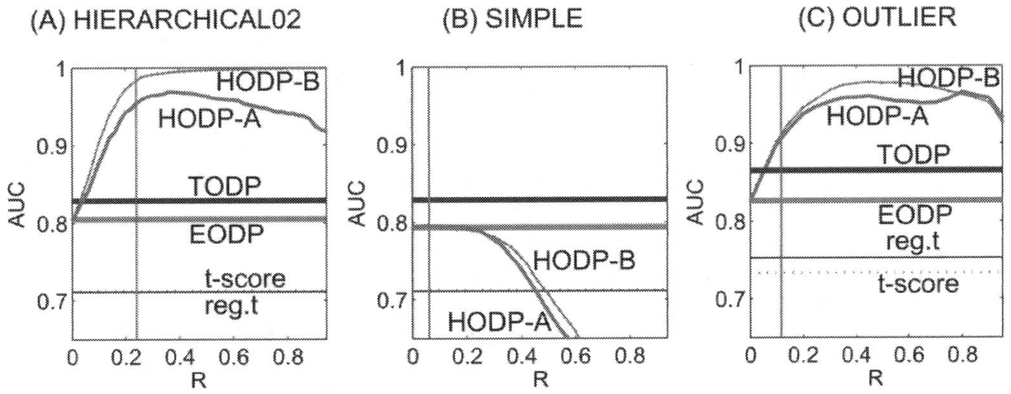

Figure 4. AUC comparisons for three artificial datasets, (A) HIERARCHICAL02, (B) SIMPLE, and (C) OUTLIER. See the caption of Fig.3(C) for details.

with certain numbers of other genes.

For the SIMPLE dataset, the HODPs no longer outweighed the ODPs, and its AUC monotonically decreased with larger R values (Figure 4(B)). This is reasonable, because this dataset includes no structural information which the HODPs can incorporate. On the other hand, as can be seen by the fact that the estimated parameter R_{est} was nearly zero, the non-structural nature of this data matrix was detected, so that the HODP with the estimated R_{est} showed a comparable performance to the standard ODP.

The HODPs also showed good performance for the OUTLIER dataset (Fig.4(C)). Although the estimated R_{est} did not correspond to the peak of the performance curve, the HODPs with the estimated parameter $R = R_{\text{est}}$ still outweighed the EODP and TODP substantially.

5.2. Demonstration on real datasets

We applied the new statistics to two real datasets, BRCA and LYMPHOMA. The former FDR analyses reported both of the two datasets include as many as thousands of DE genes [6]; to examine how the statistics behave, we compared lists of top 200 genes based on five statistics: t-score, regularized t-score, EODP, HODP-A, and HODP-B. Note that the estimated R values, $R_{\text{est}} = 0.87$ and $R_{\text{est}} = 0.53$ for BRCA and LYMPHOMA, respectively, were used for calculating both the HODP-A and HODP-B scores. We counted the numbers of genes which were commonly extracted as DE genes by each pair of scores (Table 1), and found that the lists of top 200 genes included large numbers of common members between conventional scores, t-score, regularized t-score, and EODP. In contrast, the lists extracted by the new scores, HODP-A and HODP-B, were largely different from those by the conventional scores.

Figure 5 visualizes the difference between the ODP and the HODP-A. Both of the two real datasets have prominent cluster structures in the top 200 genes, which are obvious from the color plots of the expression matrices. The estimated parameter values, $R_{\text{est}} = 0.87$ and $R_{\text{est}} = 0.53$, also indicate that the hierarchical noise model is more plausible than the non-hierarchical simple noise model. Consequently, these real datasets seemed closer to the HIERARCHICAL than to the SIMPLE situations. We found that the HODP-A tends to extract genes belonging to clear gene clusters in the top-ranked genes' list, even though the typical expression pattern of the cluster is not the same as the class label pattern. To confirm this observation, we calculated the Pearson's correlation coefficient for every pair of genes in the lists of 200 genes, $c_{ij}, i = 1, \cdots, 200, j = 1, \cdots, 200$, and took maximum correlation within all genes from each

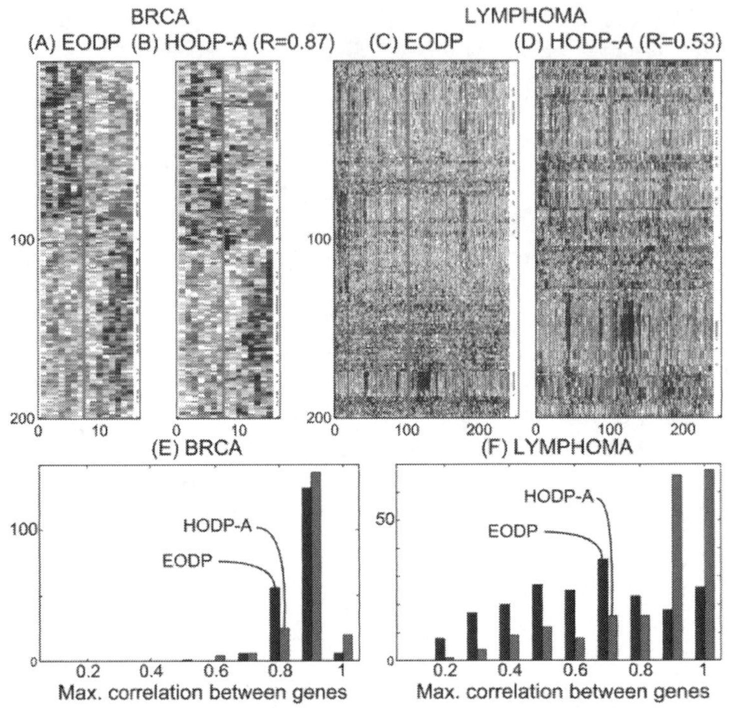

Figure 5. Comparisons between the ODP and HODP scores, when applied to the two real datasets, BRCA ((A) and (B)), and LYMPHOMA ((C) and (D)). In each panel, 200 top genes were selected based on the ODP ((A) and (C)) or HODP-A ((C) and (D)) score, and expression matrices over those genes are illustrated by color diagrams (so called, Eisen plots). Samples are aligned according to the class labels BRCA1 vs. BRCA2 (favorable vs. unfavorable) for the BRCA (LYMPHOMA) dataset. The green vertical line separates the two classes. The 200 genes in each panel are aligned according to a hierarchical clustering result in order for visibility of background patterns (cluster structures). A red mark on the right of each panel signifies a gene which has been commonly selected as within top 200 by both the ODP and HODP scores. (E) and (F) are histograms of maximum correlation coefficients for the BRCA and LYMPHOMA datasets, respectively. In each figure, we calculated correlation coefficients between all pairs of 200 top-ranked genes extracted by the EODP (or HODP-A), took the maximum value for each gene, and drew a histogram of the maximum values over all listed genes.

gene i, i.e., $c_i^{\mathrm{MAX}} \stackrel{\mathrm{def}}{=} \max_{j:i\neq j} c_{ij}$. Figures 5(E) and (F) show the histograms of maximum correlations of the genes in the top-ranked genes' list for the BRCA and LYMPHOMA datasets, respectively. The maximum correlations tend to be larger by the HODP-A than by the ODP for both of the datasets, suggesting that each top-ranked gene selected by the HODP-A is likely to have another top-ranked gene whose expression pattern is quite similar. This is a desired character of the HODP because a gene is more likely to be a DE gene if its expression pattern resembles that of a distinct cluster of genes.

6. Concluding remarks

In this study, we proposed a new statistical thresholding function for DE gene detection tasks based on the HODP framework, which showed prominent detection power of DE genes when gene expression patterns have structures, such as background patterns and cancer outlier genes.

Table 1. The number of common genes in each pair of the lists of 200 top-significant genes

BRCA

	t-score	reg.t	EODP	HODP-A
reg.t	159			
EODP	176	182		
HODP-A	63	74	70	
HODP-B	42	53	51	117

LYMPHOMA

	t-score	reg.t	EODP	HODP-A
reg.t	137			
EODP	176	132		
HODP-A	37	31	38	
HODP-B	33	20	38	73

The general framework of the HODP has further potential fields of applications because latent variable models constitute a model class of a quite wide range. Probably the most important contribution of our current study is the introduction of the concept of *sharing information between simultaneous tests via estimation of latent variables*, which was realized by the introduction of CMCL terms to the ODP statistics. We showed that there exist situations where simultaneous testing based on a hierarchical noise model can provide higher performance than that based on a simple model. Similar situations will be found in brain imaging, DNA sequence analysis, visual image processing, and other fields of research.

Although we showed that detection ability of the empirically estimated thresholding function of HODP can be better than those of both the true ODP as well as the empirical estimation of ODP, the estimation of HODP could have large room to approach the true HODP because estimation of the latent variables may easily over-fit to noisy observations. When true values of the latent variables were available, on the other hand, the AUC of the statistic was found to be as high as 1.0, implying the better estimation of the latent variables as well as the parameters will lead to improvement in the statistics. The modification that discards SL terms in the ODP estimation made a large step toward better estimation; there would be a lot of ways to improve the estimation, for example, shrinkage estimation and low rank matrix factorization.

In this study, we did not discuss much about controlling FDR or other kinds of type I error because of the following two reasons. First, calculation of FDR for the thresholding function based on the HODP framework is found to be problematic. For simpler models without considering background patterns, permutation-based resampling is widely used for simulating null distribution of the thresholding function ([6] for example). With background patterns, however, we found in our preliminary study that usual random permutation of sample labels does not simulate the null distribution well possibly because randomly permuted order of samples do not simulate well the null distribution of the background patterns. Second, type I error control may not be important when a thresholding function is used in realistic screening studies. For example, DE gene detection by microarrays is often exhibited for obtaining a certain pre-fixed number of candidate genes that can be the subjects of further examination by other more laborious but more trustworthy procedures such as RT-PCR and immunostaining. In such a case, a good thresholding function is enough for obtaining a set of top significant genes for further experiments even if type I error control is not available. On the other hand, appropriate type I error control is an important remaining issue in our future work.

Abbreviations

DE: differential expression, EB: empirical Bayes, ODP: optimal discovery procedure, HODP: hidden variables' ODP, EODP: estimated ODP, TODP: true ODP, CL: cross-likelihood, SL: self-likelihood, SMCL: self-marginalized cross-likelihood, CMCL: cross-marginalized cross-likelihood, FDR: false discovery rate, FDP: false discovery proportion, ROC: receiver operational characteristic, AUC: area under ROC curve.

Acknowledgments

This work was supported by Grant-in-Aid for Young Scientists 19710172 and Grant-in-Aid for Scientific Research on Priority Areas "Deepening and expansion of statistical mechanical informatics", both from the MEXT, Japan.

The MATLAB codes for calculating the HODP and synthetic datasets used in this study are freely available at our website `http://hawaii.sys.i.kyoto-u.ac.jp/~oba`.

References

[1] Tusher V G, Tibshirani R and Chu G 2001 *Proc. Natl. Acad. Sci. USA* **98** 5116–21
[2] Efron B and Tibshirani R 2002 *Genet. Epidemiol.* **23** 70–86
[3] Storey J 2007 *J. R. Stat. Soc.* B **69** 347–68
[4] Storey J, Dai J and Leek J 2007 *Biostatistics* **8** 414–32
[5] Oba S and Ishii S 2009 *The International Journal of Biostatistics* **5** Article 20
[6] Perelman E, Ploner A, Calza S and Pawitan Y 2007 *BMC Bioinformatics* **8** 28+
[7] Cao J, Xie X, Zhang S, Whitehurst A and White M 2009 *BMC bioinformatics* **10** 5
[8] Bair E and Tibshirani R 2004 *PLoS Biology* **2** 511–22
[9] Tibshirani R and Hastie T 2007 *Biostatistics* **8** 2–8
[10] Wu B 2006 *Biostatistics* **8** 566–75
[11] Leek J and Storey J 2008 *Proc. Natl. Acad. Sci. USA* **105** 18718–23
[12] Oba S and Ishii S 2006 *BMC Bioinformatics* **7** 414+
[13] Neyman J and Pearson E S 1933 *Philos. T. R. Soc.* A **231** 289–337
[14] Oba S, Sato M, Takemasa I, Monden M, Matsubara K and Ishii S 2003 *Bioinformatics* **19** 2088–96
[15] Alter O, Brown P and Botstein D 2000 *Proc. Nat. Acad. Sci. USA* **97** 10101–6
[16] Hedenfalk I *et al.* 2001 *N. Engl. J. Med.* **344** 539–48
[17] Rosenwald A *et al.* 2002 *N. Engl. J. Med.* **346** 1937–47

Metric learning for DNA microarray data analysis

Ichiro Takeuchi

Nagoya Institute of Technology, Gokiso-cho, Showa-ku, Nagoya 466-8555, Japan

E-mail: takeuchi.ichiro@nitech.ac.jp

Masao Nakagawa and Masao Seto

Aichi Cancer Center, 1-1 Kanokoden, Chikusa-ku, Nagoya 464-8681, Japan

E-mail: mnakagawa@aichi-cc.jp, mseto@aichi-cc.jp

Abstract. In many microarray studies, gene set selection is an important preliminary step for subsequent main task such as tumor classification, cancer subtype identification, etc. In this paper, we investigate the possibility of using *metric learning* as an alternative to gene set selection. We develop a simple metric learning algorithm aiming to use it for microarray data analysis. Exploiting a property of the algorithm, we introduce a novel approach for extending the metric learning to be adaptive. We apply the algorithm to previously studied microarray data on malignant lymphoma subtype identification.

1. Introduction

Microarray technology enables simultaneous monitoring of the expression levels of tens of thousands of genes. A wide range of biological and medical problems are being addressed with microarray studies, and these studies are expected to yield critical insights. One of the most fundamental tasks in microarray data analysis is to identify differentially expressed genes between different experimental conditions. In many studies, gene set selection is performed as a preprocessing step prior to the execution of more challenging tasks such as tumor classification or cancer subtype identification.

The selected genes are sometimes used for the succeeding classification task. In [1], for instance, a set of differentially expressed genes between two types of leukemia samples was first selected based on a simple univariate two-sample statistic, and a linear classifier was then trained with the selected genes. Gene set selection is also performed as a preliminary step of cancer subtype identification. In [2], for example, the authors selected a set of differentially expressed genes between two different types of normal B cells. The selection was performed based on their biological insight that these B cells are normal counterparts [3] of the different subtypes of malignant lymphoma. Using the selected genes they conducted a hierarchical clustering analysis of malignant lymphoma sample, and identified molecularly and clinically-distinct lymphoma subtypes (see section 5 for detail). Selected sets of genes are also used as *signatures* in biological databases. For example, the *Connectivity Map (CMap)* database [4] was developed to find functional connections between biological perturbations and drug (chemical compound) treatments. To construct CMap, a series of treatment-control microarray experiments were conducted to identify a differentially expressed set of genes for each chemical compound treatment. A user can upload their own treatment-control microarray data to CMap for pattern matching, i.e., identifying chemical compound reactions that have similar effects on gene expression as the biological process of their own interest.

In many current microarray studies, however, a naive and sometimes even subjective approach has been adopted for gene set selection. In this study, we use metric learning as an alternative to gene set selection. Metric learning is a kind of supervised learning algorithm which has been extensively studied in the machine learning literature [5, 6, 7]. In a two sample (binary classification) problem, for example, it allows us to find a metric by which the two samples are more distinctively separated than the original (e.g., Euclidean) metric.

In this paper, we develop a simple metric learning algorithm aiming to use as an alternative to gene set selection. By exploiting a property of the algorithm, we introduce a novel approach for making metric learning adaptive. We introduce the metric learning algorithm in section 2, and extend it into an adaptive algorithm in section 3. After providing a simple illustrative example in section 4, we apply our algorithm to a lymphoma subtype identification problem [2] in section 5.

2. Metric learning for microarray

In this section, we develop a simple metric learning algorithm aiming to apply it to microarray data analysis. In what follows, we assume that we have n maicroarrays with m genes. The expression level of i-th array is represented as $\boldsymbol{x}_i \in \mathbb{R}^m$ and the label (e.g., control or treatment) of i-th array is denoted as y_i for $i = 1, \cdots, n$.

2.1. Maharanobis metric learning

Metric learning has been intensively studied in machine learning literature [5, 6, 7]. In most existing studies, the focus is restricted to Maharanobis distance metric. Maharanobis distance metric is represented by a positive semi-definite matrix $M \in \mathbb{R}^{m\times m}$. The distance between two microarrays $\boldsymbol{x}_i, \boldsymbol{x}_j \in \mathbb{R}^m$ is defined as $d(\boldsymbol{x}_i, \boldsymbol{x}_j) = (\boldsymbol{x}_i - \boldsymbol{x}_j)^\top M(\boldsymbol{x}_i - \boldsymbol{x}_j)$. If we represent the square-root of M as A, i.e., $M = A^\top A$, the distance is rewritten as $d(\boldsymbol{x}_i, \boldsymbol{x}_j) = (A\boldsymbol{x}_i - A\boldsymbol{x}_j)^\top(A\boldsymbol{x}_i - A\boldsymbol{x}_j)$. The latter formulation allows us to interpret the metric learning as the problem of learning linear transformation $A : \mathbb{R}^m \to \mathbb{R}^m$ [1]. Let $S \in \mathbb{R}^{n\times n}$ be a matrix to represent the similarity between training examples. The similarity matrix S is defined in several ways. For example, in two-sample (binary classification) problem, the similarity matrix may be simply defined as

$$S_{i,j} = \begin{cases} +1 & \text{if } i\text{-th and } j\text{-th arrays are in the same class,} \\ -1 & \text{if } i\text{-th and } j\text{-th arrays are in different classes.} \end{cases} \tag{1}$$

A general form of Maharanobis metric learning for the former representation is formulated as: $\min_{M\succeq 0} \frac{1}{2}\sum_{i,j=1}^{n}(\boldsymbol{x}_i - \boldsymbol{x}_j)^\top M(\boldsymbol{x}_i - \boldsymbol{x}_j)S_{i,j}$, where $M \succeq 0$ represent M is positive semi-definite. If we use the latter formulation, metric learning is formulated as: $\min_A \frac{1}{2}\sum_{i,j=1}^{n}(A\boldsymbol{x}_i - A\boldsymbol{x}_j)^\top(A\boldsymbol{x}_i - A\boldsymbol{x}_j)S_{i,j}$.

When the sample size n is much smaller than the dimension m (as microarray studies), it is intractable to learn the entire $m \times m$ positive semi-definite matrix M. To make it feasible, we need to incorporate our prior knowledge on M. For example, if we prefer sparse Maharanobis matrix M, we may introduce ℓ_1 regularization term $\sum_{i,j}^{n}|M_{i,j}|$ [8]. If we have a particular metric M_0 as a prior knowledge, we may incorporate regularization term $||M - M_0||$, where $||\cdot||$ is any matrix norm.

[1] If we consider $A : \mathbb{R}^m \to \mathbb{R}^{m'}$ with $m' < m$, we can learn low rank distance metric for dimensionality reduction.

2.2. Regularized diagonal metric learning

In this study, we consider a very simple approach. We restrict ourselves to learn a diagonal matrix M, i.e., we only adjust the weight of each variable (gene) to find a desired metric. Let $\boldsymbol{w} = [w_1, w_2, \cdots, w_m]^\top$ with $w_k \geq 0,\ k = 1, 2, \cdots, m$, be the weights of m genes. Our simple metric learning problem is formulated as

$$\min_{\boldsymbol{w}} \quad \sum_{i,j}^{n} S_{i,j} \sum_{k=1}^{m} w_k (x_{i,k} - x_{j,k})^2 + \frac{1}{2\theta}(\boldsymbol{w} - \frac{1}{m}\mathbf{1}_m)^\top (\boldsymbol{w} - \frac{1}{m}\mathbf{1}_m) \tag{2a}$$

$$\text{s.t.} \quad \mathbf{1}_m^\top \boldsymbol{w} = 1,\ \boldsymbol{w} \geq \mathbf{0}_m, \tag{2b}$$

where $\theta \in \mathbb{R}^+$ is a regularization parameter [2], $\mathbf{1}_m$ and $\mathbf{0}_m$ are m-dimensional vectors with all 1 and 0, respectively. The first term of the objective function represents the weighted metric, and the second one is the regularization term on our prior knowledge that all the variables (genes) are equally weighted, i.e., Euclidean metric. The first equality constraint states that the sum of the weights $\sum_{k=1}^{m} w_k$ is 1, and the second constraint restricts the weights to be non-negative.

Note that the first term of the objective function is written as a linear combination of the weights:

$$\sum_{i,j}^{n} S_{i,j} \sum_{k=1}^{m} w_k (x_{i,k} - x_{j,k})^2 = \boldsymbol{c}^\top \boldsymbol{w}, \text{ where } c_k = \sum_{i,j}^{n} S_{i,j}(x_{i,k} - x_{j,k})^2,\quad k = 1, 2, \cdots, m.$$

When $\theta \to 0$, the second term of the objective function dominates the first term and the solution to (2) represents Euclidean metric:

$$w_1 = w_2 = \cdots = w_m = \frac{1}{m} \tag{3}$$

On the other hand, when $\theta \to \infty$, it is easy to verify that

$$w_k = \begin{cases} 1, & \text{if } k = \arg\min_\ell c_\ell, \\ 0, & \text{otherwise.} \end{cases} \tag{4}$$

is the solution to (2). In this case, only one variable (gene) is used and others are ignored when computing distance. If we go through the solution path with respect to θ from 0 to ∞, we obtain a series of diagonal metrics between (3) and (4).

Interpretability of the result is very important aspect of microarray study because the main goal is to understand the underlying biological mechanism rather than performing prediction or classification using the trained machine. The interpretation of the solution of the problem (2) is quite straightforward because we have only one weight parameter per each gene. The weight parameter indicates the *importance* of the corresponding gene for the learned metric. In addition, when θ is large, the solution is sparse, i.e., only a fraction of genes have positive weights and others have zero weights. Thus, the algorithm serves as a way of gene (variable) selection [3].

[2] It is customary to formulate the regularization term as $\frac{\lambda}{2}(\boldsymbol{w} - \frac{1}{m}\mathbf{1}_m)^\top (\boldsymbol{w} - \frac{1}{m}\mathbf{1}_m)$ with $\lambda \in \mathbb{R}^+$, but we use the formulation using θ as (2) for notational convenience for later discussion.

[3] This gene selection algorithm is *not interesting* in the sense that the algorithm merely selects the genes with an increasing order of the corresponding elements of $\boldsymbol{c}$. On the other hand, this property can be exploited to construct a computationally efficient learning algorithm (see the next subsection). This *uninteresting* selection rule changes to *interesting* one in the extended adaptive approach presented in section 3.

2.3. Optimality conditions and learning algorithm

The optimization problem (2) is an instance of quadratic programming (QP). We can solve the problem for a particular θ using general-purpose QP solver such as active set method or interior-point method, which would generally costs $\mathcal{O}(m^3)$. However, if we exploit the structure of the problem (2), we can construct an algorithm which can compute the entire path of solutions for all $\theta \in [0, \infty)$ in $\mathcal{O}(m^2)$ and can find a single solution for a particular θ in $\mathcal{O}(m \log m)$.

To obtain the algorithm, let us rewrite (2) as

$$\min_{\boldsymbol{w}} \quad (\theta \boldsymbol{c} - \frac{1}{m}\mathbf{1}_m)^\top \boldsymbol{w} + \frac{1}{2}\boldsymbol{w}^\top \boldsymbol{w} \tag{5a}$$

$$\text{s.t.} \quad \mathbf{1}_m^\top \boldsymbol{w} = 1, \ \boldsymbol{w} \geq \mathbf{0}_m. \tag{5b}$$

The Lagrangian of (5) is represented as $L = (\theta \boldsymbol{c} - \frac{1}{m}\mathbf{1}_m)^\top \boldsymbol{w} + \frac{1}{2}\boldsymbol{w}^\top \boldsymbol{w} + \alpha(\mathbf{1}_m^\top \boldsymbol{w} - 1) - \boldsymbol{\gamma}^\top \boldsymbol{w}$, where $\alpha \in \mathbb{R}$ and $\boldsymbol{\gamma} \in \mathbb{R}^{m+}$ are Lagrange multipliers for the first equality and the second inequality constraints, respectively. The Karush-Khun-Tucker optimality conditions are

$$\partial L / \partial \boldsymbol{w} = 0 \iff (\theta \boldsymbol{c} - \frac{1}{m}\mathbf{1}_m) + \boldsymbol{w} + \alpha \mathbf{1}_m - \boldsymbol{\gamma} = \mathbf{0}. \tag{6a}$$

$$\mathbf{1}_m^\top \boldsymbol{w} = 1, \tag{6b}$$

$$\boldsymbol{w} \geq 0, \tag{6c}$$

$$\boldsymbol{\gamma} \geq 0, \tag{6d}$$

$$\gamma_k w_k = 0, \ k = 1, 2, \cdots, m. \tag{6e}$$

Let us divide m weight parameters into those with positive values and those with zeros. We denote the former as an index set $\mathcal{P}$ and the latter as an index set $\mathcal{Z}$, and the size of them as $p = |\mathcal{P}|$ and $z = |\mathcal{Z}|$, respectively. In addition, if we use these index sets as subscript of a vector with length m, e.g., $\boldsymbol{c}_\mathcal{P}$ or $\boldsymbol{c}_\mathcal{Z}$, it indicates the sub-vector whose elements are only those in the index set.

Using the complementary condition (6e), the optimality condition (6a) is divided as

$$(\theta \boldsymbol{c}_\mathcal{P} - \frac{1}{m}\mathbf{1}_p) + \boldsymbol{w}_\mathcal{P} + \alpha \mathbf{1}_p \qquad = \mathbf{0}_p, \tag{7a}$$

$$(\theta \boldsymbol{c}_\mathcal{Z} - \frac{1}{m}\mathbf{1}_z) \qquad + \alpha \mathbf{1}_z - \boldsymbol{\gamma}_\mathcal{Z} = \mathbf{0}_z. \tag{7b}$$

Similarly, (6b) is represented as

$$\mathbf{1}_p^\top \boldsymbol{w}_\mathcal{P} = 1. \tag{8}$$

Multiplying (7a) by $\mathbf{1}_p^\top$ and subtracting (8) yields

$$\alpha = \frac{1}{m} - \frac{\mathbf{1}_p^\top \boldsymbol{c}_\mathcal{P}}{p}\theta - \frac{1}{p} \tag{9}$$

and substituting it back into (7a) yields

$$\boldsymbol{w}_\mathcal{P} = (\frac{\mathbf{1}_p^\top \boldsymbol{c}_\mathcal{P}}{p}\mathbf{1}_p - \boldsymbol{c}_\mathcal{P})\theta + \frac{1}{p}\mathbf{1}_p. \tag{10}$$

For a given θ, if we have information on the allocation of m variables into $\mathcal{P}$ or $\mathcal{Z}$, we can compute the optimal weight parameters using (10).

When we go through the solution path of (2) with respect to θ from 0 to ∞, the genes moves from $\mathcal{P}$ to $\mathcal{Z}$ one by one in decreasing order of the corresponding elements of $\boldsymbol{c}$. We initialize the algorithm with $\theta = 0$, $\mathcal{P} = \{1, 2, \cdots, m\}$, $p = m$, and $\boldsymbol{w} = \frac{1}{m}\mathbf{1}_m$ [4]. Then we increase the θ until the weight corresponding to the largest element of $\boldsymbol{c}$ moves from $\mathcal{P}$ to $\mathcal{Z}$. Let the index of the largest $\boldsymbol{c}$ be k_1. From (10), w_{k_1} becomes 0 (k_1 moves from $\mathcal{P}$ to $\mathcal{Z}$) at

$$\theta^{(1)} = (pc_{k_1} - \mathbf{1}_p^\top \boldsymbol{c}_\mathcal{P})^{-1}. \tag{11}$$

In addition, it can be confirmed that w_{k_1} never become positive again in $\theta > \theta^{(1)}$ because, from (7b), (9) and (11), the corresponding Lagrange multiplier γ_{k_1} satisfies

$$\gamma_{k_1} = \theta c_{k_1} - \frac{1}{m} + \alpha = (c_{k_1} - \frac{\mathbf{1}_p^\top \boldsymbol{c}_\mathcal{P}}{p})\theta^{(1)} - \frac{1}{p} > 0, \text{ if } \theta > \theta^{(1)}. \tag{12}$$

Let $k_1, k_2, \cdots, k_{m-1}$ be the sequence of indices such that $c_{k_1} \geq c_{k_2} \geq \cdots \geq c_{k_{m-1}}$. When we go through θ from 0 to ∞, the weights $w_{k_1}, w_{k_2}, \cdots, w_{k_{m-1}}$ becomes zero one by one in this order at $\theta^{(1)}, \theta^{(2)}, \cdots, \theta^{(m-1)}$, respectively, where

$$\theta^{(t)} = (pc_{k_t} - \mathbf{1}_p^\top \boldsymbol{c}_\mathcal{P})^{-1}, \quad t = 1, 2, \cdots, m-1.$$

The algorithm to compute the solution to (2) for the entire $\theta \in [0, \infty)$ is summarized as follows.

Algorithm 1: Regularized Diagonal Metric Learning Path

- **Input**: Given m-genes microarray expression data $\boldsymbol{x}_i \in \mathbb{R}^m$, class label $y_i \in \{1, 2\}$ for $i = 1, 2, \cdots, n$, and similarity matrix S.
- **Preprocess**: Compute $c_k = \sum_{i,j}^n S_{i,j}(x_{i,k} - x_{j,k})^2$ for $k = 1, 2, \cdots, m$, and sort $\{c_k\}_{k=1}^n$ in decreasing order and obtain the ordered indices $k_1, k_2, \cdots, k_m$ such that $c_{k_1} \geq c_{k_2} \geq \cdots \geq c_{k_m}$.
- **Initialize**: Set $\mathcal{P} = \{1, 2, \cdots, m\}$, $p = m$, $\boldsymbol{w} = \frac{1}{m}\mathbf{1}_m$ and $t = 1$.
- **Step1**: Compute $\theta^{(t)} = (pc_{k_t} - \mathbf{1}_p^\top \boldsymbol{c}_\mathcal{P})^{-1}$.
- **Step2**: Compute $w_k^{(t)} = I(k \in \mathcal{P})((\mathbf{1}_p^\top \boldsymbol{c}_\mathcal{P} - pc_k)\theta^{(t)} + 1)/p$ for $k = 1, 2, \cdots, m$ [5].
- **Step3**: Update $t \leftarrow t + 1$. If $t = m$, terminate the algorithm, otherwise, update $\mathcal{P} \leftarrow \mathcal{P} - \{k_t\}$, $p \leftarrow p - 1$ and return to **Step1**.
- **Output**: Obtain $\theta^{(1)} \leq \theta^{(2)} \leq \cdots \leq \theta^{(m-1)}$, and $\boldsymbol{w}^{(1)}, \boldsymbol{w}^{(2)}, \cdots, \boldsymbol{w}^{(m-1)}$.

Using this algorithm, we obtain a sequence of regularization parameters $\theta^{(1)}, \theta^{(2)}, \cdots, \theta^{(m-1)}$ and corresponding weight parameter sequence $\boldsymbol{w}^{(1)}, \boldsymbol{w}^{(2)}, \cdots, \boldsymbol{w}^{(m-1)}$. Since the weight parameters are linear in any intervals $[\theta^{(t)}, \theta^{(t+1)}]$, $t = 1, 2, \cdots, m-2$, the entire path of weight parameter has piecewise-linear form. In the optimization literature, $\theta^{(1)}, \theta^{(2)}, \cdots, \theta^{(m-1)}$ are called *breakpoints*. If we want the weight parameters between any two breakpoints $\theta^{(t)}$ and $\theta^{(t+1)}$, $t = 1, 2, \cdots, m-2$, we can use linear interpolation:

$$w_k = \kappa w_k^{(t)} + (1 - \kappa) w_k^{(t+1)}, \text{ where } \kappa = \frac{\theta^{(t+1)} - \theta}{\theta^{(t+1)} - \theta^{(t)}}, \quad k = 1, 2, \cdots, m. \tag{13}$$

Note that the computational costs of step1 and step2 are $\mathcal{O}(1)$ and $\mathcal{O}(m)$, respectively. Thus, the total computational cost of computing the entire solution path for all $\{\theta^{(t)}, \boldsymbol{w}^{(t)}\}_{t=1}^{m-1}$ is $\mathcal{O}(m^2)$. If we want to know a weight vector only at a particular θ, we can skip step2 in the above algorithm. After searching for the nearest $\theta^{(t)}$ (with linear time), we set $\mathcal{P} = \{1, 2, \cdots, m\} - \{k_1, k_2, \cdots, k_t\}$, $p = m - t$, and directly compute the weight parameters using (10). In this case, the total computational cost is $\mathcal{O}(m \log m)$ for sorting $\{c_k\}_{k=1}^m$ in the preprocess.

[4] Note that, when $\theta = 0$, the first term of (10) diminishes and $\boldsymbol{w} = \boldsymbol{w}_\mathcal{P} = \frac{1}{p}\mathbf{1}_p = \frac{1}{m}\mathbf{1}_m$.

[5] $I(\cdot)$ is an indicator function such that $I(z) = 1$ if z is true, and $I(z) = 0$ otherwise.

3. Adaptive metric learning

In this section, we extend the above metric learning algorithm to adaptive one[6].

3.1. Similarity measure

As mentioned above, similarity matrix S is defined in several ways. A simple approach is to use the similarity matrix in (1). With the use of this similarity, we aim to find a metric in which all the arrays in the same class come close and conform a single cluster. This simple similarity is useful for homogeneous sample with unimodal distribution.

On the other hand, if equally labeled arrays are heterogeneous entity and the distribution is multi-modal, local learning approach is useful. One popular local learning approach is to define *K-target neighbors* for each array. K-target neighbors of an array $\boldsymbol{x}_i$ is K other arrays in the same class that we want to have small distance to $\boldsymbol{x}_i$ after metric learning. If we have no prior knowledge, we can define the K-target neighbors to be K-nearest neighbor (using Euclidean metric) in the same class. Let $\eta_{i,j} \in \{0,1\}$ represents whether $\boldsymbol{x}_j$ is one of the target neighbors of $\boldsymbol{x}_i$ ($\eta_{i,j} = 1$) or not ($\eta_{i,j} = 0$) [7]. The similarity using the target neighbors is defined as

$$S_{i,j} = \begin{cases} \eta_{i,j} & \text{if } i\text{-th and } j\text{-th arrays are in the same class,} \\ -1 & \text{if } i\text{-th and } j\text{-th arrays are in different classes.} \end{cases} \tag{14}$$

There will be many possible extensions on similarity matrix. For example, we may allow $\eta_{i,j}$ to take a continuous value in $[0, 1]$.

3.2. Adaptive regularized diagonal metric learning

In local learning approach, the target neighbors must be a priori specified and they are fixed during metric learning. However, the actual neighbor assignment may vary in the course of metric learning. This observation suggests an iterative approach. First, we initialize the target neighbor using Euclidean metric, followed by the metric learning using the fixed target neighbors. Then target neighbors are re-assigned using the previously learned metric and metric learning is re-applied using the new target neighbors. This process is repeated many times. The iterative approach can alleviate the bias stemming from the target neighbor assignment changes during learning process. Although the advantage of the iterative approach was experimentally demonstrated [7], more precise treatment to this problem is desired.

In our regularized diagonal metric learning algorithm, we have the entire path of solutions w.r.t. the regularization parameter $\theta \in [0, \infty)$. Exploiting this advantage, we can construct an algorithm for adaptive target neighbor assignment. We start from the solution at $\theta = 0$ representing Euclidean metric. When we continuously increase θ, we encounter a point at which the target neighbor assignment changes, which we call *target neighbor change point*, and denote it as $\theta_{[1]}$. At target neighbor change point $\theta_{[1]}$, we renew the similarity matrix S and restart the path-following in the increasing direction of θ from $\theta_{[1]}$. We repeat this process until $\theta \to \infty$, during which, we will have a sequence of target neighbor change points: $\theta_{[1]} \leq \theta_{[2]} \leq \cdots \leq \theta_{[L]}$, where L is the number of target neighbor change points.

To explain how to find the target neighbor change points, let us first assume that a target neighbor change point $\theta_{[\ell]}$ is in $[\theta^{(t)}, \theta^{(t+1)})$ for a certain $t \in \{1, 2, \cdots, m-2\}$, and one of the i-th array's target neighbor j is replaced with h at $\theta_{[\ell]}$. This target neighbor change happens when the distance between i and j-th arrays $d(\boldsymbol{x}_i, \boldsymbol{x}_j)$ equals to the distance between i and h-th arrays $d(\boldsymbol{x}_i, \boldsymbol{x}_h)$. Remembering that the weights are linear in θ in the interval $[\theta^{(t)}, \theta^{(t+1)})$, we

[6] Let us focus on two sample problem here although most of the discussion below can be analogously applied to the problems with more than two samples.

[7] Note that $\eta_{i,j}$ is not necessarily equals to $\eta_{j,i}$.

can easily derive the target neighbor change point $\theta_{[\ell]}$ as:

$$d(\boldsymbol{x}_i, \boldsymbol{x}_j) = d(\boldsymbol{x}_i, \boldsymbol{x}_h) \iff \theta_{[\ell]} = -\frac{\sum_{k\in\mathcal{P}}((x_{ik}-x_{jk})^2-(x_{ik}-x_{hk})^2)}{\sum_{k\in\mathcal{P}}(\mathbf{1}_p^\top \boldsymbol{c}_\mathcal{P} - pc_k)((x_{ik}-x_{jk})^2-(x_{ik}-x_{hk})^2)}, \tag{15}$$

where we used (10) with $\theta = \theta_{[\ell]}$. In order to find the interval $[\theta^{(t)}, \theta^{(t+1)})$ such that $\theta_{[\ell]} \in [\theta^{(t)}, \theta^{(t+1)})$, we only have to check the target neighbor changes at finite breakpoints $\theta^{(1)}, \theta^{(2)}, \cdots, \theta^{(m-1)}$ because the distance between any two arrays is linear in θ in any interval.

The entire algorithm roughly works as follows. Starting from Euclidean metric ($\theta = 0$), we compute a similarity matrix $S_{[0]}$. Using $S_{[0]}$, we obtain a sequence of breakpoints $\theta_{[0]}^{(1)}$, $\theta_{[0]}^{(2)}$, $\cdots$, $\theta_{[0]}^{(m-1)}$, as described in 2.3, where the subscript $_{[0]}$ indicates that the breakpoint is computed using the similarity matrix $S_{[0]}$. At each breakpoint, we check if there is a target neighbor change or not. If there is, we compute the target neighbor change point $\theta_{[1]}$, using (15) [8]. At the target neighbor change point $\theta_{[1]}$, we update the similarity matrix from $S_{[0]}$ to $S_{[1]}$ and re-compute the breakpoints $\theta_{[1]}^{(1)}$, $\theta_{[1]}^{(2)}$, $\cdots$, $\theta_{[1]}^{(m-1)}$. Then we restart the path following w.r.t. θ from $\theta_{[1]}$. We repeat these processes until we arrive at the last breakpoint $\theta_{[L]}^{(m-1)}$. In this way, we obtain a mixed sequence of breakpoints and target neighbor change points.

For example, the mixed sequence looks like this:

$$\underbrace{\theta_{[0]}^{(1)} \to \theta_{[0]}^{(2)} \to \theta_{[0]}^{(3)} \to}_{S_{[0]}} \theta_{[1]} \underbrace{\to \theta_{[1]}^{(5)} \to \theta_{[1]}^{(6)} \to}_{S_{[1]}} \theta_{[2]} \underbrace{\to \theta_{[2]}^{(5)} \to}_{S_{[2]}} \theta_{[3]} \underbrace{\to}_{S_{[3]}} \theta_{[4]} \underbrace{\to \theta_{[4]}^{(8)} \to}_{S_{[4]}} \cdots \underbrace{\to \theta_{[L]}^{(m-1)}}_{S_{[L]}},$$

where $S_{[0]}$ indicates the similarity matrix based on Euclidean metric, and $S_{[\ell]}$, $\ell = 1, 2, \cdots, L$, represents the similarity matrix based on the weighted metric at ℓ-th target neighbor change point $\theta_{[\ell]}$. The underbrace notation $\underbrace{\qquad}_{S_{[\ell]}}$ indicates that the similarity matrix $S_{[\ell]}$ is used in that interval of θ. Furthermore, $\theta_{[\ell]}^{(t)}$, $t = 1, 2, \cdots, m-1$, $\ell = 0, 1, \cdots, L$, denotes the t-th breakpoint computed based on the similarity matrix $S_{[\ell]}$. Note that, for $\ell \neq \ell'$, $\theta_\ell^{(t)} \neq \theta_{\ell'}^{(t)}$ in general. Therefore, for example, the breakpoint $\theta_{[0]}^{(3)}$ followed by $\theta_{[1]}^{(5)}$ (not $\theta_{[1]}^{(4)}$) when the similarity matrix is updated from $S_{[0]}$ to $S_{[1]}$ in the above example. The adaptive algorithm is summarized as follows:

Algorithm 2: Adaptive Reglarized Diagonal Metric Learning Path

- **Input**: Given m-genes microarray expression data $\boldsymbol{x}_i \in \mathbb{R}^m$, class label $y_i \in \{-1, +1\}$ for $i = 1, 2, \cdots, n$, and the number of target neighbors K.
- **Initialize**: Set $\mathcal{P} = \{1, 2, \cdots, m\}$, $p = m$, $\boldsymbol{w} = \frac{1}{m}\mathbf{1}_m$ and initialize the similarity matrix $S_{[0]}$ based on K-target neighbor assignment using Euclidean metric (i.e., the metric at $\theta = 0$). Set $\theta_{[0]} = 0$ and $u = 1$.
- **Step1**: Compute $c_k = \sum_{i,j}^n w_k S_{[u-1]i,j} d_{ijk} - \frac{2}{m}, \forall k$, and sort them in decreasing order and obtain an ordered indices $k_1, k_2, \cdots, k_m$ such that $c_{k_1} \geq c_{k_2} \geq \cdots \geq c_{k_m}$. Compute $\theta_{[u]}^{(t)} = \{p(-c_{k_t} - \frac{\mathbf{1}_p^\top \boldsymbol{c}_\mathcal{P}}{p})\}^{-1}$ for $t = 1, 2, \cdots, m-1$. If $\theta_{[u-1]} >= \theta_{[u]}^{(m-1)}$, set $L = u - 1$ and terminate the algorithm. Let $t_{[u]}$ be the smallest t such that $\theta_{[u]}^{(t)} > \theta_{[u-1]}$. Set $\tau = 1$, $P = \{1, 2, \cdots, m\} - \{k_1, k_2, \cdots, k_{t_{[u]}-1}\}$, and $p = m - t_{[u]} + 1$.

[8] Many target neighbor changes can occur between two breakpoints. In such case, we compute all possible target neighbor change points using (15), and select the smallest one as the next target neighbor change point.

Journal of Physics: Conference Series **197** (2009) 012008 doi:10.1088/1742-6596/197/1/012008

- **Step2**: If $t_{[u]} + \tau = m$, set $M_{[u]} = \tau - 1$, $L = u - 1$ and terminate the algorithm. Compute $w_k^{(t_{[u]}+\tau)} = I(k \in \mathcal{P})((\frac{\mathbf{1}_p^\top \boldsymbol{c}_\mathcal{P}}{p}\mathbf{1}_p - c_k)\theta^{(t_{[u]}+\tau)} + \frac{1}{p})$ for $k = 1, 2, \cdots, m$, and compute the similarity matrix $S_{[u]}$ based on K-target neighbor assignment using the weight $\boldsymbol{w}^{(t_{[u]}+\tau)}$.
- **Step3**: If the similarity matrix $S_{[u]}$ is identical to $S_{[u-1]}$, $\mathcal{P} \leftarrow \mathcal{P} - \{k_{t_{[u]}+\tau}\}$, $p \leftarrow p - 1$ and $\tau \leftarrow \tau + 1$ and go back to **Step2**.
- **Step4**:
Find the pair of matrix elements (i, j) and (i, h) in $\mathbb{R}^{n \times n}$ such that $S_{[u]i,j} = 1$, $S_{[u]i,h} = 0$, $S_{[u+1]i,j} = 0$, $S_{[u+1]i,h} = 1$, and compute $\theta_{[u]}$ using (15). Compute $\boldsymbol{w}$ at $\theta_{[u]}$. Set $M[u] = \tau$, $u \leftarrow u + 1$ and go back to **Step1**.
- **output**: Obtain $\theta_{[1]} \leq \theta_{[2]} \leq \cdots \leq \theta_{[L]}$, $S_{[1]}, S_{[2]}, \cdots S_{[L]}$, and

$$
\begin{aligned}
0 = \theta_{[0]} \;\; & \leq \;\; \theta_{[1]}^{(t_{[1]})} \leq \theta_{[1]}^{(t_{[1]}+1)} \leq \cdots \leq \theta_{[1]}^{(t_{[1]}+M_{[1]}-1)} \leq \theta_{[1]} \\
& \leq \;\; \theta_{[2]}^{(t_{[2]})} \leq \theta_{[2]}^{(t_{[2]}+1)} \leq \cdots \leq \theta_{[2]}^{(t_{[2]}+M_{[2]}-1)} \leq \theta_{[2]} \\
& \;\; \vdots \\
& \leq \;\; \theta_{[L]}^{(t_{[L]})} \leq \theta_{[L]}^{(t_{[L]}+1)} \leq \cdots \leq \theta_{[L]}^{(t_{[L]}+M_{[L]}-1)} \leq \theta_{[L]} \\
& \leq \;\; \theta_{[L+1]}^{(t_{[L+1]})} \leq \theta_{[L+1]}^{(t_{[L+1]}+1)} \leq \cdots \leq \theta_{[L+1]}^{(t_{[L+1]}+M_L-1)} < \infty.
\end{aligned}
$$

4. Illustrative example

To demonstrate how the algorithm works, let us provide a very simple artificial example. We consider an extremely simple microarray study with $n = 20$ and $m = 6$. Among $n = 20$ arrays, half of arrays ($n_1 = 10$) belong to `class1`, and the other half ($n_2 = 10$) belong to `class2`. We assume that `class1` is heterogeneous entity and it actually consists of two subtypes called `subtype1` and `subtype2` each with 5 arrays, but the analyst does not have any information on these subtypes. The expressions of $m = 6$ genes were randomly generated according to the distributions in figure 1. The 1st and 2nd genes (a) are differentially expressed among `class1/subtype1`, `class1/subtype2` and `class2`. The 3rd and 4th genes (b) are differentially expressed in `class1` and `class2`, but no difference between `class1/subtype1` and `class1/subtype2`. The expressions of 5th and 6th genes (c) are equally distributed among all classes and subtypes. In microarray studies, we often encounter heterogeneous entity as `class1` [9]. Local learning approach effectively works for heterogeneous sample.

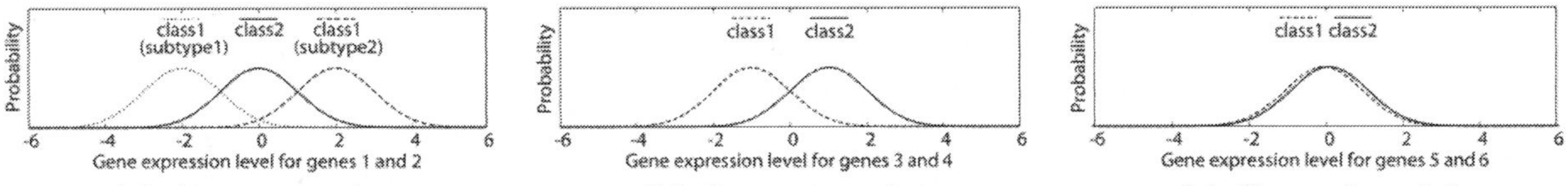

(a) Genes 1 and 2 (b) Genes 3 and 4 (c) Genes 5 and 6

Figure 1. Distribution of artificial gene expressions

Figure 2 shows the solution path w.r.t. θ for non-adaptive algorithm (a) and adaptive algorithm with $K = 4$ (b). Note that all the weight parameters have equal value $1/m \simeq 0.167$ at $\theta = 0$ (Euclidean metric), and only one gene (gene 3 in this example) has nonzero weight 1.0 at the end of the path. In non-adaptive algorithm (a), the number of breakpoints in the path is $m - 1 = 5$. On the other hand, in adaptive algorithm (b), there are in total 41 points in the

Journal of Physics: Conference Series **197** (2009) 012008 doi:10.1088/1742-6596/197/1/012008

path: 37 of them are target neighbor change points and 4 of them are breakpoints [9]. In (a), the paths of 6 genes never cross, i.e., the regularization parameter θ only controls the proportion of the importance (weight) of each gene but it never changes the order of importance. In (b), on the other hand, the order of importance of each gene changes over the regularization parameter θ because the similarity matrix S changes through the path. In (a), only the genes 3 and 4 (see figure 1(b)) have large weights. In (b), by contrast, the genes 1 and 2 (see figure 1(a)) also have significant amount of weights for a wide range of θ; this is an advantage of local learning approach.

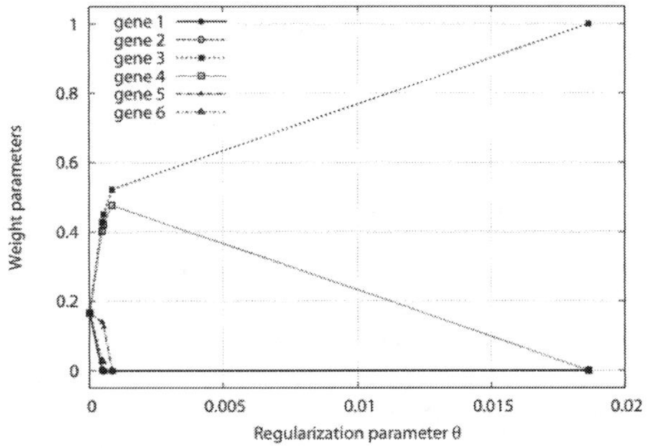

(a) Non-adaptive algorithm

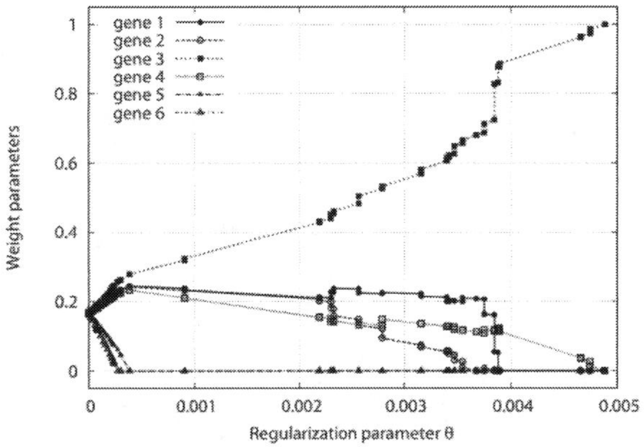

(b) Adaptive algorithm ($K = 4$)

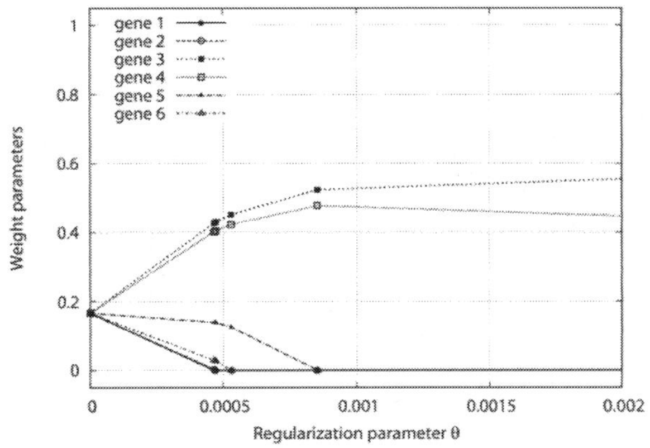

(c) Non-adaptive algorithm (enlarged for $0 \leq \theta \leq 0.002$)

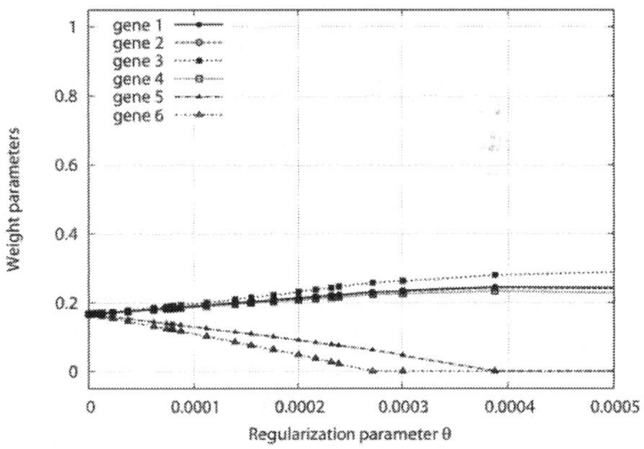

(d) Adaptive algorithm ($K = 4$) (enlarged for $0 \leq \theta \leq 0.0005$)

Figure 2. The path of weight parameters w.r.t. regularization parameter θ for non-adaptive algorithm ((a) and (c)) and adaptive algorithm with $K = 4$ ((b) and (d)).

5. Application to lymphoma subtype identification

In this section, we apply the metric learning algorithm to the microarray data on diffuse large B-cell lymphoma (DLBCL) in Lymphoma/Leukemia Molecular Profiling Project (LLMPP) [2, 10].

[9] In general, the number of target neighbor change points depends on the sample size n and the number of breakpoints depends on the number of genes m. Since $n > m$ in this artificial example, there are more target neighbor change points than breakpoints. In real microarray study (as in the next section), however, n is much smaller than m and target neighbor change points tends to be much less than breakpoints.

DLBCL is the most common lymphoma in adults. When chemotherapy is applied to DLBCL patients, 35 - 40 % of patients can be cured, but it does not work for the rest of patients. Several studies indicate that DLBCL is heterogeneous entity and actually consists of several subtypes, and it has been speculated that the variable clinical courses would reflect this heterogeneity.

Alizadeh et al. [2] investigated the expression profiles of thousands of genes for 96 samples of various types of normal and malignant lymphocytes, including 46 malignant DLBCL, 2 normal germinal center (GC) B cell, and 10 normal activated B cell lymphocytes (see figure 1 in [2]). They applied hierarchical clustering to the entire microarray data. Since they have biological knowledge that the GC B cells and activated B cells play significant roles in the development of lymphoma, they paid attention on the two branches of the hierarchical tree, each of which are highly expressed in GC B cells and activated B cells, respectively [10]. Then, they re-applied hierarchical clustering to 46 malignant DLBCL arrays using only those genes in the two branches, and identified two highly separated clusters. Since one of them have common expression pattern as normal germinal center B-cells, and the other have similar pattern as normal activated B-cells, they claimed that DLBCL has two subtypes called GC B-like DLBCL and activated B-like DLBCL (see figure 3 in [2]). Later, they investigated the clinical outcomes of these GC B-like and activated B-like DLBCL subtypes, and found that there is a statistically significant difference in their prognosis ($p = 0.01$ in logrank test). In the follow-up study by LLMPP project [10], the subtype of 240 DLBCL patients were identified and they reported that the prognosis of GC B-like DLBCL is significantly better than activated B-like DLBCL (see figure 1 in [10]). The characteristics of GC B-like DLBCL and activated B-like DLBCL subtypes have been intensively studied [11, 12, 13] and it is now recognized that they are molecularly and immunologically distinct subtypes with different clinical courses.

Although the biological and clinical impact of this work is great, their data analysis procedure is clearly immature and subjective. In this paper, we offer a more objective approach to this subtype identification problem by applying the metric learning algorithm to this DLBCL microarray data. We collected the microarray data in [2] and [10] from the website of LLMPP project. Using gene-bank accession number and unigene code, we identified 2227 common genes both in [2] and [10]. Then, we learned the metric to differentiate 2 normal GC B cells and 10 activated B cells. Using the learned metric, we applied hierarchical clustering to 46 DLBCLs in [2] and 240 DLBCLs in [10]. In both cases, we identified two significantly separated clusters and compared the prognosis of these two clusters.

Figure 3 displays the paths of 10 weight parameters with smallest c_k for non-adaptive algorithm (a) and adaptive algorithm with $K = 5$ (b). In (a), the number of breakpoints was $m - 1 = 2226$. In (b), there were 44 target neighbor change points and 2257 breakpoints. In both (a) and (b), the top 10 genes were same, but the order of importance changed in (b). In particular, the 2nd gene had larger weight than 1st one for large θ.

Figure 4 shows the hierarchical clustering analysis using learned metrics and Kaplan-Meiyer survival analysis of the identified two distinct clusters. The top 4 panels (a and b) are the results for 46 DLBCL cases in [2], and the bottom 4 panels (c and d) are those for 240 DLBCL cases in [10]. The left ones (a and c) were obtained by non-adaptive metric learning, while the right ones (b and d) were obtained by adaptive ($K = 5$) metric learning. The regularization parameter θ in both (non-adaptive and adaptive) algorithms and the number of nearest neighbor K in adaptive algorithm were selected by leave-one-out cross-validation with one-standard error rule [14]. In non-adaptive and adaptive metric learnings, the number of nonzero weights were 35 and 19 at the selected θs, respectively. In all the scenarios of (a), (b), (c), and (d), the hierarchical clustering analysis identified distinct two clusters (indicated as `cluster 1` and `cluster 2` in the plots). Tables 1 and 2 show the relationships between the two clusters and the subtypes

[10] They selected the two branches visually from the heatmap representation of the clustering result.

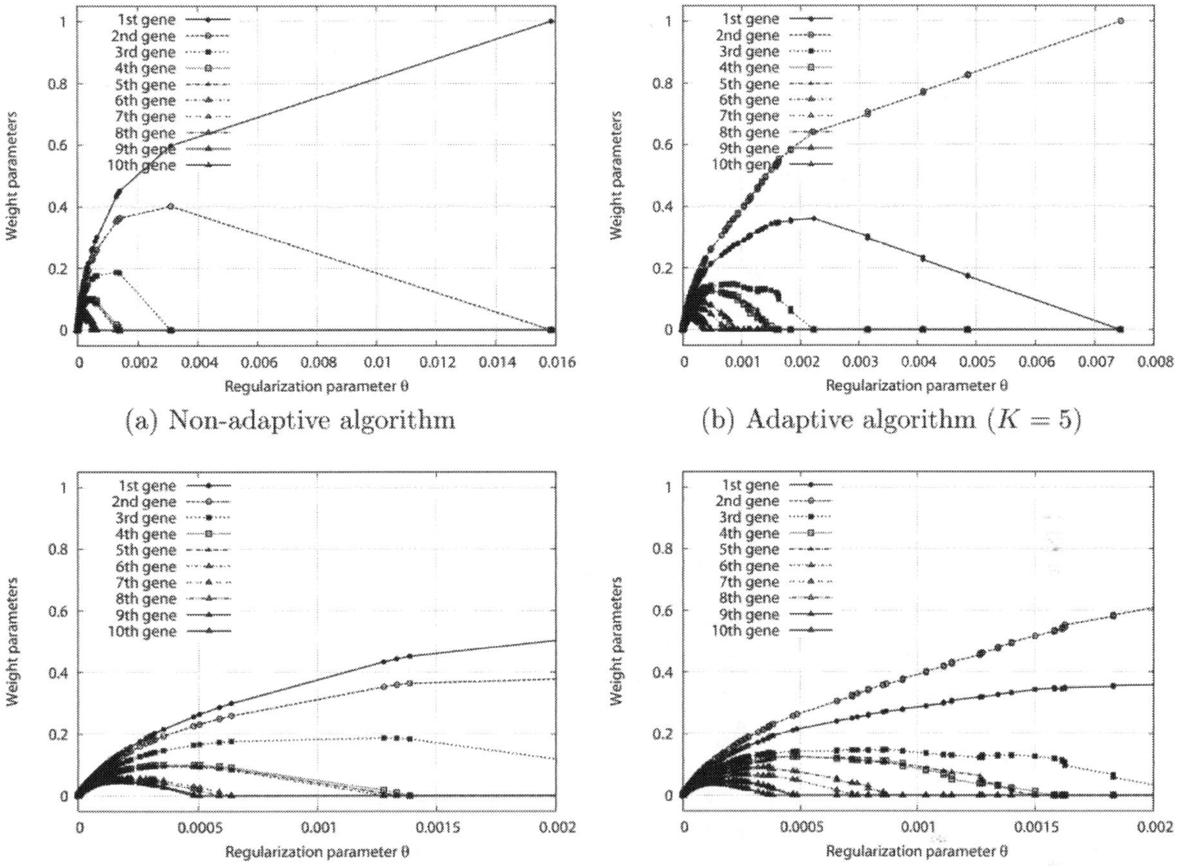

(a) Non-adaptive algorithm

(b) Adaptive algorithm ($K = 5$)

(c) Non-adaptive algorithm (enlarged for $0 \leq \theta \leq 0.002$)

(d) Adaptive algorithm ($K = 5$) (enlarged for $0 \leq \theta \leq 0.002$)

Figure 3. The path of 10 weight parameters w.r.t. regularization parameter θ for non-adaptive algorithm ((a) and (c)) and adaptive algorithm with $K = 5$ ((b) and (d)). The 1st gene, 2nd gene, ... indicates the gene with the smallest c_k, that with the 2nd smallest c_k, ..., respectively.

identified in [2] and [10], respectively. Although there are no exact correspondence between the clusters in the current study and the subtypes in the previous study, `cluster 1` and `cluster 2` seem to represent the activated B-like DLBCL and GC B-like DLBCL subtypes, respectively. We divided the DLBCL sample according to the two clusters and compared the prognosis of two groups by Kaplan-Meiyer analysis. In all the four scenarios, the prognosis of the cluster 1 (representing activated B-like DLBCL subtype) was significantly worse than that of the cluster 2 (representing GC B-like DLBCL subtype). In this application study, we did not find a significant difference in the performances between non-adaptive and adaptive algorithms. However, it has been recognized that many disease entities are biologically heterogeneous and we conjecture that the adaptive local approach presented in section 3 will be more valuable in such situation.

6. Conclusion

In this paper, we investigated the possibility of using metric learning as an alternative to gene set selection in microarray data analysis. We proposed regularized diagonal metric learning algorithm and its adaptive version. Our metric learning is extremely simple, but it has computational advantage and straightforward interpretation. We applied the algorithms to

Table 1. Relationship between the two clusters and the subtypes identified in Alizadeh et al (2000)

	non-adaptive			adaptive ($K = 5$)		
	cluster1	cluster2	total	cluster1	cluster2	total
activated B-like DLBCL	17	6	23	17	6	23
GC B-like DLBCL	0	22	22	2	20	22
other	0	1	1	0	1	1
total	17	29	46	19	27	46

Table 2. Relationship between the two clusters and the subtypes identified in Rosenwald et al (2002)

	non-adaptive			adaptive ($K = 5$)		
	cluster1	cluster2	total	cluster1	cluster2	total
activated B-like DLBCL	61	12	73	64	9	73
GC B-like DLBCL	22	93	115	38	77	115
other(typeIII)	42	10	52	27	5	52
total	125	115	240	149	91	240

previously published microarray data on malignant lymphoma subtype identification. As the future work, we plan to apply the metric learning to other microarray tasks.

References

[1] Golub T R *et al* 1999 *Science* **286** 531–7
[2] Alizadeh A A *et al* 2000 *Nature* **403** 503–11
[3] Harris N L *et al* 1994 *Blood* **84** 1361–92
[4] Lamb J *et al* 2006 *Science* **313** 1929–35
[5] Xing E P, Ng A Y, Jordan M I and Russell S 2002 *Advances in Neural Information Processing Systems* vol 14 ed T G Dietterich *et al* (Cambridge: MIT press) pp 505–12
[6] Goldberger J, Roweis S, Hinton G and Salakhutdinov R 2005 *Advances in Neural Information Processing Systems* vol 17 ed L K Saul *et al* (Cambridge: MIT press) pp 513–20
[7] Weinberger K Q, Blitzer J and Saul L K 2006 *Advances in Neural Information Processing Systems* vol 18 ed Y Weiss *et al* (Cambridge: MIT press) 1473–80
[8] Qi G J, Tang J, Zha Z J, Chua T S and Zhang H J 2009 *Proc. Int. Conf. on Machine Learning* (Montreal) vol 382 (Omnipress) p 841–8
[9] Tomlins S A *et al* 2005 *Science* **310** 644–8
[10] Rosenwald A *et al* 2002 *N. Engl. J. Med.* **346** 1937–47
[11] Wright G *et al* 2003 *Proc. Natl. Acad. Sci. USA* **100** 9991–6
[12] Shipp M A *et al* 2002 *Nat. Med.* **8** 68–74
[13] Takeuchi I, Tagawa H, Tsujikawa A, Nakagawa M, Suguro M, Guo Y and Seto M 2009 *Haematologica* **94** 51–69
[14] Hastie T, Tibshirani R and Friedman J 2001 *The Elements of Statistical Learning* (Springer)

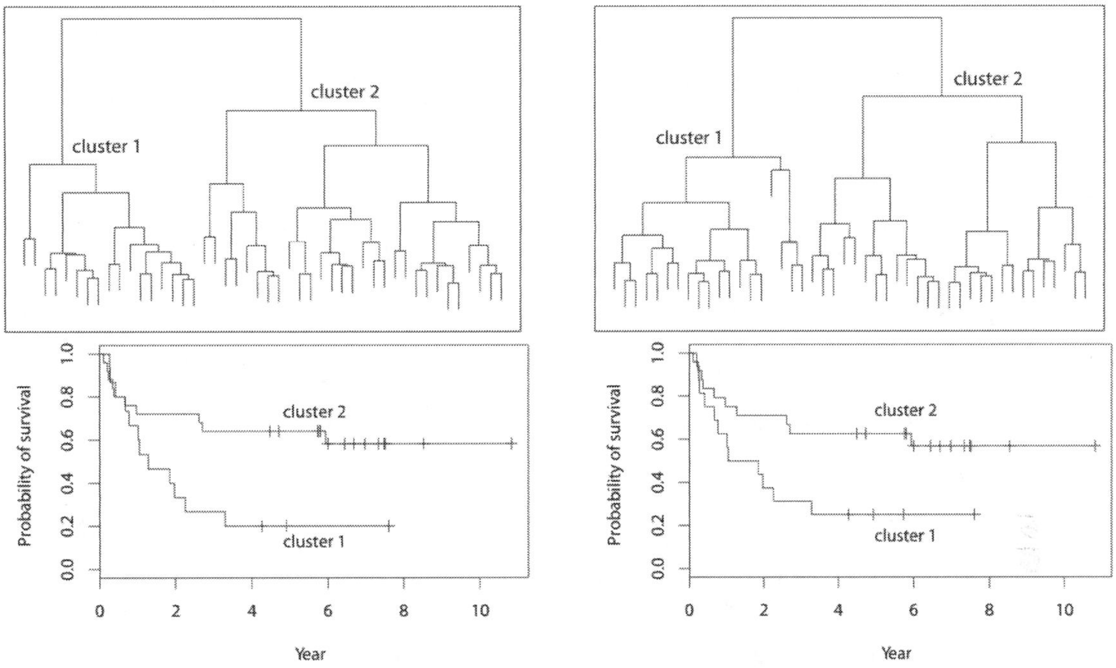

(a) Non-adaptive algorithm (b) Adaptive algorithm ($K = 5$)

46 DLBCL patients in Alizadeh et al (2000)

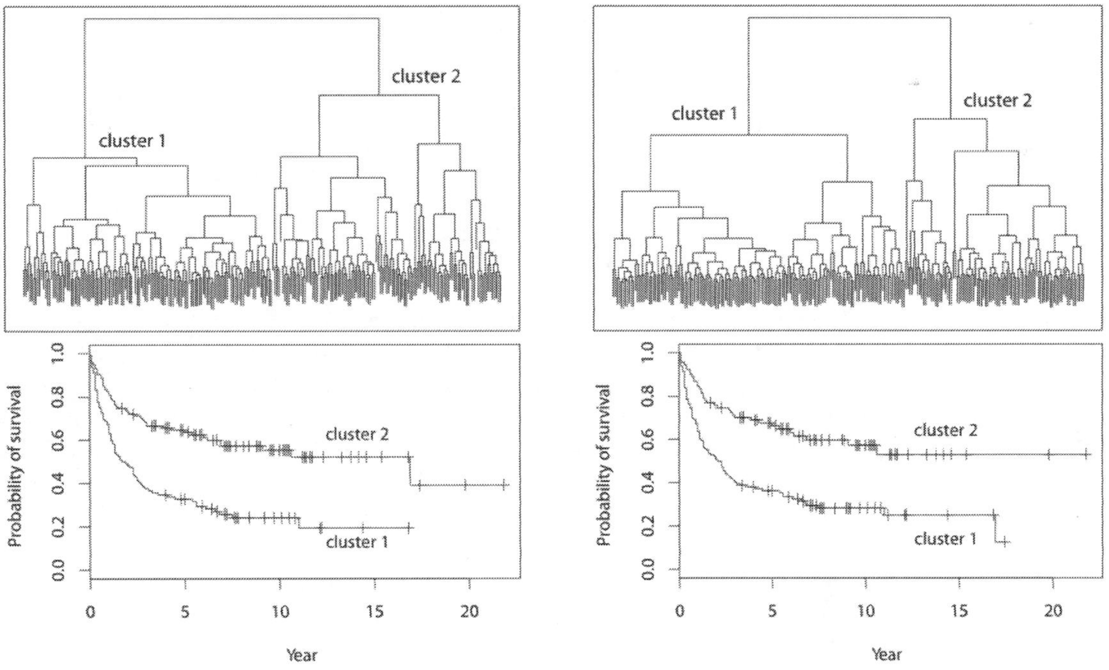

(c) Non-adaptive algorithm (d) Adaptive algorithm ($K = 5$)

240 DLBCL patients in Rosenwald et al (2002)

Figure 4. Hierarchical clustering and Kaplan-Meiyer analysis of DLBCL using learned metrics. The top 4 panels are the results of 46 DLBCL cases in [2], and the bottom 4 panels are those of 240 DLBCL cases in [10]. The left ones were obtained by non-adaptive metric learning, while the right ones were obtained by adaptive ($K = 5$) metric learning.

International Workshop on Statistical-Mechanical Informatics 2009 (IW-SMI 2009) IOP Publishing
Journal of Physics: Conference Series **197** (2009) 012009 doi:10.1088/1742-6596/197/1/012009

Grouping preprocess for haplotype inference from SNP and CNV data

Hiroyuki Shindo[1], **Hiroshi Chigira**[1], **Tomoyo Nagaoka**[1], **Naoyuki Kamatani**[2] **and Masato Inoue**[1]

[1]Department of Electrical Engineering and Bioscience, School of Advanced Science and Engineering, Waseda University, 3–4–1, Okubo, Shinjuku-ku, Tokyo 169–8555, Japan
[2]Institute of Rheumatology, Tokyo Women's Medical University, 10–22, Kawada-cho, Shinjuku-ku, Tokyo 162–0054, Japan

E-mail: masato.inoue@eb.waseda.ac.jp

Abstract. The method of statistical haplotype inference is an indispensable technique in the field of medical science. The authors previously reported Hardy-Weinberg equilibrium-based haplotype inference that could manage single nucleotide polymorphism (SNP) data. We recently extended the method to cover copy number variation (CNV) data. Haplotype inference from mixed data is important because SNPs and CNVs are occasionally in linkage disequilibrium. The idea underlying the proposed method is simple, but the algorithm for it needs to be quite elaborate to reduce the calculation cost. Consequently, we have focused on the details on the algorithm in this study. Although the main advantage of the method is accuracy, in that it does not use any approximation, its main disadvantage is still the calculation cost, which is sometimes intractable for large data sets with missing values.

1. Introduction

Genome-wide association studies have been revealing the relationships between genetic variations and phenotypic traits. Current high-throughput genotyping technologies have provided us with a number of single nucleotide polymorphisms (SNPs), which could be associated with common diseases such as type II diabetes or rheumatoid arthritis [1]. In addition to SNPs, it has been reported that structural variations such as copy-number variations (CNVs) affect phenotypic traits. CNVs are genetic variations in the copy number of DNA segments in comparison with a reference genome, a copy unit of which ranges from 1 kilobase to several megabases. CNVs are caused by genetic deletion or duplication, and they cover more than 12% of the human genome [2]. Since SNPs and CNVs are in linkage disequilibrium [3, 4], we need to do association analyses of integrated SNP-CNV sequences.

Haplotype information is crucial for association tests and other genetic studies because it is more powerful than information about the alleles of a single marker. (A haplotype is a single genetic constituent of an individual chromosome inherited from the father or mother.) Since experimental genotype data are observed as unphased pairs of alleles in SNPs and the total number of segment copies in CNVs of over two chromosomes, the haplotype phase is usually determined by statistical and computational methods.

Although various methods of haplotype phasing have been proposed [5, 6, 7, 8, 9, 10, 11], they cannot be directly applied to SNP-CNV haplotype phasing. One of the most difficult problems is that the number of possible haplotypes exponentially increases for the number of heterozygous loci. One of the standard methods used for haplotype inference, the EM algorithm [6], also suffers from huge calculation costs.

In this paper, we propose a haplotype-phasing algorithm based on the grouping preprocess (GrEM) [12], which can be applied to SNP and CNV genotypes. The GrEM involves a method of efficiently exploring haplotypes whose frequency in population is exactly zero in the maximum likelihood solution without using any approximation to reduce the number of possible haplotypes. From the mathematical aspect, SNPs appear as logical additions of alleles on two chromosomes, while CNVs appear as arithmetical additions. The proposed method can handle mixed SNP and CNV data through an integrated framework, which utilizes the common properties that both SNP and CNV have. We tested our algorithm by applying it to real and artificial unphased genotypes and found that it relaxed limitations on the calculation cost.

2. Model

Here, we define the probabilistic model for the data. We assume that the observed data are unphased multilocus genotypes. For example, [A/T, A/C, G/G, 3] represents 4 polymorphic genotypes composed of 3 SNPs and 1 CNV. 'A/T' denotes the heterozygous genotype, i.e., one of the alleles is 'A' and the other is 'T'. '3' denotes the total number of allelic copies over two chromosomes, i.e., the combinations of the numbers can be '0' and '3' or '1' and '2'. Here, we assume that the number of the alleles of each SNP locus is two; this is the ordinary situation. As to the copy number of each CNV site, we made no assumptions, so any non-negative integer is acceptable. For simplicity, we shall use 'A' to express a major allele or homozygous genotype of major alleles for each SNP locus. Similarly, 'C' is used for minor allele(s). Incidentally, as our method does not distinguish whether a certain allele is major or minor, 'A' and 'C' are absolutely exchangeable throughout this paper. 'B' is also used for briefly expressing a heterozygous genotype 'A/C'. According to these notations, [A/T, A/C, G/G, 3] is expressed as [B, B, A, 3] or briefly as [BBA3].

First, we define $\boldsymbol{\theta}$ as the haplotype frequencies of the population. For example, in the 3 SNPs and 1 CNV case, where the maximum copy number of the CNV locus is 3, $\boldsymbol{\theta} = (\theta_{[\mathrm{AAA0}]}, \theta_{[\mathrm{AAA1}]}, ..., \theta_{[\mathrm{CCC3}]})$ is a $2 \times 2 \times 2 \times 4$ dimensional vector. As $\boldsymbol{\theta}$ are frequencies, the sum of each element is normalized to 1 ($\sum_h \theta_h \equiv 1$), where h denotes every possible haplotype. The diplotype configuration for each individual is assumed to follow the Hardy-Weinberg equilibrium (HWE),

$$P(\boldsymbol{d}_i|\boldsymbol{\theta}) \equiv \begin{cases} \theta_{d_{i,1}}^2 & \text{if} \quad d_{i,1} = d_{i,2} \\ 2\theta_{d_{i,1}}\theta_{d_{i,2}} & \text{if} \quad d_{i,1} \neq d_{i,2} \end{cases}, \tag{1}$$

where $\boldsymbol{d}_i \equiv (d_{i,1}, d_{i,2})$ $(i = 1, 2, ..., I)$ denotes the pair of haplotypes for the ith individual and $\theta_{d_{i,1}}$ denotes the haplotype frequency for haplotype $d_{i,1}$. To prevent permutation symmetry, we impose a constraint, $d_{i,1} \leq d_{i,2}$, on $\boldsymbol{d}_i$ by defining a certain order. The probability for each individual is assumed to be independent. Also, measurement of the genotypes is assumed to include no errors:

$$P(\{g_i\}, \{\boldsymbol{d}_i\}|\boldsymbol{\theta}) \equiv \prod_i^I \delta_{\boldsymbol{d}_i \in g_i} P(\boldsymbol{d}_i|\boldsymbol{\theta}), \tag{2}$$

where g_i denotes the observed unphased genotype data for the ith individual, $\{g_i\}$ denotes the set of data for all individuals, I denotes the number of individuals in a data set, and δ denotes an indicator function, which yields 1 when the given condition is true and 0 otherwise. $\boldsymbol{d}_i \in g_i$ denotes that a diplotype, $\boldsymbol{d}_i$, is consistent with the genotype data, g_i.

Our final target is to determine the optimal diplotype configurations, $\{\hat{\boldsymbol{d}}_i\}$, such as $\arg\max_{\{\boldsymbol{d}_i\}} P(\{\boldsymbol{d}_i\}|\{g_i\})$ by assuming a certain prior distribution, $P(\boldsymbol{\theta})$, but this discrete optimization problem is usually hard to solve. Instead, we take a two-step approach; we first determine the maximum likelihood estimator of $\boldsymbol{\theta}$:

$$\hat{\boldsymbol{\theta}} \equiv \arg\max_{\boldsymbol{\theta}} P(\{g_i\}|\boldsymbol{\theta}). \tag{3}$$

Then, we infer

$$\{\hat{\boldsymbol{d}}_i\} \equiv \arg\max_{\{\boldsymbol{d}_i\}} P(\{\boldsymbol{d}_i\}|\{g_i\}, \hat{\boldsymbol{\theta}}) \tag{4}$$

using the determined $\boldsymbol{\theta}$.

Equation (4) is easy to solve, but Eq. (3) is not. The EM algorithm is an effective approach to solving this problem; it is an iterative procedure to estimate the maximum likelihood estimator by starting randomly with set $\boldsymbol{\theta}^{(0)}$ and sequentially calculating

$$\boldsymbol{\theta}^{(t+1)} \equiv \arg\max_{\boldsymbol{\theta}} \sum_{\{\boldsymbol{d}_i\}} P(\{\boldsymbol{d}_i\}|\{g_i\}, \boldsymbol{\theta}^{(t)}) \ln P(\{g_i\}, \{\boldsymbol{d}_i\}|\boldsymbol{\theta}) \tag{5}$$

under the expectation that $\boldsymbol{\theta}^{(t)}$ will converge to the maximum likelihood estimator. The EM algorithm guarantees convergence to a locally optimal but not globally optimal solution. Therefore, several trials with different initial values are carried out, and the best solution obtained from these is employed. Equation (4) can be reduced to

$$\theta_h^{(t+1)} \propto \sum_i^I \sum_{\boldsymbol{d}_i \in g_i} \theta_{d_{i,1}}^{(t)} \theta_{d_{i,2}}^{(t)} (\delta_{d_{i,1}=h} + \delta_{d_{i,2}=h}) \tag{6}$$

in this haplotype-inference problem. Thus, the calculation cost is roughly on the order of the total number of possible diplotype configurations satisfying $\boldsymbol{d}_i \in g_i$ over all individuals. This calculation cost occasionally becomes intractable because it increases exponentially with the number of heterozygous loci. For example, if the genotype data of a certain individual are [BB...B] with 50 SNP loci, the calculation cost is $\mathcal{O}(2^{50-1})$.

3. Theory

The idea behind the GrEM we previously proposed is simple. Our direct task here is to find one of the optimal solutions, $\hat{\boldsymbol{\theta}}$, which maximizes the following likelihood function under given genotype data $\{g_i\}$,

$$P(\{g_i\}|\boldsymbol{\theta}) = \prod_i^I \sum_{\boldsymbol{d}_i \in g_i} P(\boldsymbol{d}_i|\boldsymbol{\theta}). \tag{7}$$

By focusing on the symmetrical structure of this function, the GrEM reduces the calculation cost. Here, we show several simple examples composed of only SNP loci, but this theory is also valid for CNV only data and the mixed data of SNP and CNV.

For the first example, let us assume one individual has [ABBB], as shown in figure 1(a). In this case, the likelihood function is given as

$$P(\{g_i\}|\boldsymbol{\theta}) = (\theta_{[\mathrm{AAAA}]}\theta_{[\mathrm{ACCC}]} + \theta_{[\mathrm{AAAC}]}\theta_{[\mathrm{ACCA}]} + \theta_{[\mathrm{AACA}]}\theta_{[\mathrm{ACAC}]} + \theta_{[\mathrm{AACC}]}\theta_{[\mathrm{ACAA}]}) \tag{8}$$

There are four haplotype pairs in this equation, and obviously, assigning non-zero values to one of them and zero values to the others produces the optimal solution. Actually,

$\theta_{[\mathrm{AAAA}]} = \theta_{[\mathrm{ACCC}]} = 1/2$ is one of the optimal solutions (note that other θs are zero because $\sum_h \theta_h \equiv 1$). In other words, these four haplotype pairs are equivalent to one another because of the symmetry, and only one pair has non-zero probabilities that satisfy the optimality of the likelihood function. We can thus set the probabilities of three arbitrary haplotype pairs to zero while running the EM algorithm. If there are numerous heterozygous loci, this type of manipulation greatly reduces the calculation cost.

For the second example, let us assume two individuals have [ABBB] and [BAAB], as shown in figure 1(b). In this case, the likelihood function is given as

$$\begin{aligned} P(\{g_i\}|\boldsymbol{\theta}) = {} & (\theta_{[\mathrm{AAAA}]}\theta_{[\mathrm{ACCC}]} + \theta_{[\mathrm{AAAC}]}\theta_{[\mathrm{ACCA}]} + \theta_{[\mathrm{AACA}]}\theta_{[\mathrm{ACAC}]} + \theta_{[\mathrm{AACC}]}\theta_{[\mathrm{ACAA}]}) \\ & \times (\theta_{[\mathrm{AAAA}]}\theta_{[\mathrm{CAAC}]} + \theta_{[\mathrm{AAAC}]}\theta_{[\mathrm{CAAA}]}). \end{aligned} \tag{9}$$

As a result, there are two optimal solutions, the first is $\theta_{[\mathrm{AAAA}]} = 1/2$, $\theta_{[\mathrm{ACCC}]} = \theta_{[\mathrm{CAAC}]} = 1/4$, and the second is $\theta_{[\mathrm{AAAC}]} = 1/2$, $\theta_{[\mathrm{ACCA}]} = \theta_{[\mathrm{CAAA}]} = 1/4$. As we can easily guess that [AAAA] and [AAAC] are included in both g_1 and g_2, they have advantages. [ACCC], [ACCA], [CAAC], and [CAAA] also have advantages because each of them can be paired with an advantageous one. As the remaining four haplotypes, [ACAC], [ACAA], [AACC], and [AACA] are less advantageous, we can set their probabilities to zero. This type of manipulation also greatly reduces the calculation cost. Similar to the first example, two haplotype combinations { [AAAA], [ACCC], [CAAC] } and { [AAAC], [ACCA], [CAAA] } are symmetrical and equivalent to each other. Thus, we can set the probabilities of one of them to zero.

Such optimality can easily be demonstrated through simple equations. Assume $\boldsymbol{\theta}$ such that every θ in Eq. (9) is greater than zero. Also, suppose $\boldsymbol{\theta}'$ such that

$$\begin{aligned} \theta'_{[\mathrm{AAAA}]} &= \theta_{[\mathrm{AAAA}]} + \theta_{[\mathrm{AAAC}]} + \theta_{[\mathrm{AACA}]} + \theta_{[\mathrm{AACC}]} \\ \theta'_{[\mathrm{ACCC}]} &= \theta_{[\mathrm{ACCC}]} + \theta_{[\mathrm{ACCA}]} + \theta_{[\mathrm{ACAC}]} + \theta_{[\mathrm{ACAA}]} \\ \theta'_{[\mathrm{CAAC}]} &= \theta_{[\mathrm{CAAC}]} + \theta_{[\mathrm{CAAA}]} \\ \theta'_{[\mathrm{AAAC}]} &= \theta'_{[\mathrm{AACA}]} = \theta'_{[\mathrm{AACC}]} = \theta'_{[\mathrm{ACCA}]} = \theta'_{[\mathrm{ACAC}]} = \theta'_{[\mathrm{ACAA}]} = \theta'_{[\mathrm{CAAA}]} = 0. \end{aligned} \tag{10}$$

Then, as the likelihood of $\boldsymbol{\theta}$ is always less than that of $\boldsymbol{\theta}'$,

$$P(\{g_i\}|\boldsymbol{\theta}) < P(\{g_i\}|\boldsymbol{\theta}'), \tag{11}$$

such $\boldsymbol{\theta}$ is necessarily suboptimal. Similarly, we can prove that $\boldsymbol{\theta}$ is suboptimal except for two cases such that only $\theta_{[\mathrm{AAAA}]}$, $\theta_{[\mathrm{ACCC}]}$, and $\theta_{[\mathrm{CAAC}]}$ have positive values or only $\theta_{[\mathrm{AAAC}]}$, $\theta_{[\mathrm{ACCA}]}$, and $\theta_{[\mathrm{CAAA}]}$ have positive values. Therefore, we can assert that the left θs should be exactly zero before running the EM algorithm.

4. Algorithms

Next, how can we perform such manipulations? Doing so manually is not realistic because there are usually hundreds of individuals. Our approach consists of three steps. (1) The first determines all *borders* that correspond to ellipses in the Venn diagram of haplotypes in figure 1. Because a haplotype paired to some advantageous haplotype is also advantageous, not only the *borders* corresponding to individuals but also new *borders* are necessary to distinguish these advantageous haplotypes. Intuitively, a new *border* is a mirror image of a shared haplotype set reflected by some individual. According to our experience, the number of *borders* usually greatly exceeds that of individuals. (2) The second step locally determines mostly nested territories that correspond to all intersections of arbitrary *borders* but that cannot be separated by any *borders*. We call such haplotype sets (territories) *leaves*. Figure 2 shows examples of *leaves*. The number

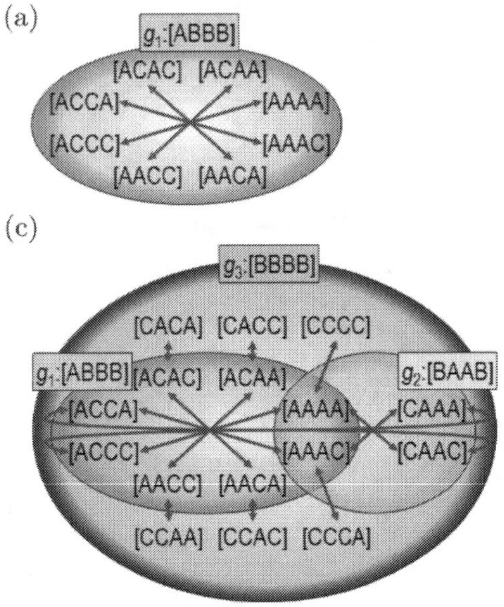

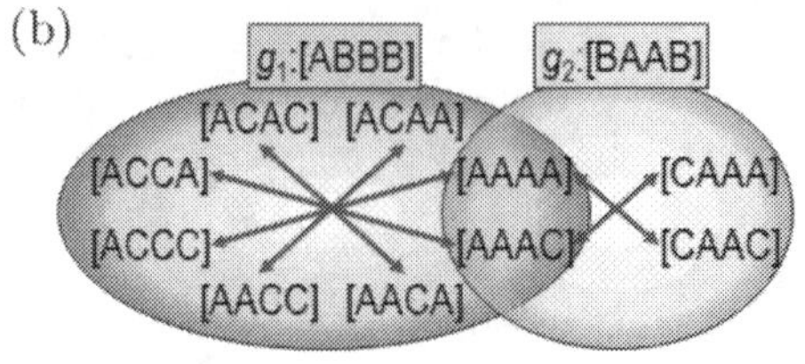

Figure 1. Venn diagrams of haplotypes for (a) first, (b) second, and (c) third examples. g_1: [ABBB] denotes multilocus genotype data from the 1st individual. 'A' denotes major and 'C' denotes minor alleles, while 'B' denotes a heterozygous genotype. [ACAC], [ACAA],... denote haplotypes. The bidirectional arrows denote diplotype configurations for g_i. Each ellipse labeled g_i includes all haplotypes this individual may have.

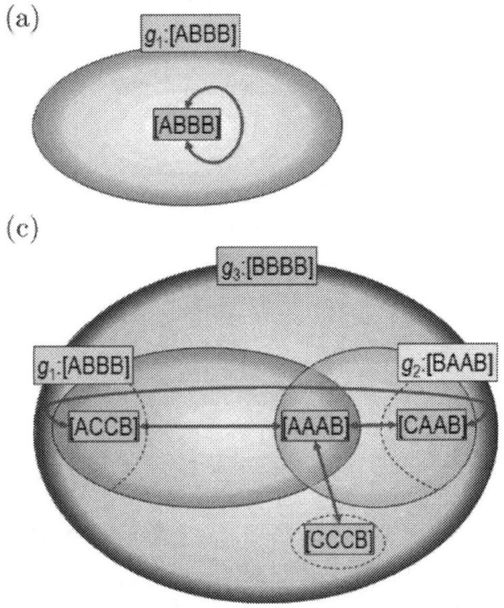

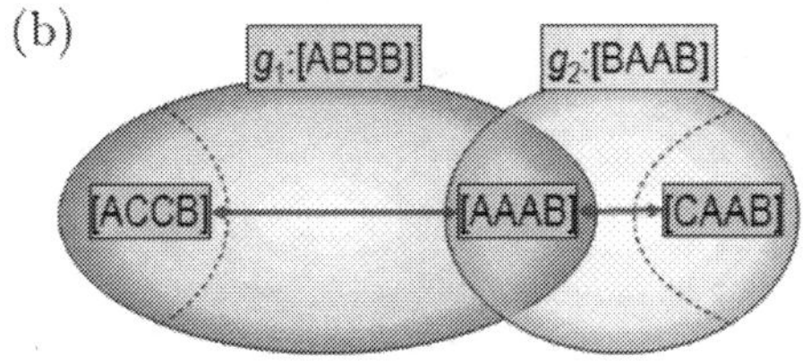

Figure 2. Venn diagrams of *leaves* after Step 2 of GrEM for (a) the first, (b) second, and (c) third examples. *Borders* determined by Step 1 are (a) {[ABBB]}, (b) {[ABBB], [BAAB], [ACCB], [CAAB]}, (c) {[ABBB], [BAAB], [BBBB], [BCCB], [CBBB]}. For (b), [AAAB] is temporarily created as a candidate for a *border* but is dropped because it intersects some *borders*, i.e., [AAAB] = [ABBB] ∩ [BAAB]. Same as [ACCB] and [CAAB] for (c).

of leaves is usually much smaller than 2^N when N is large. (3) The third step detects loopy graphs and replaces some *leaves* with haplotypes if necessary. Briefly, the *leaves* and individuals can be expressed using a graph (figure 3). Here, graph vertices and edges correspond to *leaves* and individuals. The graph is usually disconnected, forming several separated subgraphs. For each subgraph, we check whether it has any odd-step loops such as Möbius strips, and if it has, each vertex (*leaf*) is replaced with some haplotypes, which are also called *leaves*. Empirically, such replacements rarely occur but are necessary for an exact maximum likelihood estimator. GrEM is mainly composed of these three steps of manipulation.

After these steps, we run the EM algorithm not for all possible haplotypes (haplotypes such as those in figure 1) but for *leaves* (leaves such as those in figure 4). Although the number of all possible haplotypes may be not tractable, we expect that the number of *leaves* will be much

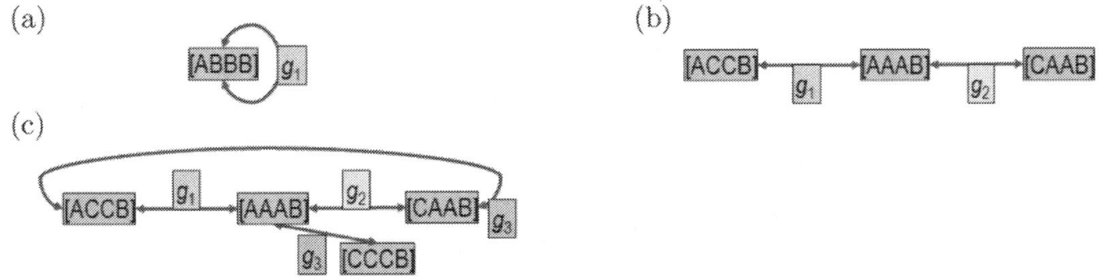

Figure 3. Graphical expression of *leaves* after Step 2 of GrEM for (a) the first, (b) second, and (c) third examples. [ABBB], [ACCB],... denote vertices of *leaves*. g_i denotes the edge of the ith individual. In each example, the number of subgraphs is only one by chance, but there are usually several disconnected subgraphs for larger data sets. For (a), there is a 1-step loop: [ABBB]-g_1-[ABBB], which is expanded with Step 3 of GrEM. For (b), there are no loops. For (c), there is a 3-step loop: [ACCB]-g_1-[AAAB]-g_2-[CAAB]-g_3-[ACCB], which is also expanded with Step 3.

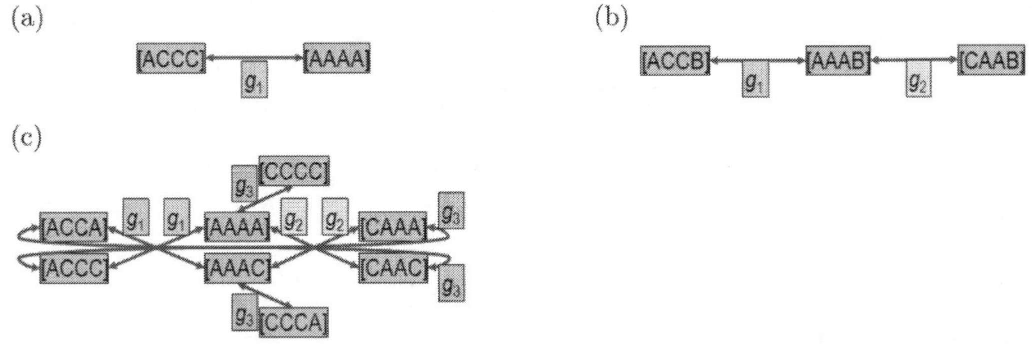

Figure 4. Graphical expression of *leaves* after Step 3 of GrEM for (a) the first, (b) second, and (c) third examples. For (a) and (c), the original *leaves* are replaced by new *leaves* with the loop expansion in Step 3, while loop expansion did not occur in (b). For the graph in figure 3(c), we can start from [ACCB], go around though [AAAB] and [CAAB], and return back to [ACCB]. However, if we check the haplotype connectivity more closely, this path is in fact impossible. As shown in (c), we start from [ACCA], go around through [AAAC] and [CAAA], and cannot return back to [ACCA]. We have to go further through [ACCC], [AAAA], and [CAAC] to reach the starting haplotype. We can find such a Möbius strip and expand it with Step 3 of GrEM. The calculation cost of the following EM algorithm is roughly in proportion to the number of diplotypes, which has been indicated by the bidirectional arrows here.

smaller and tractable. More precisely, the calculation cost of the EM algorithm is in proportion to the number of possible diplotypes, as shown in Eq. (6); we evaluate the proposed method by comparing the numbers of possible diplotypes before and after these steps.

Before going into details on the algorithms for the GrEM, let us define some operators, as listed in table 1. Again, h denotes a single haplotype, e.g., [AACC]. g denotes multilocus genotype data from an individual, and we simply call g an individual; e.g., [ABCB] means this individual may have [AACA] and [ACCC] or [AACC] and [ACCA]. b denotes a *border*; e.g., [ABCB] means the set of haplotypes, { [AACA], [AACC], [ACCA], [ACCC] }. The set of haplotypes a *border* can express is limited to the direct product of each genotype at each locus.

Table 1. Operation tables of $\oplus$, $\odot$, $\cap$, and $\ominus$ for SNP data. Note: b denotes a *border*, and g denotes genotype data from a certain individual. All of these operations are performed by locus-wise processes according to the tables above. Superscript (n) denotes the identity of a locus. 'A' denotes major and 'C' denotes minor alleles, while 'B' denotes a heterozygous genotype and 'Z' denotes a missing datum of a genotype; this may be 'A', 'C', or 'B'.

$g := h \oplus h'$

	$h'^{(n)}$	
$h^{(n)}$	A	C
A	A	B
C	B	C

$b := \odot g$

$g^{(n)}$	A	C	B	Z
$b^{(n)}$	A	C	B	B

$b'' := b \cap b'$

	$b'^{(n)}$		
$b^{(n)}$	A	C	B
A	A	-	A
C	-	C	C
B	A	C	B

$b' := g \ominus b$

	$b^{(n)}$		
$g^{(n)}$	A	C	B
A	A	-	A
C	-	C	C
B	C	A	B
Z	B	B	B

Algorithm 1 DetermineBorders($\mathcal{G}$)

1: initialize $\mathcal{B} := \phi$, $\mathcal{C} := \phi$, and $\mathcal{D} := \phi$
2: **foreach** $g \in \mathcal{G}$, calculate $b := \odot g$, and add b to $\mathcal{C}$ and $\mathcal{D}$
3: **foreach** $g \in \mathcal{G}$, **if** g has homozygous b **then** add b to $\mathcal{C}$ and $\mathcal{D}$
4: **if** $\mathcal{C} = \phi$ **then return** $\mathcal{B}$
5: remove largest b^* from $\mathcal{C}$
6: **if** $b^* = b' \cap b'' \cap b''' \ldots$ $({}^{\exists}b', b'', b''' \ldots \in \mathcal{B})$ **then goto** 4
7: **foreach** $g \in \mathcal{G}$ and $g \supset b^*$, calculate $b := g \ominus b^*$, **if** $b \notin \mathcal{D}$ **then** add b to $\mathcal{C}$ and $\mathcal{D}$
8: add b^* to $\mathcal{B}$
9: **goto** 4

We also define operation tables for these items as shown in table 1. $\oplus$ is a binary operator that produces the observed result, g, from two given haplotypes, e.g., [AACC] $\oplus$ [ACAC] is [ABBC]. $\odot$ is a unary operator that produces a *border* from a given g, e.g., $\odot$ [ABCA] is [ABCA]. The result is the same as for the operand for SNP data, but differs for missing values of SNP and CNV data. We will explain the CNV case later. $\cap$ is a binary operator that produces the intersection of two given *borders*, e.g., [ABBC] $\cap$ [BABB] is [AABC]. $\ominus$ is a binary operator that produces a set of paired haplotypes to a given b for g, e.g., [ABBB] $\ominus$ [AAAB] is [ACCB]. Note that if b includes unpaired haplotypes for g while $g \ominus b$ being calculated, such haplotypes are simply neglected, e.g., [ABBB] $\ominus$ [BAAB] is also [ACCB] (refer to figure 2(b)).

The first step of the GrEM is as follows, and its specific algorithm is algorithm 1. This is a function whose argument is $\mathcal{G}$, the set of all individuals, and it returns $\mathcal{B}$, the set of determined *borders*. This function intuitively produces mirror images sequentially. ϕ denotes the empty set. $\mathcal{C}$ denotes the set of candidates for *borders* that are worth keeping. $\mathcal{D}$ denotes the set of processed *borders* that prevents a *border* from being added to $\mathcal{C}$ more than once at line 7. On line 2, $\odot g$ denotes an ellipse in the Venn diagram corresponding to each individual g_i. 'Homozygous b' on line 3 will be explained later in connection to CNV data. The 'largest b^*' on line 5 denotes the *border* with the largest number of 'B's in its sequence, e.g., the first is the largest of [ABBB], [ABCB], and [AACB], because the first has three 'B's while the others have less than three 'B's. Line 6 is a cost saver. This means that if a new candidate coincides with any intersection of an arbitrary number of already determined *borders*, such a candidate is not worth keeping. Line 7 represents the main operation for this function. A mirror image of a *border* for each individual is added to $\mathcal{C}$. These manipulations are repeated until $\mathcal{C}$ has no elements.

Journal of Physics: Conference Series **197** (2009) 012009 doi:10.1088/1742-6596/197/1/012009

Algorithm 2 DetermineLeaves($\mathcal{B}$)

1: initialize $\mathcal{L} := \phi$ and $\mathcal{B}' := \phi$
2: **if** $\mathcal{B} = \phi$ **then return** $\mathcal{L}$
3: remove largest b^* from $\mathcal{B}$
4: **foreach** $b \in \mathcal{B}$, **if** $b \subset b^*$ **then** remove b from $\mathcal{B}$, and add b to $\mathcal{B}'$
5: **foreach** $b \in \mathcal{B}$ and $b \cap b^* \neq \phi$, calculate $b' := b \cap b^*$, and add b' to $\mathcal{B}'$
6: **if** $\mathcal{B}' = \phi$ **then** add b^* to $\mathcal{L}$ **else** add (**call** DetermineLeaves($\mathcal{B}'$)) to $\mathcal{L}$
7: **goto** 2

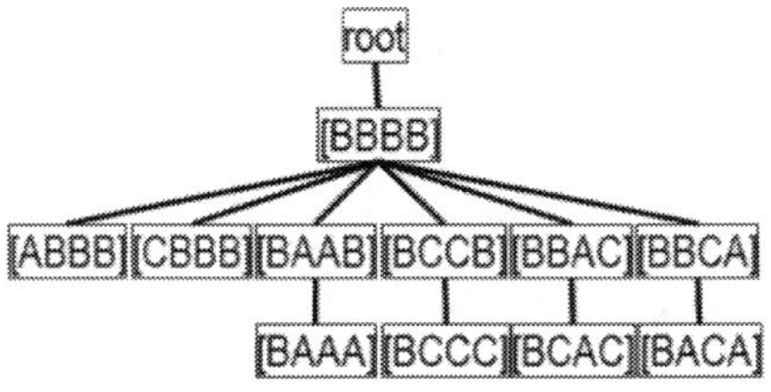

Figure 5. Example of a tree of *borders*. A tree is usually more powerful than a set when we check whether a certain *border* is already included or not. If we introduce a certain order for siblings in a tree, it will be more efficient.

The second step of the GrEM is as follows, and its specific algorithm is algorithm 2. This is a recursive function whose argument is $\mathcal{B}$, the result of the first step, and it returns $\mathcal{L}$, the set of determined *leaves*. *Leaves* are defined using the set of *borders*. A *leaf* must coincide with either some *border* or the intersection of an arbitrary number of *borders*, and it must not be separated by any *borders*. Recall that shared haplotypes and their mirror images have advantages. *Leaves* coincide with such advantageous haplotype sets. If line 6 in algorithm 1, the cost saver, is removed, the resulting $\mathcal{B}$ necessarily includes all *leaves* because applying the operator, $\ominus$, twice is equivalent to the simple intersection, $g \ominus (g \ominus b) = g \cap b$ (except for missing values). The approach without the cost saver seems more straightforward, but it usually suffers from numerous *borders* resulting in memory overflow. On line 3, one of the *borders* is selected, and the intersection is only calculated inside this *border*. Such a step is hierarchically repeated on line 6. This approach suppresses the peak number of *borders* and results in a faster calculation. When all the intersections are checked, this function returns the set of all determined *leaves*.

These two algorithms can be made more efficient if we introduce a tree of *borders* as shown in figure 5. The construction of this tree is intuitive. *Borders* are added to the tree in descending order of their cardinalities. We start from the root and check all its children to find whether each child is a proper superset of a new *border* or not. If there is such a child, we move to the child and check all its children again. We repeat this check until there are no such children. Then, we add a new *border* to the current *border* as a child. We also construct such a tree for $\{\odot g_i\}$. This is because these two algorithms contain numerous inclusion checks, e.g., $g \supset b^*$ on line 7 of algorithm 1, or $b \supset b^*$ on line 4 of algorithm 2. Also, the tree of *borders* is efficient for checking the if-clause on line 6 of algorithm 1.

The third step of the GrEM is as follows. First, we group all *leaves* into several disconnected subgraphs. Each subgraph is checked as to whether it has any odd-step loops or not. More specifically, we arbitrarily select a *leaf*, l, from the subgraph and assign a specific haplotype, h, to l where $h \in l$. Then, we propagate h to the whole subgraph; we find the adjacent vertex l' to l by following edge g where $l' = g \ominus l$, and assign $h' := g \ominus h$ to l'. If l' has already been assigned the same haplotype, we do not assign it twice, whereas when h' is different from the assigned haplotype, we assign h' to l' in addition to the already assigned one(s). This assignment is repeated throughout the whole subgraph. Finally, we check whether every *leaf* in the subgraph is only assigned one haplotype or not. If so, we do not have to do anything else. If not, we replace all *leaves* in the subgraph with assigned haplotypes, which become new *leaves* instead.

Table 2. Operation tables of $\oplus$, $\odot$, $\cap$, and $\ominus$ for CNV data. b denotes a *border* and g denotes genotype data from a certain individual. All these operations are performed by locus-wise processes according to the tables above. Superscript (n) denotes the identity of a locus. A *border* has two values, i.e., minimum and maximum numbers of copies, for each locus, which is expressed as '$0..x$' when the minimum and maximum numbers correspond to 0 and x.

$g := h \oplus h'$	
	$h'^{(n)}$
$h^{(n)}$	y
x	$x + y$

$b := \odot g$	
$g^{(n)}$	x
$b^{(n)}$	$0..x$

$b'' := b \cap b'$	
	$b'^{(n)}$
$b^{(n)}$	$x..y$
$w..z$	$\max(w, x)$ $..\min(z, y)$

$b' := g \ominus b$	
	$b^{(n)}$
$g^{(n)}$	$x..y$
z	$\max(0, z - y)$ $..\max(0, z - x)$

This algorithm successfully detects odd-step loops in the subgraph. Intuitively, this step detects Möbius strips and expands them. Three examples before this step are shown in figure 3, and those after this step are shown in figure 4. Through Steps 1 and 2, we obtain advantageous sets of haplotypes, but their connectivities sometimes form Möbius strips. Note that if all the cardinalities of *leaves* are one in a subgraph, this replacement never occurs. Empirically, this replacement rarely occurs. Also, this step does not take a lot of time.

We will only give a short explanation here of missing values of SNP data. The missing value denoted by 'Z' may be 'A', 'C', or 'B'. The proposed approach successfully manages missing values. More specifically, missing values are dealt with in the same manner as the other genotypes, as listed in table 1. Empirically, missing values greatly increase the number of *borders* and *leaves*. Thus, the number of missing values is critical to the calculation cost of the proposed method.

Each copy number for each polymorphic locus of a certain individual appears as an arithmetical addition of two copy numbers from each haplotype. In our approach, SNP and CNV have two common properties. These properties are related to the introduced operators, $\odot$, $\ominus$, and $\cap$. Operator $\odot$ produces a set of haplotypes that a given individual may have. We can define this operator for CNV as listed in table 2, e.g., $\odot$[0, 1, 2, 3] is [0..0, 0..1, 0..2, 0..3], where '$x..y$' denotes a set of integers from x through y. Similar to SNP data, the set of haplotypes corresponding to a certain individual can also be expressed as the direct product of each locus. This operation can also be calculated with a locus-wise process. Although these properties are trivial, they are important for the proposed method. Operator $\ominus$ produces a set of haplotypes that can be paired with shared haplotypes. We can also define this operator for CNV as seen in table 2, e.g., [1, 2, 3, 4] $\ominus$ [0..2, 1..2, 0..1, 2..3] is [0..1, 0..1, 2..3, 1..2]. An operation result can also be expressed as a direct product of each locus, and this operation can be performed by using a locus-wise process. Operator $\cap$ produces new *borders* that coincide with the intersection of two given *borders*. This operator also retains these properties.

The property of the direct product guarantees that a *border* can be expressed using the size of the $\mathcal{O}(N)$ memory, where N denotes the number of polymorphic loci. The property of the locus-wise process guarantees that each operation can be performed with the calculation cost of $\mathcal{O}(N)$ if a locus-wise process requires a calculation cost of $\mathcal{O}(1)$. Without these properties, we would incur a memory requirement and calculation cost of $\mathcal{O}(2^N)$, which is intractable. These properties also make haplotype inference from mixed SNP and CNV data possible. For example, we can manage such individual data as [A, B, C, 1, 2, 3].

For CNV, a set of copy numbers for each loci of a *border* is convex and we can express it in the form of '$x..y$' by using two integers, i.e., minimum and maximum numbers. Although this property is not crucial, it is advantageous. Without this convexity, we would have to hold a set

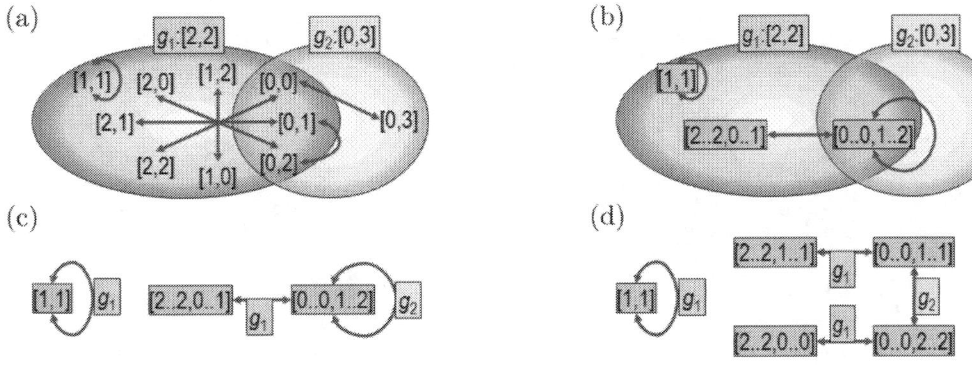

Figure 6. Example of CNV data. (a) Venn diagram of haplotypes. (b) Venn diagram of *leaves*. (c) Graphical expression of *leaves* after Step 2. (d) Graphical expression of *leaves* after Step 3.

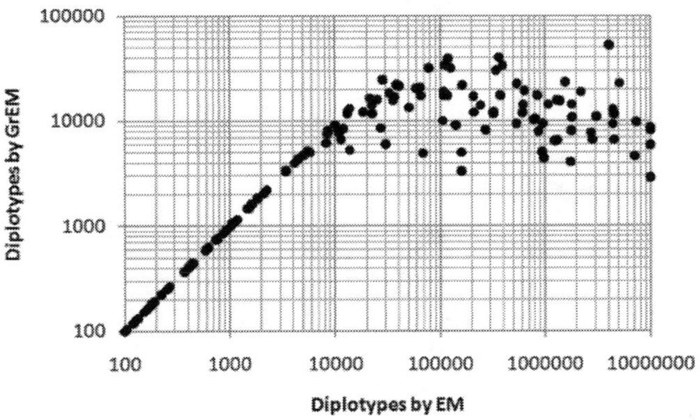

Figure 7. EM and GrEM results for the number of diplotypes. Each closed circle denotes the result from an artificial data set: 100 individuals of 1–30 SNP loci including 1% missing values. GrEM successfully reduced the number of diplotypes. EM could not handle 7 out of 150 data sets because of memory overflow. Such data sets are plotted on the extreme right of the graph.

of copy numbers for each loci, e.g, [{0, 1, 3}, {2}, {2, 3}, {0, 2, 4}]. If the average cardinality of each set was five, the memory requirement and calculation cost would increase five times.

One significant difference in the properties of SNP and CNV is that regarding the set of haplotypes a certain individual may have, $\odot g$; each haplotype in the set is symmetrical for SNP (figure 1(a)) but not for CNV. Let us consider an individual of $g = [2, 2]$. $\odot g$ is {[0,0], [0,1], [0,2], [1,0], [1,1], [1,2], [2,0], [2,1], [2,2]}. Of the elements in this set, only [1,1] can be a homozygous haplotype (figure 6(a)). This type of homozygous haplotype exists if g does not contain any homozygous genotype 'B' or odd copy numbers, e.g., [A, C, 2, 4]. The expression of 'g has homozygous b' in algorithm 1 means such a situation. Except for this, CNV data or mixed SNP and CNV data can be dealt with by using the same algorithms. Figure 6 shows such one example.

5. Results

We first show how much GrEM reduces the calculation cost by using artificial data sets. Figure 7 compares the numbers of diplotypes, which are roughly in proportion to the calculation cost of the EM algorithm. Memory overflow often occurred with the original EM when there were more than 28 SNP loci. However, GrEM successfully reduced the number of diplotypes when there were more than 12 SNP loci.

The effect of the cost saver on line 6 in algorithm 1 is plotted in figure 8. The cost saver reduced about 90% of the *borders* (figure 8(a)), which roughly corresponds to the calculation cost and memory requirement of Step 1. Also, fewer *borders* reduced the calculation time of

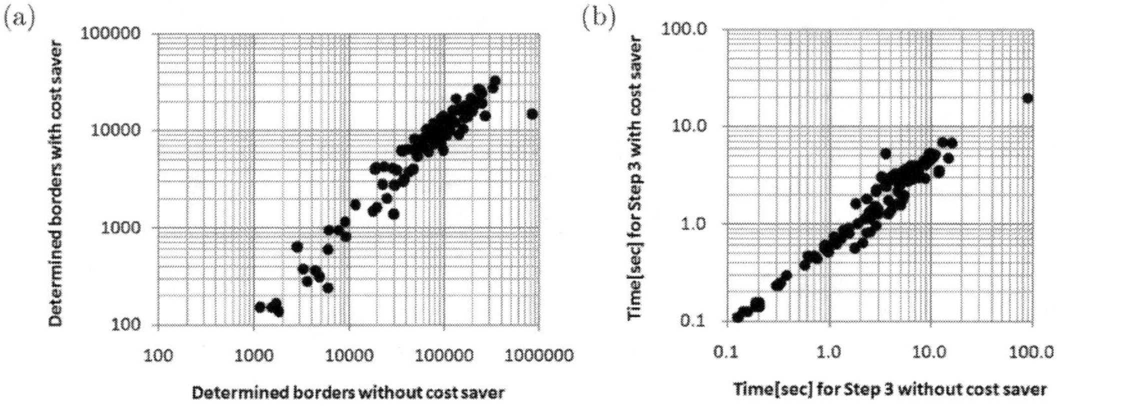

Figure 8. Effect of cost saver on line 6 in algorithm 1. The data sets are the same as those used in figure 7. The cost saver reduced about 90% of the *borders* and thus helped to prevent memory overflow.

Table 3. Results for real data sets. Loci denotes the number of polymorphic loci, where 1 means 1 CNV locus, while 12 means 1 CNV and 11 SNP loci. Log-likelihood denotes $\max_{\boldsymbol{\theta}} \ln P(\{\hat{\boldsymbol{d}}_i\}|\{g_i\}, \boldsymbol{\theta})$. '–' denotes intractable cases.

dataset	loci	number of individuals	missing rate %	log-likelihood GrEM	EM	MOCSphaser
MRGPRX1(CEU)	1	60	0.00	−0.7999	−0.7999	−0.7999
MRGPRX1(YRI)	1	60	0.00	−0.7872	−0.7872	−0.7872
CYP2D6(CEU)	1	60	0.00	−0.2303	−0.2303	−0.2303
CYP2D6(YRI)	1	60	0.00	−0.5447	−0.5447	−0.5447
MRGPRX1(CEU)+SNP	12	60	4.44	−1.9656	−1.9656	−1.9668
MRGPRX1(YRI)+SNP	28	60	3.04	−2.3971	–	–
CYP2D6(CEU)+SNP	58	60	1.35	−1.4572	–	–
CYP2D6(YRI)+SNP	76	60	1.23	−2.6574	–	–

Step 2, as can be seen from figure 8(b). This effect was especially clear when there were many *borders*.

The accuracy of GrEM was tested and validated using real data sets provided by Kato et al. [13]. They were measured by quantitative polymerase chain reaction (PCR) on *CYP2D6* and *MRGPRX1* genes for individuals of Northern and Western European descent from Utah, USA (CEU) and for individuals of the Yoruba people from Ibadan, Nigeria (YRI) in the HapMap populations. We also prepared mixed data for these experimental CNVs and neighboring SNPs. As shown in table 3, GrEM worked well on both small and large data sets. In regard to the small data sets, the log-likelihoods of the three methods were the same. In regard to the large data sets, memory overflow occurred in the original EM algorithm and the conventional mixture-of-CNV-SNP phaser (MOCSphaser) [13]. These results demonstrated that the proposed method could handle SNP-CNV data and greatly reduce the calculation cost.

Concerning the calculation time, GrEM was much faster than MOCSphaser. We prepared 7 middle-sized data sets composed of one CNV locus and 9–14 SNP loci, for which both GrEM and MOCSphaser could perform the inference. GrEM was 1.4–108 times faster than MOCSphaser, the median case was 2.5 sec and 28 sec, respectively, although calculation time greatly depends on a given data.

6. Discussion

The proposed method assumes Hardy-Weinberg equilibrium (HWE) for the generative model. Also, it does not employ recombination or single mutation for the model or the prior distribution. If the population of given data is large, this simple model is considered to be adequate, and it avoids overfitting to the given data. The EM algorithm is a standard method that is used to solve this model. The EM algorithm has serious problem in that it exponentially increases the calculation cost. The proposed method focuses on the symmetrical structure of the model and successfully reduces the calculation cost by using such symmetry. If the population is small or, more extremely, if pedigree data are given, the HWE model is not an adequate way of representing the data. One useful extension of the proposed method would be the ability to impose a blood relationship on the model.

Missing values create realistic problems for haplotype inference. The proposed method can manage the missing values for SNP data. More precisely, completely missing values are acceptable, but partially missing values are not acceptable, e.g., for a certain SNP locus, such a case would be where one of the alleles is major and the other allele is unknown. This is because such data are asymmetric and GrEM does not work well with these. The proposed method has currently not been implemented to manage CNV data from ambiguously determined copy numbers. We would like to extend it to handle such data.

Single-nucleotide variation in a copy unit (SNVC) is also a big problem with haplotype inference [14]. As CNV involves multiple copies of DNA segments longer than 1 kilobase, and the frequency of SNP is about 0.1%, the CNV region usually contains multiple SNP loci. These SNP data cannot be distinguished from one another but the quantity of each allele can be measured. We would also like to extend the method to SNVC data in the future.

Acknowledgments

This work was partially supported by a Grant-in-Aid for Scientific Research on Priority Areas (No. 18079012) and by the "High-Tech Research Center" Project of the Japanese Ministry of Education, Culture, Sports, Science and Technology.

References

[1] The Wellcome Trust Case Control Consortium 2007 *Nature* **447** 661–78
[2] Redon R *et al* 2006 *Nature* **444** 444–54
[3] Locke D P *et al* 2006 *Am. J. Hum. Genet.* **79** 275–90
[4] McCarroll S *et al* 2008 *Hum. Mol. Genet.* **17** R135–42
[5] Clark A G 1990 *Mol. Biol. Evol.* **7** 111–22
[6] Excoffier L and Slatkin M 1995 *Mol. Biol. Evol.* **12** 921–7
[7] Stephens M, Smith N J and Donnelly P 2001 *Am. J. Hum. Genet.* **68** 978–89
[8] Niu T H, Qin Z H S, Xu X P and Liu J S 2002 *Am. J. Hum. Genet.* **70** 157–69
[9] Qin Z H S, Niu T H and Liu J S 2002 *Am. J. Hum. Genet.* **71** 1242–7
[10] Stephens M and Donnelly P 2003 *Am. J. Hum. Genet.* **73** 1162–9
[11] Xing E P, Jordan M I and Sharan R 2007 *J. Comput. Biol.* **14** 267–84
[12] Shindo H, Chigira H, Tanaka J, Kamatani N and Inoue M 2008 *J. Hum. Genet.* **53** 747–56
[13] Kato M, Nakamura Y and Tsunoda T 2008 *Bioinformatics* **24** 1645–6
[14] Kato M, Nakamura Y and Tsunoda T 2008 *Am. J. Hum. Genet.* **83** 157–69

Journal of Physics: Conference Series **197** (2009) 012010 doi:10.1088/1742-6596/197/1/012010

How to measure genetic heterogeneity

Ryo Yamada

Professor, Unit of Statistical Genetics, Center for Genomic Medicine, Graduate School of Medicine, Kyoto University, Yoshida-Konoe-cho, Sakyo-ku, Kyoto 606-0815, JAPAN

E-mail: ryamada@genome.med.kyoto-u.ac.jp

Abstract. Genetic information of organisms is coded as a string of four letters, A, T, G and C, a sequence in macromolecules called deoxyribonucleic acid (DNA). DNA sequence offers blueprint of organisms and its heterogeneity determines identity and variation of species. The quantitation of this genetic heterogeneity is fundamental to understand biology. We compared previously-reported three measures, covariance matrix expression of list of loci (pair-wise r^2), the most popular index in genetics, and its multi-dimensional form, Ψ, and entropy-based index, ϵ. Thereafter we proposed two methods so that we could handle the diplotypic heterogeneity and quantitate the conditions where the number of DNA sequence samples is much smaller than the number of possible variants.

1. Introduction

Parents hand down the blueprint of life to children. The blueprint consists of macromolecules, deoxyribonucleic acid (DNA), coding information in a string of four letters, A, T, G and C. The sequence is copied before the inheritance with two types of modifications, mutations and recombinations. These modifications are the origin of genetic heterogeneity among biological population. The accumulation of the modifications is believed to result in speciation. The genetic heterogeneity increases the phenotypic heterogeneity and it is believed that the phenotypic heterogeneity enables species to survive against environmental changes because a fraction of the population has higher chance to be resistant to the changes.

The evolution and speciation is one of the most important topics in genetics.[1] The study to investigate relation between genetic and phenotypic heterogeneity among individuals of a species, particularly of *homo sapience*, is in the spotlight of medical genetics.[2] The former and the latter are the studies of inter-species and intra-species genetic heterogeneity, respectively. In this paper, I characterize the conventional measures of DNA sequence heterogeneity and introduce new measures thereafter. I discuss these measures from mathematical and genetic standpoints to prompt further studies to develop better methods to quantitate genetic heterogeneity.

2. Genetics

Although species take various reproduction strategies, I focus on one reproduction system, sexual reproduction of diploid organisms, to which *homo sapience* belongs. DNA in such system is classified into autosomal and sex chromosomes. The majority of DNA is autosomal. Every individual has two copies of autosomal chromosomes regardless of sex and the number of copy of sex chromosomes depends on sex. In this paper I handle autosomal heterogeneity.

See Figure 1 describing the reproduction system with mutations and reproductions. An individual G2 has a pair of DNA sequences, one of which comes from its mother (G1-1) and the other from its father (G1-2). When G2 gives DNA sequence to its offspring (G3), the maternal and paternal DNA sequences of G2 are copied and a segmental mosaic of them is produced. Therefore DNA sequence G2 gives to its offspring is different from G2's two sequences. The phenomenon creating mosaic is called recombination. Besides recombinations, mutations also make descending DNA sequence different from its parental sequences. Mutations are the stochastic events to exchange letters of DNA sequence among A, T, G and C.

Mutations create genetic variants and recombinations exchanges combinations of variants. Because mutations are not uncommon events, DNA sequence of a species would get saturated with so many variations, if all variants persist in a population, which is not the case. Because the selection between maternal and paternal sequences to make descending mosaic is stochastic and because population size is finite, fraction of each variant fluctuates over time and sometimes variants extinct in the population. This phenomenon is called genetic drift. Actually the majority of new-born variants do not survive more than a couple of generations due to genetic drift.

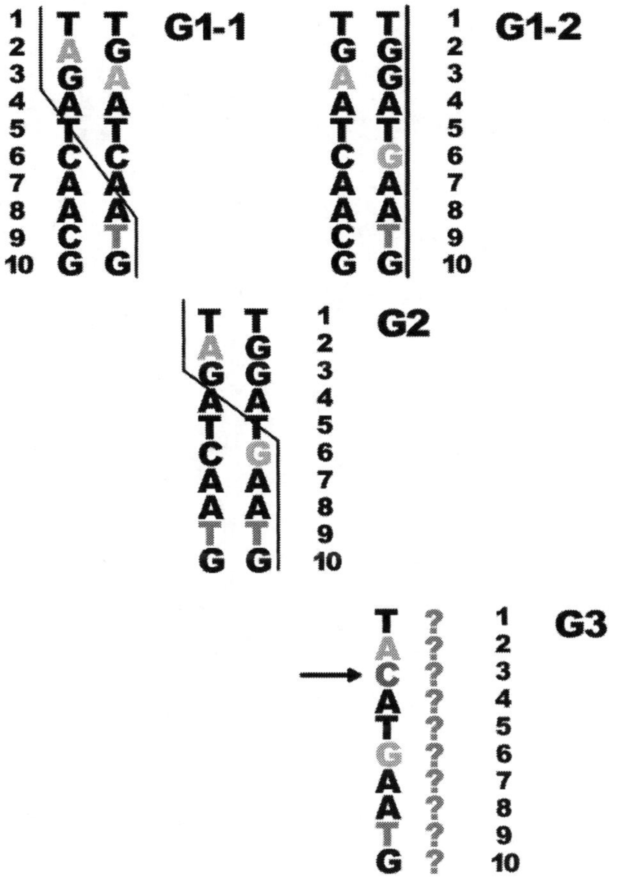

Figure 1. Mutation and Recombination. G1-1 and G1-2 represent the parents of G2 and G2 gives one of DNA sequence to G3. The sequence from G1-1 to G2 is a mosaic of two sequences of G1-1. A recombination occurs somewhere between the positions 3 and 9. No recombination occurs for the sequence from G1-2 to G2. For the sequence from G2 to G3, a recombination happens between the positions 2 and 6. "C" at the third position in the sequence that G2 gives to G3, indicated by an arrow is a mutation.

3. Space of DNA sequence

Although length of DNA molecules among species vary a lot, I set its length at a fixed value, for example $L = 3 \times 10^9$, that is an approximate length of human genome. As a chemical macromolecule, 4^L types of sequence are possible. This is the space of DNA sequence with the length of L. Because genes code functionally important information, such as particular amino acid sequence and regulation rules of their expression, a part of L loci have to be a particular letter. Therefore the space is biologically restricted to $4^l < 4^L$ where $l < L$. As far as the studies of genome of *homo sapience* have revealed, almost all the variable sites are dichotomous, or have only two of four letters. Therefore 2^l rather than 4^l is sufficient as a model. In this model the sequence strings can be expressed with 0 and 1 (Figure 2).

For a region with l loci with two allelels, there are 2^l potential variants. The 2^l variants can be represented as vertices of l-dimensional hypercube in discrete mathematics and geometry[3]. The hypercube can be drawn in a lattice as shown in Figure 3. The heterogeneity of DNA sequence is determined by frequency vector of 2^l elements.

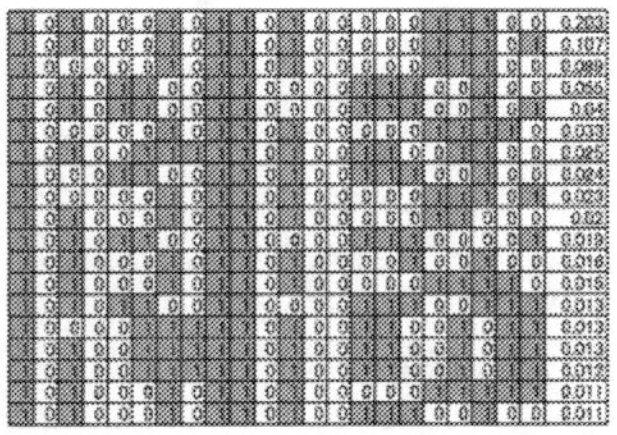

Figure 2. Example of 0 and 1 expression of sequence strings. The left on the right is fraction.

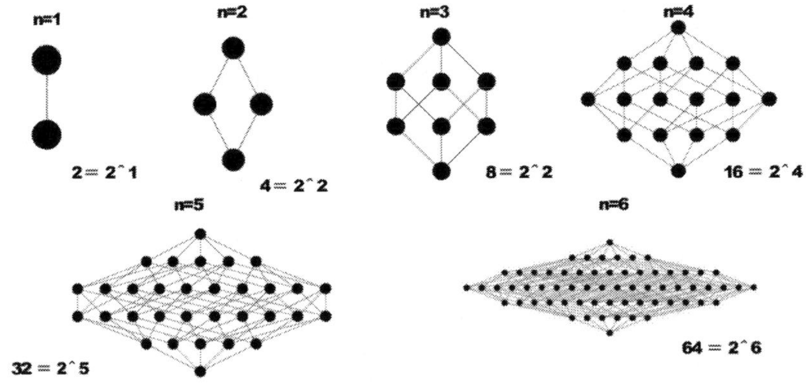

Figure 3. l-dimensional hypercube or lattice. l-dimensional hypercubes are drawn as graphs in 2-dimensional space. The number of nodes is 2^l and each node has l edges. The power set of a set S with l elements consists of 2^l subsets including the empty set and S itself. Inclusion relations among the elements of power set is drawn as a lattice, in which the subsets with the same number of elements are arranged on the same level. The number of horizontal levels is $l + 1$. The top level has only one node that is the empty set and the bottom level has also only one node, representing S itself.

4. Notation

- Sequence is a string of two letters with length l.
- Locus is an individual position in sequence: $L_i, i = 1, 2, ..., l$.
- Set of loci is a set with l elements: $S = \{L_1, L_2, ..., L_l\}$.
- Subsets of loci is a set of loci with 0 to l elements: $sS_j = \{L_{j1}, L_{j2}...\}$.
- N_{sS_j} represents the number of elements of sS_j.
- Power set of S has 2^l elements: $PS = \{sS_1, sS_2, ..., sS_{2^l}\}$.
- $Nt_{sS_j} = 2^{N_{sS_j}}$ types are defined for sS_j: $T_{sS_j} = \{t_{sS_j}(1), t_{sS_j}(2), ..., t_{sS_j}(Nt_{sS_j})\}$.
- Fraction is defined for types: $F_{sS_j} = \{f_{sS_j}(1), f_{sS_j}(2), ..., f_{sS_j}(Nt_{sS_j})\}$. $\sum_{k=1}^{Nt_{sS_j}} f_{sS_j}(k) = 1$

- Haploid and Diploid: Diploid organisms have a pair of DNA sequences, each of which comes from one of parents. Haploid means a half set of DNA and Diploid means a set of two DNA sequences.
- Nt_{dip,sS_j} is the number of diploid types of sS_j. When the diplotype of $(t_{sS_j}(x), t_{sS_j}(y))$ and the diplotype of $(t_{sS_j}(y), t_{sS_j}(x))$ are considered the same, $Nt_{dip,sS_j} = \binom{Nt_{sS_j}}{1} + \binom{Nt_{sS_j}}{2}$. When they are considered discriminable, $Nt_{dip,sS_j} = (Nt_{sS_j})^2$. The former definition is useful usually but the latter is necessary for particular biological phenomena, which needs to discriminate maternal and paternal sequences.
- $T_{dip,sS_j} = \{t_{dip,sS_j}(1), t_{dip,sS_j}(2), ..., t_{dip,sS_j}(Nt_{dip,sS_j})\}$.
- Fraction of T_{dip,sS_j} is $F_{dip,sS_j} = \{f_{dip,sS_j}(1), f_{dip,sS_j}(2), ..., f_{dip,sS_j}(Nt_{dip,sS_j})\}$.

5. Methods to quantitate haploid sequence heterogeneity

The methods to quantitate haploid sequence heterogeneity can be divided into two types. One type quantitates sequence heterogeneity with multiple parameters. The other type gives only one or a few values to describe the sequence heterogeneity as a whole.

5.1. Variance-covariance matrix and r or r^2

The most popular method to describe sequence heterogeneity in genetic studies is similar to $l \times l$ variance-covariance matrix of l loci.

Consider two subsets of one locus, $sS_{j1} = \{L_{j1}\}$ and $sS_{j2} = \{L_{j2}\}$ and their joint subset $sS_{j3} = \{L_{j1}, L_{j2}\}$. They have $F_{sS_{j1}} = \{f_{sS_{j1}}(1), f_{sS_{j1}}(2)\}$ and $F_{sS_{j2}} = \{f_{sS_{j2}}(1), f_{sS_{j2}}(2)\}$ and $F_{sS_{j3}} = \{f_{sS_{j3}}(1), f_{sS_{j3}}(2), f_{sS_{j3}}(3), f_{sS_{j3}}(4)\}$. The covariance between L_{j1} and L_{j2}, $r_{j1,j2}$, is given as,

$$r_{j1,j2} = \frac{f_{sS_{j3}}(1)f_{sS_{j3}}(4) - f_{sS_{j3}}(2)f_{sS_{j3}}(3)}{\sqrt{f_{sS_{j1}}(1)f_{sS_{j1}}(2)f_{sS_{j2}}(1)f_{sS_{j2}}(2)}}, \quad (1)$$

where

$$\begin{aligned} f_{sS_{j1}}(1) &= f_{sS_{j3}}(1) + f_{sS_{j3}}(2), \\ f_{sS_{j1}}(2) &= f_{sS_{j3}}(3) + f_{sS_{j3}}(4), \\ f_{sS_{j2}}(1) &= f_{sS_{j3}}(1) + f_{sS_{j3}}(3), \\ f_{sS_{j2}}(2) &= f_{sS_{j3}}(2) + f_{sS_{j3}}(4). \end{aligned}$$

Actually $r^2_{j1,j2}$ is used regularly in genetic studies and is called linkage disequilibrium index and it ranges from -1 to 1. $\binom{l}{2}$ pair-wise indices are plotted to display the heterogeneity visually as shown in Figure 4. [4][5] r has a good reason to be used in genetic studies as follow.

Assume a study to find genetic locus that is associated with a phenotype and assume χ^2 test is performed to test the association between markers and the phenotype. When a marker itself is functionally related with the phenotype, χ^2 value for the marker is expected as the strength of the association indicates. Let $\chi^2(L_i)$ denote the expected value. In many cases of the studies to identify such locus, markers close to the true locus, that are indirectly associated with the phenotype, are tested. The chi^2 value of those markers are expected to be

$$\chi^2(L_j) = r^2_{i,j}\chi^2(L_i),$$

where L_j is not functionally associated with the phenotype and L_i and L_j are mutually dependent with $r_{i,j}$. This relation is important because regular genetic studies to identify phenotype-causing loci have to make a part of loci represent all loci without acceptable loss of power by a limited number of loci. This method summarizes the genetic heterogeneity which requires $2^l - 1$ parameters to be fully described, with $\binom{l}{2}$ parameters instead.

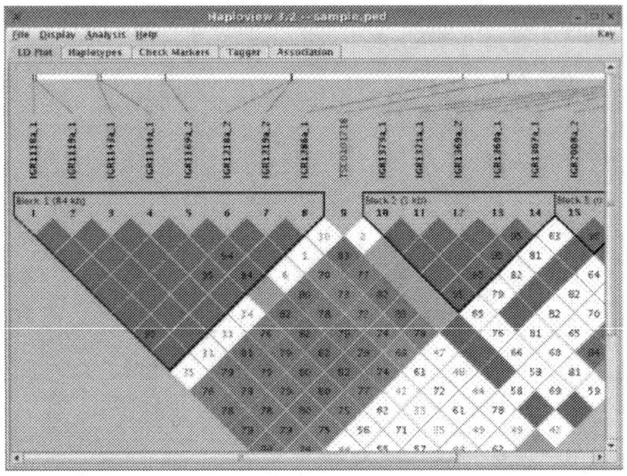

Figure 4. $\binom{l}{2}$ values are plotted in a triangle with color-intensity scale. The values of neighboring loci are plotted on the top horizontal line and the values of loci in longer distance are plotted in the lower area.

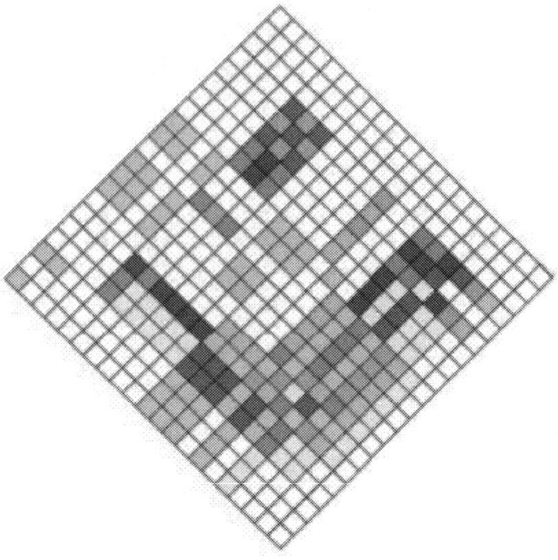

Figure 5. The upper half is the triangle of pair-wise r^2 in gray scale. The lower half is the triangle of subsets with elements in tandem. The cells on the central and horizontal line represent the pairs of loci side-by-side. The cells above the central line represent the pairs of loci separated by one locus in-between. The cells above next are for the pairs separated by two loci. The cells below the central line represent the trios of loci in tandem. The cells below next for the quartets of loci in tandem and so on.

5.2. Power set LD index and their partial display, Ψ

This method extracts information of 2^l types with $2^l - 1$ parameters. This method uses the fact that the power set of a set of l elements consists of 2^l subsets as,

$$\sum_{i=0}^{l} \binom{l}{i} = 2^l.$$

This equation is demonstrated by the fact that the elements of power set can be drawn in a lattice as shown in Figure 3.

Ψ consists of $2^l - 1$ parameters, each of which is the value for an element of power set except for the empty set as $\Psi = \{\psi_{sS_j}\}$ and they are defined as [6]

$$\psi_{sS_j} = \sum_{k=1}^{Nt_{sS_j}} v_{sS_j}(k) \times f_{sS_j}(k), \tag{2}$$

where

$$v_{sS_j}(k) = \prod_{x=1}^{N_{sS_j}} \sigma_{sS_j}(k)_x = 1 \; or -1,$$

Journal of Physics: Conference Series **197** (2009) 012010 doi:10.1088/1742-6596/197/1/012010

when the x-th locus of $t_{sS_k}(k)$ is one arbitrarily determined type

$$\sigma_{sS_j}(k)_x = 1,$$

and when the x-th locus of $t_{sS_k}(k)$ is the other

$$\sigma_{sS_j}(k)_x = -1.$$

Another simple way to express fraction distribution of 2^l types is to give one parameter for each type except the last type. Compared to this way, the parameters of Ψ are hierarchically structured and the parameters for smaller subsets tend to represent wider picture of its structure than ones for larger subsets. Although this method expresses all information of sequence heterogeneity, the major drawback is that the number of parameters is too big to display all the information. Therefore it might be the palliative answer to chose $2 \times \binom{l}{2}$ parameters among $2^l - 1$, which are displayed in a square and one half of them is identical to the triangle plot of r^2 and the other half represents subsets whose elements are in tandem. Figure 5 is the drawing for sequence fraction in Figure 2. [6]

5.3. *Entropy-based measures of LD or heterogeneity, ϵ*

This measure describes sequence heterogeneity with one value. It is based on the information theory. The definition of entropy H is

$$H = -\sum_{i=1}^{2^l} f_i ln(f_i) \;\; (f_i \neq 0). \tag{3}$$

H takes the minimum when only one type exists and the value is

$$H_{min} = 0. \tag{4}$$

H takes the maximum when all types exist with identical fraction and the value is

$$H_{max} = -\sum_{i=1}^{2^l} \frac{1}{2^l} \times ln(\frac{1}{2^l}) = l \times ln2. \tag{5}$$

Nothnagel *et al.* defined ϵ using H. They supposed that l loci had variants and the fraction of variants were given.

Actually when fraction of all types is the expected value under the assumption of independence among all loci,

$$H = \sum_{k=i}^{l} H_i,$$

where H_i is entropy of L_i. On the other hand when only two types exist, entropy of all individual loci are identical as H_i and

$$H = H_i.$$

Under the condition where fractions of individual loci were given, ϵ was standardized so that it gave the minimum value, 0, for the state where fraction of all types is the expected value under the condition all loci are mutually independent and it gave the maximum value, 1, for the state where only two out of 2^l types exist. [7] Two major benefits of this measure is:

- One value represents the whole.
- It ranges in the fixed range, from 0 to 1, regardless of l.
- Its calculation only uses fraction of existing types. Because the number of existing types is limited, this feature is useful.

5.4. An extension of r^2, r^l

ϵ has very favorable features as an index of heterogeneity as listed in the previous section. But ϵ for two loci is not identical with r or r^2. Because r or r^2 are closely related with measures for statistical evaluation such as correlation and test of independence, an index that was extended from r or r^2 and that had the characters of ϵ listed above would be useful. Following descriptions will develop such index, r^l from r^2.

$r^2_{i,j}$ can be expressed as below;

$$r^2_{i,j} = \sum_{k=1}^{4} \frac{(f_{sS_{j3}}(k) - E(f_{sS_{j3}}(k)))^2}{E(f_{sS_{j3}}(k))},$$

or

$$r^2_{i,j} = \sum_{k=1}^{4} \frac{(f_{sS_{j3}}(k))^2}{E(f_{sS_{j3}}(k))} - 1, \tag{6}$$

where

$$E(f_{sS_{j3}}(1)) = f_{sS_{j1}}(1) * f_{sS_{j2}}(1),$$
$$E(f_{sS_{j3}}(2)) = f_{sS_{j1}}(1) * f_{sS_{j2}}(2),$$
$$E(f_{sS_{j3}}(3)) = f_{sS_{j1}}(2) * f_{sS_{j2}}(1),$$
$$E(f_{sS_{j3}}(4)) = f_{sS_{j1}}(2) * f_{sS_{j2}}(2).$$

Of note r^2 is χ^2 statistics of test of independence between two loci standardized for sample size. Let $\chi^2(st)$ denote χ^2 statistics for independence standardized for samples size. Then

$$r^2 = \chi^2(st).$$

For sS_i with loci more than 2, independence of fraction under the condition where fraction of variants of individual loci is given, can be quantitated with

$$\chi^2(st) = \sum_{k=1}^{Nt_{sS_i}} \frac{(f_{sS_i}(k) - E(f_{sS_i}(k)))^2}{E(f_{sS_i}(k))},$$

or

$$\chi^2(st) = \sum_{k=1}^{Nt_{sS_i}} \frac{(f_{sS_i}(k))^2}{E(f_{sS_i}(k))} - 1.$$

This is a simple extension of r^2 for 2 loci. However this $\chi^2(st)$ ranges from 0 to 1 when $l = 2$ but it ranges from 0 to an unfixed value when $l > 2$. This unfixed range makes $\chi^2(st)$ unfavorable as an index. Therefore I defined r^l by modifying $\chi^2(st)$ so that it ranges from 0 to 1 regardless of l as

$$r^l = \sum_{k=1}^{Nt_{sS_i}} \frac{(f_{sS_i}(k))^{\frac{1}{l-1}+1}}{(E(f_{sS_i}(k)))^{\frac{1}{l-1}}} - 1. \tag{7}$$

See Appendix for further description. Because $l = 2$, equations (6) and (7) are identical, r^l is an extension of r^2 for two loci and is a modification of χ^2 statistics for independence overall and satisfies the favorable features of ϵ. The further statistical properties of r^l have yet to be investigated.

Journal of Physics: Conference Series **197** (2009) 012010 doi:10.1088/1742-6596/197/1/012010

6. Further requirements for diploids and many loci

6.1. From haploids and diploids

The methods described above, r, r^2, Ψ, ϵ and r^l, were defined to quantitate frequency data of haplotypes but not of diploids, even when the phases of diploid genotypes are known . It is sufficient to compare sequences of various species, because one sequence representing each species is treated. However it is necessary to quantitate genetic heterogeneity among diploid individuals who possess a pair of sequences. In genetics when individuals randomly mate, combination of sequences in diploid individuals is expected independent. When combination of sequences in individuals is not independent, non-random mating is usually indicated. The randomly-mating condition is genetically called "in Hardy-Weinberg equilibrium (HWE)" and the condition that is most remote from "HWE" is the condition where the population is completely subdivided into two homogeneous subpopulations ("complete subdivision"). Therefore genetic heterogeneity of diploids is heterogeneity of haploid sequence and mating. Although r or Ψ is not easily extended to handle diploids data, ϵ and r^l can be. Entropy which is the base of ϵ is calculable by fraction of diploid types.

$$H(diploid) = -\sum_{k=1}^{Nt_{dip,sS_j}} f_{fip,sS_j}(k) \times ln(f_{dip,sS_j}(k)). \tag{8}$$

It would be reasonable to define the "most" and "least" heterogeneous states of diploids as below. When haploid sequence entropy is the maximum and when diploids are in HWE, the state is "most" heterogeneous. When haploid sequence entropy is the minimum and when diploids are completely subdivided, the state is "least" heterogeneous. The values of H for both conditions are expressed as,

$$H(diploid, max) = 2\sum_{i=1}^{l} H_i,$$

$$H(diploid, min) = H_i.$$

It would be appropriate to calculate ϵ for diploids by standardizing entropy of diploid fraction data with the maximum and minimum entropy values for diploids as given above.

In order to make $r^l(diploids)$ range from 0 to 1, the exponent is changed from $\frac{1}{l-1}$ to $\frac{1}{2l-1}$ as

$$r^l(diploids) = \sum_{k=1}^{Nt_{dip,sS_j}} \frac{(f_{dip,sS_j}(k))^{\frac{1}{2l-1}+1}}{(E(f_{dip,sS_j}(k))^{\frac{1}{2l-1}}} - 1. \tag{9}$$

The calculation of H and r^l for diploids are affected by Nt_{dip,sS_j} which can vary as described in Notation section. Briefly when maternal and paternal sequences are discriminated, information of data is more than the condition without the discrimination. Details on this issue are beyond this paper.

6.2. Sparsity and "Biological" heterogeneity measure

l is quite large, such as 3×10^9 for *homo sapience*, therefore all individuals are genetically different even though human population is more than 6×10^9 worldwide. sSs with much less loci are still able to distinguish all individuals. When it is the case, $f_{sS_j}x$ or $f_{dip,sS_j}y$ is $\frac{1}{N}$ or 0 where N is sample size. Because both H and r^l are functions of $f_{sS_j}x$ or $f_{dip,sS_j}y$, fixed values of the fraction are not appropriate. Actually $H = ln(N)$ regardless of existing types.

Suppose any one haploid or diploid individual is different from all others. Then the population can be considered as a discrete distribution in l-dimensional hypercube or a lattice graph with 2^l nodes, in which only N nodes are occupied by one individual and all other nodes are vacant, where the space is sparsely occupied. This sparse distribution would be the subject of geometry or graph theory.

Here reconsider from biological and genetic standpoint whether it is appropriate to locate 2^l types in l-dimensional hypercube or its lattice graph counterpart. When DNA sequences among various species are compared, distance between any two sequences is accumulation of mutations. Therefore distance in the space or in the lattice corresponds to biological distance. However when DNA sequences among a population of same species are compared, their relations are determined both by mutations and recombinations. This means the distribution in l-dimensional hypercube or the lattice might not be appropriate for sequences in the same species. Two types of distance can be defined for sequences in the same species. One distance is defined between any two sequences as the number of mutations, that changes one sequence to the other. The other distance is defined between a sequence (offspring sequence) and a pair of sequences (parent sequences) as the number of recombinations, which creates the offspring sequence from parent sequences (Figure 6).

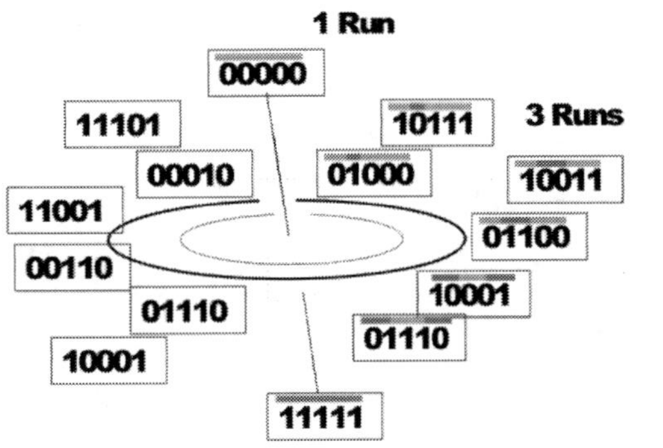

Figure 6. Distance by recombination. Sequence pair of 1 run creates a sequence of k runs with k-1 recombinations. The recombination distance of this offspring sequence from the parents as k-1. A given pair of sequences that are different at l locus has $2 \times \binom{l-1}{k}$ offsprings that are k recombination distance from them $(0 \leq k \leq l-1)$.

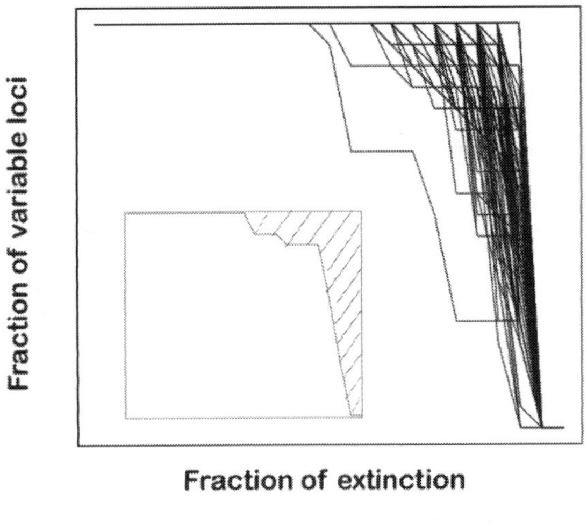

Figure 7. Each line represents an order of samples, in which the samples are removed from the population, and the decrease of heterogeneous loci is drawn. The inner plot in the lower left is one of 100 lines and the slashed area was calculated to extract representing values from 100 lines.

This indicates that straightforward mathematical approaches based on geometry or graph-theory might not be relevant for combination of mutations and recombinations. Because possession of some amount of mutations makes a species survive through environmental stresses, it is beneficial to have variants in the way where genetic drift will not easily lose the variants excessively. When some fraction of population is removed from the whole, some loci become homogeneous in the remnant. Figure 7 shows the relation between the fraction of population extinction and the fraction of loci remaining heterogeneous, for a population that consists of 19 samples, each of which corresponds to one of the sequences shown in Figure 2. One hundred permutations of 19 samples were randomly selected and samples were withdrawn from the total one by one in the order of each permutation and the fraction of withdrawn samples and the fraction of heterogeneous loci in the remnant were calculated and plotted. When the fraction of extinction was 0, all loci remained heterogeneous and when only one individual remained, all loci are homogeneous. Therefore all lines in Figure 7 started at the top left and ended at the bottom right. The orders, in which individuals were removed from the population, determined the pathway from the top left down to the bottom right. Based on this idea, I randomly obtained the extinction lines in Monte-Carlo methods and quantitated the lines of multiple trials with average and variance of area above the extinction lines. Some conditions have similar average value of the extinction lines but different variance value. Therefore at least two values from the extinction simulations, average and variance, will be necessary to quantitate their heterogeneity. The example shown in Figure 7 was one of haploid data but it is simple and easy to have its diploid counterpart.

7. Brief summary

In this paper, I described three methods previously reported and proposed two new methods. Two of five, variance-covariance matrix, and Ψ, describe heterogeneity of haploids but not diploids with multiple values and they are not for sparse data. Another two of five, entropy-based ϵ and χ^2-related r^l, describe data with one value and are applicable to diploid data, but not to sparse data. The last method is simulation-based and two descriptive values from simulation results seem to represent haploid and diploid heterogeneity and it is applicable to sparse data. (Table 1) Only a few simple rules, mutations, recombinations and genetic drifts, can produce very complex discrete data and further studies are necessary to develop better methods to quantitate the complexity.

Table 1. Control sequences to describe lines and symbols in figure captions.

Methods	No. parameters	Sparse data	Diploid
Variance-covariance matrix of r or r^2	$\binom{l}{2}$	+	-
Power set system, Ψ	$2^l - 1$	+	-
Entropy-based ϵ	1	-	+
Standardized χ^2, r^l	1	-	+
Extinction-resistance	2	+	+

8. Appendix

When there are l loci, the condition where all loci have the same allele frequency and all loci are mutually in absolute linkage disequilibrium, is the most deviate state from the equilibrium. Let p denote their allele frequency, then only two haplotypes exist in the population and their frequencies are p and $1-p$. Their expected frequencies in linkage equilibrium are p^l and $(1-p)^l$, respectively.

$$r^l = \frac{p^{\frac{1}{l-1}+1}}{(p^l)^{\frac{1}{l-1}}} + \frac{(1-p)^{\frac{1}{l-1}+1}}{((1-p)^l)^{\frac{1}{l-1}}} - 1$$

$$r^l = p + (1-p) - 1$$

$$r^l = 1.$$

Journal of Physics: Conference Series **197** (2009) 012010 doi:10.1088/1742-6596/197/1/012010

References

[1] Reznick D N and Ricklefs R E 2009 Darwin's bridge between microevolution and macroevolution *Nature* **457** 837–42

[2] Altshuler D, Daly M J and Lander E S 2008 Genetic Mapping in Human Disease *Science* **322** 881–8

[3] Henk M, Richter-Gebert J and Ziegler G M 2004 Basic properties of convex polytopes *Handbook of Discrete and Computational Geometry (2nd)* ed J E Goodman and J O'Rourke (Boca Raton, FL: Chapman & Hall/CRC) chapter 16

[4] Devrin B and Risch N 1995 A comparison of linkage disequilibrium measures for fine-scale mapping *Genomics* **29** 311–322

[5] Barrett J C, Fry B, Maller J and Daly M J 2005 Haploview: analysis and visualization of LD and haplotype maps *Bioinformatics* **21** 263–5

[6] Yamada R and Matsuda F 2007 A novel method to express SNP-based genetic heterogeneity, Ψ, and its use to measure linkage disequilibrium for multiple SNPs, D(g), and to estimate absolute maximum of haplotype frequency *Genet. Epidemiol.* **31** 709–26

[7] Nothnagel M, Furst R and Rohde K 2002 Entropy as a measure for linkage disequilibrium over multilocus haplotype blocks *Hum. Hered.* **54** 186–98

Journal of Physics: Conference Series **197** (2009) 012011 doi:10.1088/1742-6596/197/1/012011

Discovering large network motifs from a complex biological network

Aika Terada and Jun Sese

Department of Computer Science, Ochanomizu University, 2-1-1 Ohtsuka, Bunkyo-ku, Tokyo 112-8610, Japan

E-mail: terada@sel.is.ocha.ac.jp, sesejun@is.ocha.ac.jp

Abstract. Graph structures representing relationships between entries have been studied in statistical analysis, and the results of these studies have been applied to biological networks, whose nodes and edges represent proteins and the relationships between them, respectively. Most of the studies have focused on only graph structures such as scale-free properties and cliques, but the relationships between nodes are also important features since most of the proteins perform their functions by connecting to other proteins. In order to determine such relationships, the problem of network motif discovery has been addressed; network motifs are frequently appearing graph structures in a given graph.

However, the methods for network motif discovery are highly restrictive for the application to biological network because they can only be used to find small network motifs or they do not consider noise and uncertainty in observations. In this study, we introduce a new index to measure network motifs called AR index and develop a novel algorithm called ARIANA for finding large motifs even when the network has noise. Experiments using a synthetic network verify that our method can find better network motifs than an existing algorithm. By applying ARIANA to a real complex biological network, we find network motifs associated with regulations of start time of cell functions and generation of cell energies and discover that the cell cycle proteins can be categorized into two different groups.

1. Introduction

Complex network analysis is an important part of statistical analysis and is applied to various networks, such as biological networks, social networks, and the Web. Most network studies are classified into three types. The first type includes analysis of scale-free property [1, 2], from which we can understand the development of such networks. The second type includes analysis of cliques [3, 4, 5, 6], from which we find highly associated vertices from a given network. The last type includes determination of network motifs [7, 8, 9], which find frequently appearing network structures.

The issue of network motif discovery has been addressed in the last few years, and many biologically important results have been obtained [8]. Network motif discovery is different from the analysis of scale-free networks and cliques, and it is used to detect frequently appearing network structures and also to determine the role of vertices in a network. In biological networks, vertices and edges in a graph represent genes and the relationships between them, respectively. Since most genes perform functions by cooperating with other genes, it is very important to establish the the functional relationships between neighboring genes in order to understand the functions of the genes.

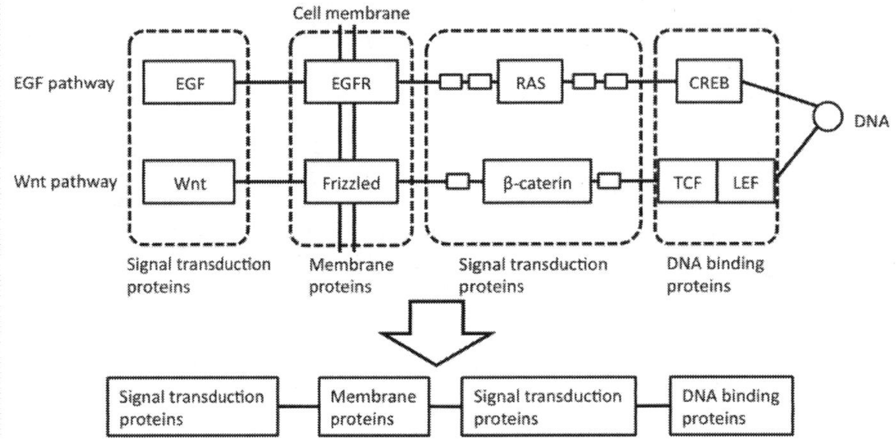

Figure 1. Example of a network motif over different two pathways.

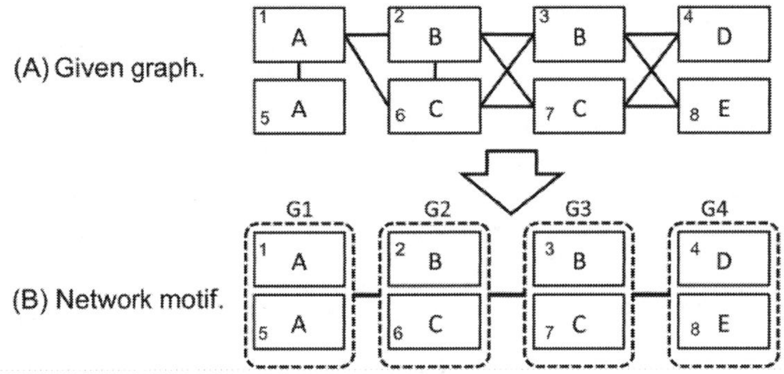

Figure 2. Motivating example.

Figure 1 shows an example of a small network motif in a molecular biological system. The squares and lines indicate proteins and the relationships between two proteins, respectively. The upper sequence starting from the protein EGF is one of the well-known signal transduction pathways called the EGF pathway, which is closely related to cell growth. The lower sequence starting from the Wnt is the other well-known pathway associated with cancer. In this study, we have determined the network motifs by finding frequently appearing structures in these networks. The frequently appearing structures are shown at the bottom of Figure 1. The beginning of both pathways is marked by signal transduction proteins, that are connected to membrane proteins. The membrane proteins are linked to signal transduction proteins within cells, and finally, the signal transduction proteins are connected to DNA binding proteins. From this network motif structure, we can establish the general pathway of information flow within cells.

Figure 2(A) shows a motivating example. The vertices represent proteins, and the edges represent the relationships between two vertices connected to the edge. Each number indicates their ID number. Each letter in the vertex indicates an attribute, i.e., the function, of the vertex. This example shows five types of attributes: A-E. Two vertices having the attribute "A" have the same function.

Here, we find the network motifs from Figure 2(A). Figure 2(B) shows a network motif of the example shown in Figure 2(A). In the figure, the squares with dotted lines represent groups of vertices (G1, G2, G3, and G4). Vertices 2 and 6 belong to G2, and the members of this group connect to the members of G1 and G3 because both vertices 2 and 6 connect to 1, 3, and 7. Therefore, G2 connects to G1 and G3 in the network motif. Further, vertices in G2 do not connect to vertices in G4, and hence, there are no edges between G2 and G4. Vertices in G3 also connect to vertices in G2 and G4 but do not connect to G1, and hence, in the network motif, G3 connects to G2 and G4 and does not connect to G1.

In this example, we can determine a network motif with no ambiguity. However, most graph structures representing real biological networks are subject to noise and uncertainty, and hence network motif cannot represent whose given network. In this study, we introduce a novel index to measure the network motifs even when the network includes noise and develop an algorithm to find network motifs by maximizing this index.

2. Related work

In recent years, graph mining has received increased attention from researchers. The frequent subgraph discovery methods [10, 11, 12, 13] can be used to enumerate all frequently occurring common-structured subgraphs in a graph database. These algorithms cannot be used to find large frequently occurring subgraphs because of the noise present in biological networks. Our approach can enumerate large motifs networks from the entire protein-protein network.

Finding cliques is one of the research topics in network analysis. In the motivating example, nodes 2, 3, 6, and 7 form a clique. It is better to divide them into two groups — nodes 2 and 6 and nodes 3 and 7 — where the former two nodes connect to node 1 and the latter two connect to nodes 4 and 8. Because the two groups connect differently to the other groups, the method cannot apply to find network motifs.

Network motif discovery algorithms have been studied in the fields of both biology [8] and data mining [7, 9]. In the field of biology, only some genes or proteins have been investigate at once [8]. Chen *et al.* [7] studied networks containing less than 25 genes, and their analysis was limited to graphs whose vertices did not have labels. k-SNAP [9] can handle large networks, but it cannot find motifs from networks having a large number of labels on the vertices. In contrast, our method can handle a large number of labels.

3. Preliminaries

3.1. Attributes and relationships compatible grouping

In this section, we define a network motif model and introduce an optimization model that we attempted to maximize in this study. For finding network motifs, we define the grouping of vertices.

Definition 1 *(Definition of a Graph) Let* $G = (V(G), E(G), A(G))$ *be an undirected and unweighted graph. Let* $V(G)$ *and* $E(G)$ *be vertices and edges on* G*, respectively. Let* $A(G)$ *be the attributes on vertices in graph* G*. Each vertex* v *has an attribute* $a(v) \in L(G)$*. Edges do not have labels and weights. We refer to this graph as a labeled graph.*

On this labeled graph, we define non-overlapping groups of vertices.

Definition 2 *(Vertex Grouping of a Graph) For a group* G, $\mathcal{C} = \{C_1, C_2, \ldots, C_k\}$ *is called a vertex grouping of* G *or simply grouping of* G*, if and only if*

(i) $\forall C_i \in \mathcal{C}$, $C_i \subseteq V(G)$ *and* $C_i \neq \phi$,

(ii) $\bigcup_{C \in \mathcal{C}} C = V(G)$, *and*

(iii) for $\forall C_i, C_j \in \mathcal{C}$ *and* $i \neq j$, $C_i \cap C_j = \phi$.

Next, we define a special kind of grouping on the basis of attributes.

Definition 3 *(Attribute Compatible Grouping) Let* $\mathcal{C}$ *be a vertex grouping of* G*. Let* $c(v)$ *be a group in* $\mathcal{C}$ *to which* v *belongs. We call* $\mathcal{C}$ *A-compatible if* $\mathcal{C}$ *satisfies the following condition: for* $\forall u, v \in V(G)$, *if* $c(u) = c(v)$, *then* $a(u) = a(v)$.

Various A-compatible groupings can be generated from a graph. Here, we define a partial order relation $\preceq$ on the set of all groupings of a graph and find an A-compatible grouping that has the minimum number of groups.

Definition 4 *(Dominance Relation) For a graph G, the grouping $\mathcal{C}$ dominates the grouping $\mathcal{C}'$, denoted as $\mathcal{C}' \preceq \mathcal{C}$, if and only if $\forall C' \in \mathcal{C}', \exists C \in \mathcal{C}$ s.t. $C' \subseteq C$.*

Among all the A-compatible groupings of a graph, there exists a unique global maximum grouping with respect to the dominance relation $\preceq$ [9].

A-compatible grouping only accounts for the vertex attributes. Vertices have not only attributes but also relationships with other nodes that are represented by edges. Next, we define groupings compatible with both attributes and relationships.

Definition 5 *(Attributes and Relationships Compatible Grouping) Let $\mathcal{N}(v)$ be $\{c(u) \mid (u,v) \in E(G)\}$. $\mathcal{N}(v)$ indicates a set of groups to which the neighbors of v belong. Let $\mathcal{C}$ be a group of vertices. When the grouping satisfies*

(i) $\mathcal{C}$ is A-compatible, and

(ii) $\forall u, v \in V(G)$, if $c(u) = c(v)$, then $\mathcal{N}(v) = \mathcal{N}(u)$,

we call $\mathcal{C}$ to be (A,R)-compatible.

(A,R)-compatible grouping is that in which the nodes of each group are homogeneous in terms of attributes and relationships. Unfortunately, in practice, homogeneity is often highly restrictive because most graph representing real biological networks are subject to noise and uncertainty. (A,R)-compatible groupings with noise may increase the number of small groups. These small groups do not represent network motifs, because the network structure appears only few times in a given graph. To generate frequently appearing network structures, we introduce a novel index that relaxes the homogeneity requirement for both attributes and relationships and can be used to control the size of groups by means of a regularization term.

3.2. Network motif index

Here, we discuss the relaxation of homogeneity in attributes and relationships. First, we relax the attribute homogeneity using a relative entropy measure.

Definition 6 *(Relaxing Attribute Homogeneity Requirement) Let $p_A(a)$ and $p_A(a \mid C_i)$ be*

$$\frac{|\{v \mid a(v) = a\}|}{|V(G)|} \text{ and } \frac{|\{v \in C_i \mid a(v) = a\}|}{|C_i|}, \text{respectively.}$$

We define $Ent(C_i)$ as

$$Ent(C_i) = \sum_{a \in A(G)} p_A(a \mid C_i) \log_2 \frac{p_A(a \mid C_i)}{p_A(a)}.$$

This index can be used to measure the similarity between a grouping and an A-compatible grouping. When a grouping is A-compatible, $\sum_{C_i \in \mathcal{C}} Ent(C_i)$ is largest under the condition that $|\mathcal{C}|$ is fixed. As vertices in a group C_i consist of different attributes, $Ent(C_i)$ becomes small.

Next, we relax the homogeneity requirement for the relationships. We define a set of nodes in group C_i that participate in a group relationship (C_i, C_j) as $V_R(C_i, C_j)$ and denote the participation ratio of the group relationship as $r(C_i, C_j)$.

Definition 7 *Let $V_R(C_i, C_j)$ be $\{v \in C_i \mid (v, v') \in E(V)$ where $v' \in C_j\}$. Note that $V_R(C_i, C_j)$ is asymmetric with respect to C_i and C_j. We define $r(C_i, C_j)$ as*

$$r(C_i, C_j) = \frac{|V_R(C_i, C_j)| + |V_R(C_j, C_i)|}{|C_i| + |C_j|}.$$

As $r(C_i, C_j)$ increases, the strength of the group relationship increases. Note that in the case of an (A,R)-compatible grouping, the participation ratios between any two groups are either 0 or 1.

From the participation ratio, we measure the connectivity between two groups. When $r(C_i, C_j)$ is close to 0 or 1, the relationship between C_i and C_j is close to (A,R)-compatible grouping, and hence the grouping seems to include network motifs. On the other hand, when $r(C_i, C_j)$ is not close to 0 or 1, since the edges between groups are generated almost randomly, the grouping is undesirable. For considering this grouping, we define connectivity measure between the pair of groups.

Definition 8 *Let $c(C_i, C_j)$ be the number of vertices relating the connection between C_i and C_j.*

$$c(C_i, C_j) = \begin{cases} |V_R(C_i, C_j)| & \text{if } r(C_i, C_j) > 0.5 \\ |C_i| - |V_R(C_i, C_j)| & \text{otherwise} \end{cases}$$

In an (A,R)-compatible grouping, $c(C_i, C_j) = |C_i|$ for any pair of C_i and C_j. Note that $c(C_i, C_j)$ is asymmetric.

By treating the connectivity measure between groups as the weight of edges, we introduce an index to assess the grouping of vertices. The index is maximized when both the relative entropy of attributes and the connectivity measure between groups are large.

Definition 9 *(Definition of Attribute Relationship Index) Let $\mathcal{C}$ be a grouping of vertices. Let $w(\mathcal{C})$ be*

$$w(\mathcal{C}) = \sum_{C_i \in \mathcal{C}} \left(Ent(C_i) \times \sum_{C_j \in \mathcal{C} \text{ and } C_i \neq C_j} c(C_i, C_j) \right) - \lambda |\mathcal{C}|^2.$$

We call $w(\mathcal{C})$ an attribute relationship index, or in short, an AR index.

The last term of the AR index is a penalty term, which accounts for the increase in the index as the number of groups increases. We can regulate the cluster size by using λ. When λ is small, the group size reduces.

We use this AR index and define our optimization problem, which is to find the grouping that maximizes the index. We call this problem *Attribute Relationship Index ANAlysis* (*ARIANA*). Then, we find network motifs by using the pair of groups having a high participation ratio.

However, it is difficult to find the grouping that maximizes the AR index, and hence, we introduce a simple method to find the better grouping of a graph in the next section.

4. Method

Here, we introduce an algorithm to solve the ARIANA problem. Since it is difficult to find a grouping that maximizes $w(\mathcal{C})$, we introduce a new two-step algorithm. Then, we apply the algorithm to a real biological dataset.

Our algorithm includes an aggregation step and a division step. We start with the maximum A-compatible grouping in terms of dominance relation between groups (Definition 4). In the

Algorithm 1 : ARIANA(J, K, λ)

Require: Graph $G = (V(G), E(G), A(G))$.
1: Generate the maximum A-compatible grouping $\mathcal{C}$.
2: // Aggregation step.
3: **for** while $|\mathcal{C}| \geq J$ **do**
4:   Compute $cos(C_i, C_j)$ for all pairs of $C_i, C_j \in \mathcal{C}$.
5:   Find a pair of groups that minimizes $cos(C_i, C_j)$.
6:   Remove C_i and C_j from $\mathcal{C}$ and add a new group consisting of $C_i \cup C_j$ to $\mathcal{C}$.
7: **end for**
8: // Division step.
9: **for** while $|\mathcal{C}| \leq K$ **do**
10:   $max\_w \leftarrow 0$
11:   **for** each pair of $C_i, C_j \in \mathcal{C}$ **do**
12:     $C' = \{v \in C_i \mid (v, v') \in E(G) \text{ for } v' \in C_j\}$.
13:     Generate $\mathcal{C}' \leftarrow \mathcal{C} - \{C_i\} + \{C'\} + \{C_i - C'\}$.
14:     $max\_w \leftarrow w(\mathcal{C}')$ if $w(\mathcal{C}') > max\_w$.
15:   **end for**
16:   $\mathcal{C} \leftarrow \mathcal{C}'$ where $\mathcal{C}'$ is the grouping associated with max_w.
17: **end for**
18: return $\mathcal{C}$.

aggregation step, we merge the groups into J groups by using the following cosine distance measure:

$$cos(C_i, C_j) = \frac{\sum_{C_k \in \mathcal{C}'} |V_R(C_i, C_k)||V_R(C_j, C_k)|}{\sqrt{\sum_{C_k \in \mathcal{C}'} |V_R(C_i, C_k)|^2} \sqrt{\sum_{C_k \in \mathcal{C}'} |V_R(C_j, C_k)|^2}},$$

where $\mathcal{C}' = \mathcal{C} - \{C_i, C_j\}$. This measure gives the similarity between two groups in terms of distribution of connectivity to other groups. We merge the groups with the largest pair of cosine distance, i.e., we perform complete linkage clustering until the number of clusters is J using $cos(C_i, C_j)$. In this step, we can generate groups whose members do not have attribute homogeneity but have relationship homogeneity.

Next, we maximize the AR index by dividing the groups. In this step, we divide groups whose members have attribute homogeneity but do not have relationship homogeneity into some groups to satisfy relationship as homogeneous as possible. When we divide an attribute homogeneous group into two groups, both of them keep attribute homogeneity. Therefore, we only consider relationship homogeneity to divide a group into two groups.

Let us consider the division of group C_i into two groups C' and $C_i - C'$. To maintain the relationship as homogeneous as possible, we select members of C' to connect it to the other group. When we divide C_i into C' and $C_i - C'$ using C_j, we select C' as

$$C' = \{v \in C_i \mid (v, v') \in E(G) \text{ for } v' \in C_j\}.$$

For all pairs of C_i and $C_j \in \mathcal{C}$, we find the best division having the maximum AR index.

Algorithm 1 shows the pseudo-code of ARIANA. In the first line, we make the maximum A-compatible groups in terms of dominance relations, and $|\mathcal{C}|$ is the number of types of attributes. In lines 2 to 7, groups are merged until the number of groups is J. In line 8 to 17, the groups are divided by the grouping that maximize the AR index until the number of groups is K.

Although ARIANA requires parameters J and K, both of which are associated with the number of clusters, we can change these two values, and find the best division that maximizes the AR index on the real biological dataset.

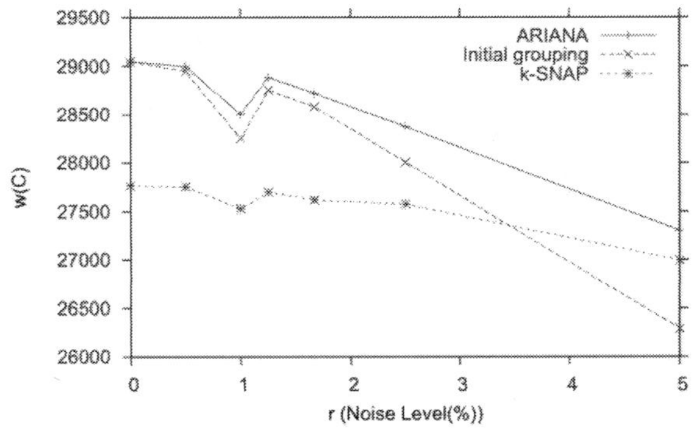

Figure 3. Changes of AR index with respect to adding noise.

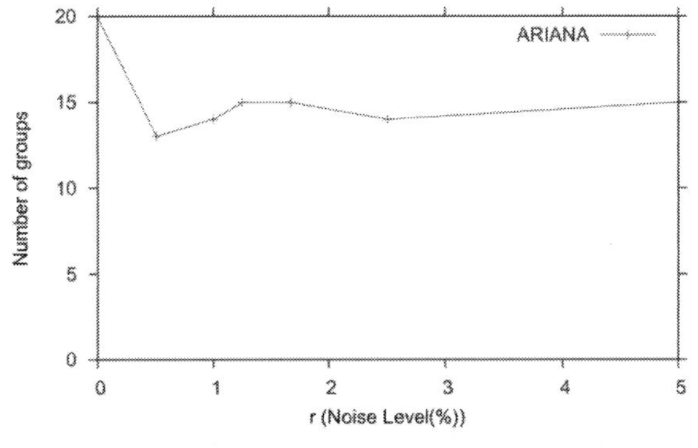

Figure 4. Changes of number of aggregated clusters when the maximum AR index is computed.

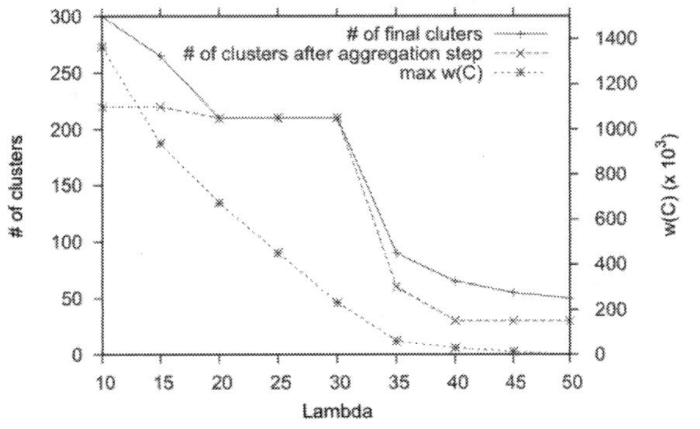

Figure 5. Changes of number of clusters and $w(\mathcal{C})$ according to λ.

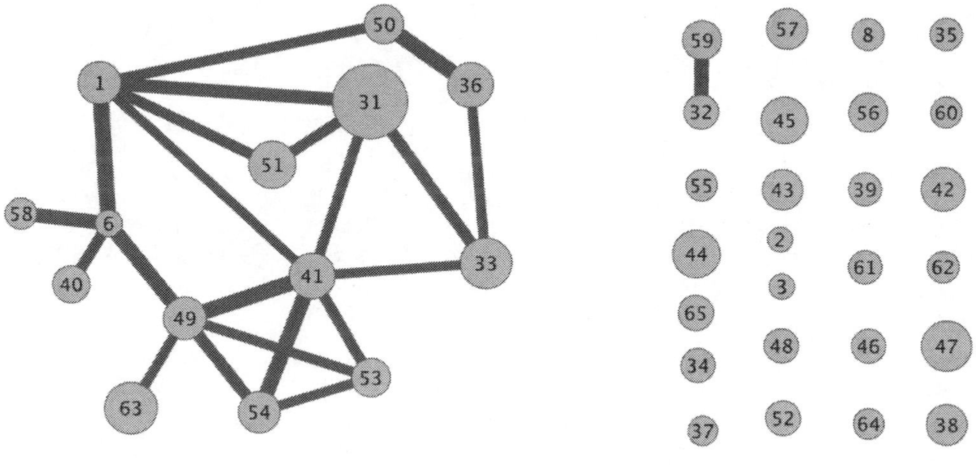

Figure 6. Network motifs discovered by ARIANA ($\lambda = 40$)

5. Experimental results

5.1. Results for a synthetic dataset

Here, we test our algorithm using a synthetic dataset. First, we generate a synthetic dataset consisting of 400 vertices, 1200 edges and 20 attributes.

(i) Generate 20 attributes, $a_1, \ldots, a_{20}$.
(ii) Generate 20 vertices whose attributes are a_i. Totally, 400 vertices are generated.
(iii) Make 10 groups, $C_1, \ldots, C_{10}$, such that $C_i (i = 1, \ldots, 10)$ contains 20 vertices and half of the vertices have attribute a_i and the rest vertices have attribute $a_{i+1 \bmod 10}$.
(iv) Make 10 groups, $C_{11}, \ldots, C_{20}$, such that $C_i (i = 11, \ldots, 20)$ contains 20 vertices and all of the vertices have attribute a_i.
(v) Generate 100 pseudo edges between the groups, randomly.
(vi) For all pairs of groups C_i and C_j connected by a pseudo edge, we generate 20 edges, each of which connect between vertices in C_i and C_j.
(vii) For $C_i (i = 0, \ldots, 20)$, generate 10 edges between two vertices selected from C_i randomly.

This procedure generate relationship compatible network consists of 20 groups. Within the groups, 10 groups consist of a single attribute; that is, attribute homogeneous. However, the other 10 groups consists of two different attributes.

Then, we add noise to the network.

(i) Let r be user-specified noise ratio.
(ii) Select $400 \times r$ nodes randomly and change their attributes randomly.
(iii) Select $1200 \times (r/5)$ edges randomly and remove the edges from the existing graph.
(iv) Add $1200 \times (r/5)$ edges between randomly selected two vertices to the existing graph.

We compare AR indexes computed by ARIANA to that by k-SNAP [9], which first make A-compatible grouping and then divide the groups until the number of groups is k. In this experiments, we set the number of groups as 20 on both algorithms. To calculate AR index, we set λ as 0.

Figure 3 shows AR index values by changes of noise. ARIANA and k-SNAP indicate AR index values calculated by the algorithms. Initial grouping indicates AR index value just after the noise is added to graph and before we do ARIANA or k-SNAP.

According to the increase of noise, AR indexes of any groupings become gradually low. In all noise levels, AR indexes computed by ARIANA are larger than those by k-SNAP, and hence, ARIANA find the better grouping than k-SNAP. When noise level is low, AR indexes of k-SNAP are lower than those of the initial groupings because attribute homogeneity in k-SNAP is too restrictive to find relationship homogeneous grouping. On the other hand, ARIANA can find the better grouping than initial grouping in terms of AR index. When noise level is high, the k-SNAP grouping is better than initial grouping. In this case, grouping of ARIANA is still better than that of k-SNAP. From these result, ARIANA is shown to be able to correct the attribute and relationship noise.

Figure 4 shows the number of clusters on the aggregation step when the best AR indexes are computed. For example, when $r = 1$, we can understand that ARIANA merge groups until the number of cluster is 14 and then divide the groups until the number of cluster is 20 when the maximum AR index is found. In all cases except $r = 0$ (no noise), ARIANA merge some groups then dividing the groups. This observation verifies the effectiveness of both merge step and division step in ARIANA.

5.2. Results for a real biological network

We apply our algorithm to real biological dataset. We use iRefIndex [14] as the target network, which is a protein network aggregated from several sources, and the attributes on vertices are pathways to which genes belong. The nodes and edges indicate proteins and the relationships between them, respectively. In this dataset, the numbers of nodes, number of edges, and number of attributes are 1,184, 9,807, and 221, respectively.

We calculate the maximum $w(\mathcal{C})$ using ARIANA as described in Section 4. ARIANA uses user-specified values: the number of clusters in the aggregation step, J; the number of final clusters, G; and coefficient of normalization term, λ. Here, we only fix the value of λ, and determine J and G by using various values of $w(\mathcal{C})$. We change the values of J and G from 10 to 300, independently.

Figure 5 shows the number of clusters and its max $w(\mathcal{C})$ with a change in λ. This figure shows that the number of clusters is large when λ is small, an vice versa. Therefore, by setting the value of λ, we can control the size of clusters. Because of the limitation that k-SNAP can find the clusters only when the number of clusters is more than the number of attributes, k-SNAP cannot find the clusters we found when $\lambda > 20$.

Figure 6 shows the network motif when $\lambda = 40$. For this value of λ, we generate 30 groups in the aggregation step, and finally generate 65 groups. The nodes and edges in the figure indicate groups and relationships with a high participation ratio, respectively. The figure shows only the nodes containing a minimum of five proteins and the edges for which the relationship has $r(C_i, C_j) \geq 0.80$. The number of edges in the figure is 22. The group size indicates the number of proteins in the group; the larger nodes contain a larger number of proteins. The numbers in the nodes represent its node ID.

Here, we check the biological functions of this network. Detail of associated pathways with the groups are described in `http://lab.se-se.jp/~sesejun/large_net_motif.html`. The figure shows a large connected component. The component contains basic biological processes: cell cycles (regulation of time of cells), ribosomes (protein generation) and purine and pyrimidine metabolisms (nucleotide metabolisms). On the other hand, single nodes include pathways working only on the specific case such as DNA replication (Replication of DNA damages) and glycan biosynthesis. From this result, we confirm that basic biological processes are highly related to each other.

Node 1 at the upper-left of the left most connected component has a hub node, which is associated with cell cycle. Cell cycle proteins are known to regulate the start time of cell functions. Moreover, the proteins are significantly associated with GO term "DNA-dependent DNA replication" (Binomial test P-value = 2.8e-6, two-sided), which means that the genes assocate with start time of cell division and connect to various cellular processes. Hence, results of this study are in agreement with common biological knowledge. Proteins in Node 51 connected to Node 1 are also associated with cell cycle but the node is only connected with Node 1 and 31. The process is associated with GO term "metabolic process" (Binomial test 5.26e-5, two-sided). This result indicates that the role of cell cycle related proteins can be categorized into more than two groups by our method: one related to DNA replication showing association with various functions and other related to metabolic process showing little association.

Node 54 is associated with both purine metabolism and pyrimidine metabolism. Both the metabolisms are known to have a similar function about metabolism of different nucleotides. Our method classified the two different but having similar function metabolisms into the same groups. The above observations are consistent with the known facts about the involved pathways.

Thus, our method can be successfully used to extract biologically significant network motifs.

6. Concluding remarks and future work

Network motif discovery is gaining importance in the field of both statistical analysis and biology. The problem with the existing methods is that they cannot handle noise and uncertainty in a given graph. In this study, we developed a novel algorithm called ARIANA to find large network motifs even when the network has noise and uncertainty. By applying ARIANA to a real biological network, we found network motifs associated with regulation of cell time and discovered that the role of ribosomes can be categorized into two different groups.

Although our result showed that our method practically succeeded in the analysis biological network, our index is no ground theory on statistics and information theory. In a future study, we intend to curry out a theoretical analysis of the AR index. In addition, we intend to develop on algorithm to find the best grouping. We introduced a simple two step algorithm, but the result may not maximize AR index. By the improvements of the algorithm, we compute the better graph in terms of AR index.

This network motif discovery problem can be applied not only to biological network but also to the Web and economic networks. Application of this study will show the new visualization and understanding of large complex networks.

References

[1] Barabasi A L and Albert R 1999 *Science* **286** 509 ISSN 0036-8075
[2] Barabási A L and Oltvai Z N 2004 *Nat. Rev. Genet.* **5** 101 ISSN 1471-0056
[3] Xiong H, He X, Ding C, Zhang Y, Kumar V and Holbrook S R 2005 Identification of functional modules in protein complexes via hyperclique pattern discovery *PSB '05* p 221
[4] Spirin V and Mirny L A 2003 *Proc. Natl. Acad. Sci. USA* **100** 12123
[5] Guimerà R and Amaral L A N 2005 *Nature* **433** 895
[6] Bachman P and Liu Y 2009 *Bioinformatics* **25** 1814
[7] Chen J, Hsu W, Lee M and Ng S K 2006 *KDD '06: Proc. of the 12th ACM SIGKDD Int. Conf. on Knowledge Discovery and Data Mining* 106
[8] Alon U 2007 *Nat. Rev. Genet.* 450
[9] Tian Y, Hankins R A and Patel J M 2008 Efficient aggregation for graph summarization *SIGMOD '08: Proc. of the 2008 ACM SIGMOD Int. Conf. on Management of Data* (New York, NY, USA: ACM) p 567
[10] Huan J, Wang W, Prins J and Yang J 2004 Spin: mining maximal frequent subgraphs from graph databases *KDD '04* p 581
[11] Inokuchi A, Washio T and Motoda H 2000 An apriori-based algorithm for mining frequent substructures from graph data *PKDD '00* p 13
[12] Kuramochi M and Karypis G 2001 Frequent subgraph discovery *ICDM* p 313
[13] Yan X and Han J 2002 gspan: Graph-based substructure pattern mining *ICDM '02* p 721 ISBN 0-7695-1754-4
[14] Razick S, Magklaras G and Donaldson I M 2008 *BMC Bioinformatics* **9** 405

Dense module enumeration in biological networks

Koji Tsuda[1] **and Elisabeth Georgii**[2,3]

[1]Computational Biology Research Center, National Institute of Advanced Industorial Science and Technology (AIST), Tokyo, Japan
[2]Max Planck Institute for Biological Cybernetics, Tübingen, Germany
[3]Friedrich Miescher Laboratory of the Max Planck Society, Tübingen, Germany

E-mail: koji.tsuda@aist.go.jp

Abstract. Analysis of large networks is a central topic in various research fields including biology, sociology, and web mining. Detection of dense modules (a.k.a. clusters) is an important step to analyze the networks. Though numerous methods have been proposed to this aim, they often lack mathematical rigorousness. Namely, there is no guarantee that all dense modules are detected. Here, we present a novel reverse-search-based method for enumerating all dense modules. Furthermore, constraints from additional data sources such as gene expression profiles or customer profiles can be integrated, so that we can systematically detect dense modules with interesting profiles. We report successful applications in human protein interaction network analyses.

1. Introduction

Today, a large number of databases provide access to experimentally observed protein-protein interactions. The analysis of the corresponding protein interaction networks can be useful for functional annotation of previously uncharacterized genes as well as for revealing additional functionality of known genes. Often, function prediction involves an intermediate step where clusters of densely interacting proteins, called modules, are extracted from the network; the dense subgraphs are likely to represent functional protein complexes [1]. However, the experimental methods are not always reliable, which means that the interaction network may contain false positive edges. Therefore, confidence weights of interactions should be taken into account.

A natural criterion that combines these two aspects is the average pairwise interaction weight within a module (assuming a weight of zero for unobserved interactions, cf. Ulitsky et al.[2]). We call this the module *density*, in analogy to unweighted networks [3]. We present a method to enumerate all modules that exceed a given density threshold. It solves the problem efficiently via a simple and elegant reverse search algorithm, extending the unweighted network approach in [4].

There is a large variety of related work on module discovery in networks. The most common group are graph partitioning methods[5, 6, 7]. They divide the network into a set of modules, so their approach is substantially different from DME, which provides an explicit density criterion for modules (Fig. 1A). Another group of methods define explicit module criteria, but employ heuristic search techniques to find the modules [8, 3]. This contrasts with complete enumeration algorithms, which form the third line of research: they give explicit criteria and return all modules that satisfy them. For example, clique search has been frequently applied [9, 10]. The enumeration of cliques can be considered as a special case of our approach, restricting it to unweighted graphs and a density threshold of 1. Further enumerative approaches use different module criteria assuming unweighted graphs [11, 12].

In recent years, many module finding approaches which integrate protein-protein interaction networks with other gene-related data have been published. One strategy, often used in the context of partitioning methods, is to build a new network whose edge weights are determined by multiple data sources [13]. Tanay et al.[14] also create one single network to analyze multiple genomic data at once; however, they use a bipartite network where each edge corresponds to one data type only. In both cases, the different data sets have to be normalized appropriately before they can be integrated. In contrast to that, other approaches keep the data sources separate and define individual constraints for each of them. Consequently, arbitrarily many data sets can be jointly analyzed without the need to take care of appropriate scaling or normalization. Within this class of approaches, there exist two main strategies to deal with profile data like gene expression measurements. In the first case, the profile information is transformed into a gene similarity network, where the strength of a link between two genes represents the global similarity of their profiles [15, 2, 16]. In the second case, the condition-specific information is kept to perform a context-dependent module analysis [17, 18, 19]. Our approach follows along this line, searching for modules in the protein interaction network that have consistent profiles with respect to a subset of conditions. In contrast to the previous methods, our algorithm systematically identifies all modules satisfying a density criterion and optional consistency constraints.

In this study, we evaluate our approach on the yeast interaction network in comparison with four other methods. Also, we discuss our results obtained from human protein interactions in the context of gene expression data.

2. Dense Module Enumeration

In this section, we describe the basic idea of dense module enumeration (DME) using the examplar graph shown in Fig. 2. Our method is based on the reverse search paradigm [20], which is quite popular in the algorithm community[1], but only in a limited degree known in the data mining community. A weighted graph is represented as a symmetric association matrix (edges that are not shown have zero weight). We denote by w_{ij} the weight between two nodes, and define the density of a node subset U as

$$\rho(U) = \sum_{i,j \in U, i<j} w_{ij} \bigg/ \frac{|U|(|U|-1)}{2}.$$

We would like to enumerate all subsets U with $\rho(U) \geq \theta$, where θ is a prespecified constant.

All subsets form a natural graph-shaped search space, where one can move downwards or upwards by adding or removing a node, respectively (Figure 3a). Here, the root node corresponds to the empty set. For efficient traversal, however, one needs a spanning tree, not a graph. When a tree is made by lexicographical ordering (Figure 3b), the search tree is not *anti-monotonic* with respect to the density. Namely, the density is not monotonically decreasing when the tree is traversed from the root to a leaf. This property disallows early pruning and makes the enumeration difficult. However, there exists indeed a search tree which is anti-monotonic (Figure 3c). It can be constructed by reverse search.

In reverse search, the search tree is specified by defining a *reduction map* $f(U)$ which transforms a child to its parent. In our case, the parent is created by removing the node with minimum degree from the child. Here, the degree of a node is defined as the sum of weights of all adjacent edges within U. If there are multiple nodes with minimum degree, the one with the smallest index is removed. In [21], it is proven that the density of a parent is at least as

[1] See a list of applications at `http://cgm.cs.mcgill.ca/~avis/doc/rs/applications/index.html`

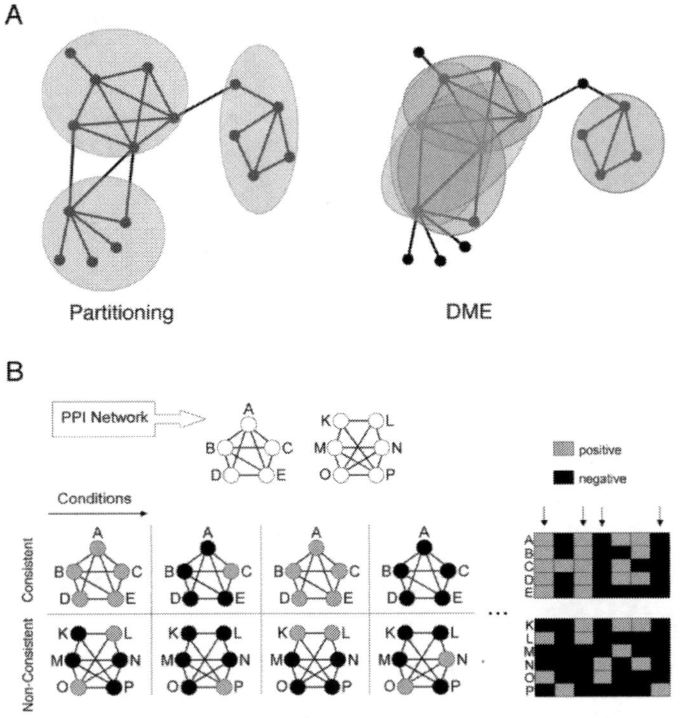

Figure 1. Dense Module Enumeration approach. A. DME versus partitioning. While partitioning methods return one clustering of the network, DME discovers all modules that satisfy a minimum density threshold. B. Combination with profile data. Integration of protein-protein interaction (PPI) and external profile data allows to focus on modules with consistent behavior of all member proteins in a subset of conditions. The top module has two conditions where all nodes are positive and one condition where all nodes are negative. The arrows in the profile shows such consistent conditions. On the other hand, the bottom module does not have such consistency.

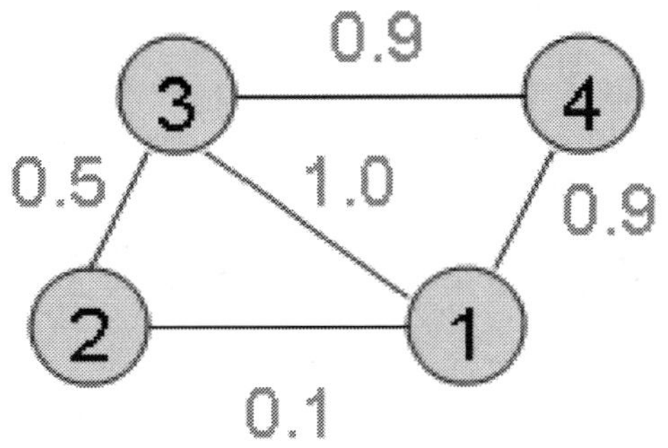

Figure 2. An examplar graph for dense module enumeration.

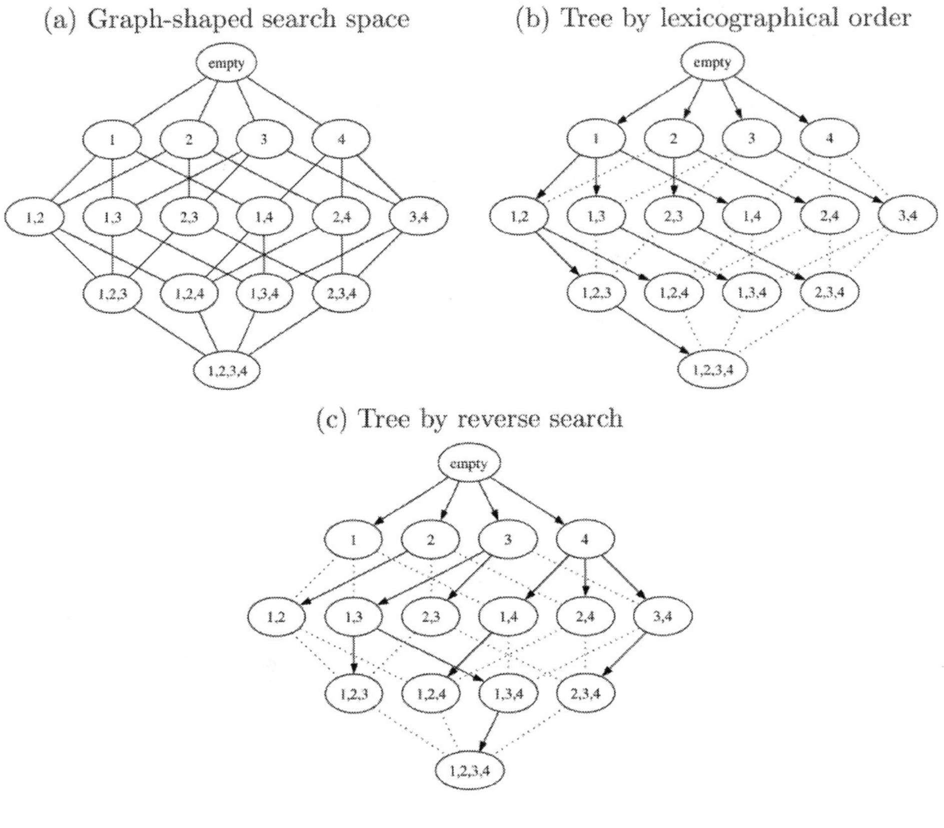

Figure 3. Illustration of reverse search.

high as the maximum density among the children, ensuring that the search tree induced by the reduction map is anti-monotonic.

In addition to the anti-monotonicity property, a valid reduction map must satisfy the following reachability condition [20]: starting from any node of the search tree, we can reach a root node after applying the reduction map a finite number of times. This condition ensures that the induced search tree is indeed spanning. For the reduction map stated above, it is trivial to show that the reachability condition is satisfied, because any cluster shrinks to the empty set by removing nodes repeatedly.

To enumerate all clusters with density $\geq \theta$, one has to traverse this implicitly defined search tree in a depth-first or breadth-first manner. During traversal, children are generated *on demand*. As the reduction map defines how to get from children to parents and not vice versa, we cannot directly derive the children from a given parent. Instead, to generate the children of a cluster U, we have to consider all candidates $U \cup \{i\}$, $i \notin U$ and apply the reduction map to every candidate (reverse search principle). Qualified candidates with $f(U \cup \{i\}) = U$ are then taken as children. A naive implementation of this child generation process can make the algorithm very slow. Thus, it is important to engineer this process well. As the search tree is anti-monotonic, one can prune the tree whenever the density goes below θ. See [22] for detailed evaluation of computational complexity.

The definition of a search tree is not an issue in the context of frequent pattern mining [23], because frequency is anti-monotonic in any tree. Reverse search is interesting because it provides a systematic way of defining an anti-monotonic tree. Notice, however, that it is not applicable to all score functions. Cluster density is an example where reverse search can be applied most effectively.

3. Integration of Additional Constraints

The DME framework makes it easy to incorporate and systematically exploit constraints from additional data sources. For illustration, consider the case where we have an additional data set which provides profiles of proteins or genes across different conditions (Fig. 1B). For simplicity, let us assume binary profiles, being 1 if the protein is positively associated with the corresponding condition, and 0 otherwise. Then, dense modules where all member proteins share the same profile across a certain number of conditions are of particular interest; we call these modules *consistent*. The problem of dense module enumeration with consistency constraints is formalized as follows.

Definition 1. *Given a graph with node set V and weight matrix W, a density threshold $\theta > 0$, a profile matrix $(m_{ij})_{i \in V, j \in C}$, and non-negative integers n_0 and n_1, find all modules $U \subset V$ with $\rho_W(U) \geq \theta$ s.t. there exist at least n_0 conditions $c \in C$ with $m_{uc} = 0 \ \forall u \in U$ and there exist at least n_1 $c \in C$ with $m_{uc} = 1 \ \forall u \in U$.*

Given such a *consistency constraint*, we can stop the module extension during the dense module mining as soon as the constraint is violated. This is due to the fact that the number of consistent profile conditions cannot increase while extending the module; more generally, this property is called *anti-monotonicity*. So we simply add to the module enumeration algorithm a condition which checks for the consistency requirements. These are then automatically taken into account in the check for local maximality. The use of additional constraints can restrict the search space considerably, so it accelerates the computation and helps to focus on biologically interesting solutions.

4. Module Ranking

The exhaustiveness of our DME approach enables us to exactly determine the uncommonness of the discovered substructures with respect to the network at hand. Let $W = (w_{ij})_{i,j \in V}$ be the matrix representation of the given network; the total number of nodes is denoted by $|V|$. Let U be a module with $|U|$ nodes and density $\rho_W(U)$. Then, the probability that a random selection of $|U|$ nodes in the network produces a module with at least the same density as U is given by

$$\left|\{U' \subset V : |U'| = |U| \wedge \rho_W(U') \geq \rho_W(U)\}\right| \Big/ \binom{|V|}{|U|} . \qquad (1)$$

The exact value of the numerator can be obtained as a side product of the DME algorithm. In the case of additional constraints, it includes only modules that satisfy them. The modules in the DME output are sorted by their probability values (in ascending order). This ranking scheme captures the intuition that the rank of a module should increase with its size and density, but from a theoretical point of view it is more principled than the ranking criterion used in [3], which is the product of size and density. Furthermore, our probability calculation refers specifically to the network at hand, in contrast to measures derived from network models [24].

5. Experimental Results

First, we validated the performance of DME on the yeast interaction network in comparison with four other methods: clique detection (Clique)[2], the clique percolation method (CPM)[2] [9], a procedure for joining cliques of a certain size to larger clusters, CPMw[2] [25], an extension

[2] implementation from `http://www.cfinder.org`

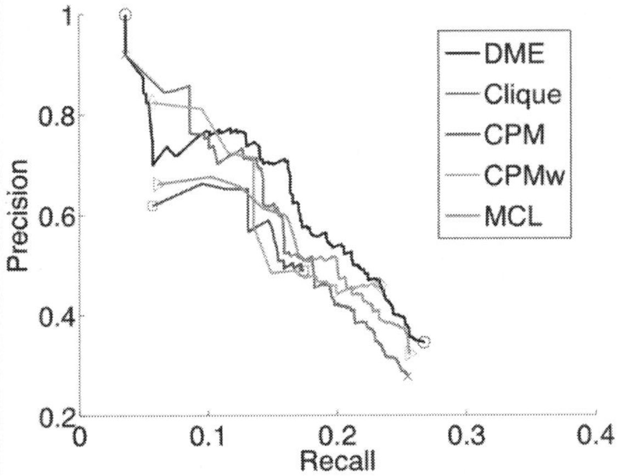

Figure 4. Comparative precision-recall analysis. To account for module overlap, the measures are based on protein pairs, see text.

of CPM which includes an additional clique filtering step, and Markov clustering (MCL)[3] [6], a popular graph clustering method simulating random walks. As a reference set of *confirmed* complexes, we used the manually curated protein complexes provided by MIPS[26]. To properly assess methods which can produce overlapping modules, we chose performance measures that are based on protein pairs rather than modules; in that way, we avoid taking the same subset of nodes several times into account even if it occurs in more than one module. Defining the intersection of pairs from predicted modules and pairs from known complexes as correctly predicted pairs, we calculated precision and recall as follows.

$$\text{Precision} = \frac{\text{nr of correctly predicted protein pairs}}{\text{nr of protein pairs in predicted modules}} \tag{2}$$

$$\text{Recall} = \frac{\text{nr of correctly predicted protein pairs}}{\text{nr of protein pairs in known complexes}} \tag{3}$$

To obtain precision-recall curves, we iteratively calculated the precision and recall values, each time extending the set of considered modules by the next highest ranking module. As the other methods do not provide a module ranking and our criterion is only applicable to enumerative approaches, we used the scoring scheme by Bader et al.[3]. In fact, it produced for our DME results almost the same ranking as our criterion; the corresponding precision-recall curves are virtually equivalent. Clique and CPM cannot handle edge weights directly, but they preselect edges according to a minimum weight threshold. Due to space constraints, we cannot describe the full details of protein interaction networks and their statistics. Please refer to [21].

The precision-recall curves are shown in Fig. 4. Overall, DME shows the best prediction performance. It has high precision with respect to the highest ranking modules and then shows a sudden drop, which is due to a big module not annotated as a known complex. Clique detects the same module, but there are some other higher ranked modules, so the drop happens later. MCL and CPM stay always below DME. Clique works quite well, however the precision drops quickly for higher recall values because edge weights are not taken into account. It seems that DME has a clear advantage compared to CPM: by explicitly using the edge weights and tuning the density parameter, it allows for more flexibility than the two-stage procedure of CPM, first

[3] implementation from `http://micans.org/mcl`

Table 1. Module statistics of the comparative analysis (see text for details). The average size of the raw modules can be larger than for the distinct modules because larger modules allow for more variants. The time measurements were performed on a 2.2 GHz processor.

	DME	Clique	CPM	CPMw	MCL
# distinct modules	1083	916	19	32	648
average size of distinct modules	3	4	16	14	3
# raw modules	24803	1971	19	33	648
average size of raw modules	10	6	16	14	3
# matched complexes	84	54	9	20	59
average complex size	5	7	19	14	7
# partially recovered complexes	133	107	20	33	117
# predicted interactions	5970	7066	2756	3935	6108
area under prec.-rec. curve (AUC)	0.183	0.166	0.107	0.153	0.148
# enriched distinct modules	112	131	18	32	69
# enriched among top-50	47	44	-	-	45
# overlapping proteins	1010	1113	12	38	1
# overlapping interactions	3664	4340	24	114	0
AUC for overlapping interactions	0.152	0.082	0.000	0.001	-
# recovered complex overlaps	18	16	0	4	0
running time (in seconds)	2667	6	5	457	4

selecting edges and subsequently joining together cliques that satisfy an overlap criterion. While CPMw allows to refine the module search, it still differs significantly from our approach. As it joins preselected cliques, it does not control directly the density of the produced modules and might also miss some dense modules. In our analysis, CPMw improved the result obtained by CPM, but is mostly inferior to Clique or DME. We deliberately determined the parameters of our method and the competing methods to make a fair comparison. Please refer to the supplementary report of [21].

Table 1 summarizes further statistics for the predicted modules. As DME and Clique produced a large number of very similar modules, we computed for better comparability the number of distinct modules. For that purpose, we grouped matching modules into clusters; each cluster was represented by its top-ranking module. To decide whether two modules match each other, we here computed the overlap score proposed by Bader et al.[3], using a stringent cutoff of 0.5. It is defined as the fraction of overlapping proteins with respect to the size of the first module multiplied by the fraction of overlapping proteins with respect to the size of the second module. The same criterion was used to determine matches between predicted modules and known complexes. While DME and Clique discovered a comparable number of distinct modules, the DME modules match many more known complexes. Among these, we also find small-sized complexes, so the overall average size of retrieved complexes is lower than for Clique. In addition, we report the number of complexes from which at least one protein pair was recovered as well as the area under the precision-recall curve from the pairwise analysis (Fig. 4). In both cases, DME is leading. Furthermore, we investigated the enrichment of the distinct modules with respect to Gene Ontology (GO) terms. For that purpose, we applied the TANGO tool [27] using the default setting with P-value threshold 0.05 after correction for multiple testing. Beside the total number of enriched modules, we also counted the number of enriched modules among the top-50 distinct modules, showing that for each method that produced more than 50 modules, most of the high-ranking modules satisfy the enrichment threshold. For small modules the enrichment test fails even if they are totally pure.

6. Tissue-Specific Modules in the Human Interaction Network

In this section, we discover tissue-specific modules of the human interaction network. As side information, we downloaded the gene expression profiles by Su et al.[28], containing measurements in 79 different human tissues and present/absent/marginal calls. For our purposes, we considered a gene to be expressed in a given tissue only if it was called as present in both of the duplicated measurements. In order to find complexes which are present in several, but not all tissues, we applied DME to enumerate all modules that are consistently expressed in at least three tissues and consistently not expressed in at least ten tissues. We used again the same procedure for selecting the density parameter and ended up with 460 distinct modules.

The two top-ranking modules cover the MCM complex (Fig. 5A). As reference, we used a manually curated set of human complexes collected by MIPS [29]. MCM is a hexameric protein complex required for the initiation and regulation of eukaryotic DNA replication. The DME modules contain two additional proteins, Ssrp1 and Orc6l. Orc6l is a member of the origin recognition complex (ORC), which plays a central role in replication initiation; in fact, the MCM and ORC complexes form the key components of the pre-replication complex [30]. This is nicely reflected by the high interaction density as well as the common expression profiles of the proteins: the module is fully expressed in three different types of bone marrow cells and fully non-expressed in 42 tissues like brain, liver, and kidney, where cells are differentiated and divide less. Ssrp1 is a member of the FACT complex, which is involved in chromatin reorganization [31].

Moreover, our analysis yields some insights about the tissue-specific reorganization of the SCF E3 ubiquitin ligase complex, which marks proteins for degradation. Fig. 5B depicts the five top-ranking modules that cover the complex (beyond those, there were three other modules covering only a single protein of the complex). One of them contains as additional component Cand1, a regulatory protein that inhibits the interaction of Cul1 with Skp1 [32]. The other four peripheral proteins are F-box proteins, which serve as substrate recognition particles for the SCF complex. Interestingly, the corresponding modules show different tissue specificities, indicating that the target proteins of SCF are selected in a tissue-dependent manner. This finding is in accordance with experimental studies [33, 34, 35]. On one hand, it has been shown that in human cells multiple variants of the SCF complex exist, each one containing a different F-box protein for substrate recognition. On the other hand, brain and blood cells have been identified as tissues of major expression for some F-box components, and expression variation of F-box components has been observed in several tissues like testis, prostate, and placenta. In our results, all detected module variants are present in natural killer (nk) cells, which play an important role in immune response [36], whereas only a few are present in B-cells and testis; in certain brain regions, for instance medulla oblongata, only the module variant with Fbxw7 is predicted to be active. As illustrated by this example, DME integrated with gene expression data can be a powerful tool to reveal functional and condition-specific variants of protein complexes.

7. Conclusion

Our algorithm, DME, extracts all densely connected modules from a given weighted interaction network. In addition to its completeness guarantee, a strength of the method lies in the possibility of transparent data integration, which is of crucial importance in biological applications. Due to its generality, we believe that DME is a useful tool in many different systems biology approaches. Our framework can also solve more general problems arising in the analysis of structured data, like dense subgraph detection in multi-partite graphs [14, 8] or in hypergraphs [37]. Moreover, module finding can assist in network comparison and classification tasks [38].

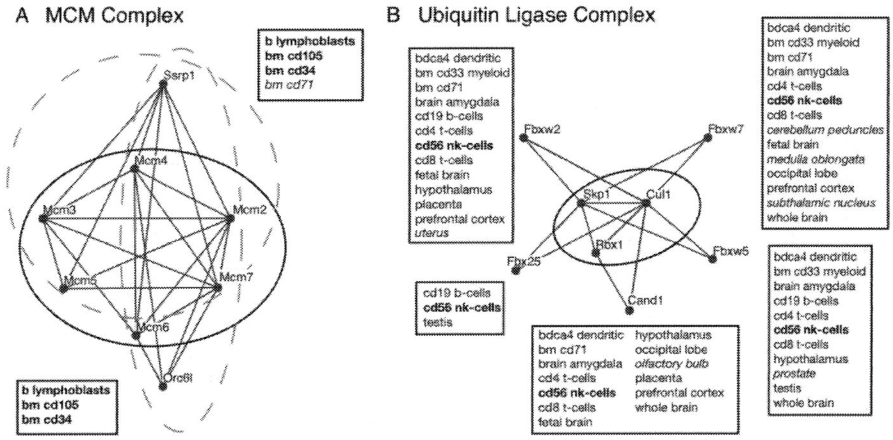

Figure 5. Tissue-specific modules in human. A. The two top-ranking modules, covering the MCM complex. Known complexes are indicated as solid ellipses, modules as dashed ellipses. B. Top-five modules around the SCF ubiquitin ligase complex, revealing its tissue-specific organization. Boxes show the tissues of consistent positive expression for the respective module. Tissues associated with all modules are marked in bold, uniquely appearing tissues in italics.

8. Acknowledgment

This proceedings paper summarizes two of our previous papers [21, 22] for the talk at IW-SMI 2009. A C++ implementation of the algorithm is available at `http://www.kyb.tuebingen.mpg.de/~georgii/dme.html`. KT is supported by MEXT Kakenhi 21680025 and the Functional RNA Project of New Energy and Industrial Technology Development Organization (NEDO).

Reference

[1] Sharan R, Ulitsky I and Shamir R 2007 *Mol. Syst. Biol.* **3** 88
[2] Ulitsky I and Shamir R 2007 *BMC Syst. Biol.* **1** 8
[3] Bader G D and Hogue C W 2003 *BMC Bioinformatics* **4** 2
[4] Uno T 2007 in *Proc. of ISAAC 2007* pp 402–14
[5] Chen J and Yuan B 2006 *Bioinformatics* **22**(18) 2283–90
[6] van Dongen S 2000 *Graph Clustering by Flow Simulation* Ph.D. thesis University of Utrecht
[7] Newman M E 2006 *Proc. Natl. Acad. Sci. USA* **103**(23) 8577–82
[8] Everett L, Wang L S and Hannenhalli S 2006 *Bioinformatics* **22**(14) e117–23
[9] Palla G, Derenyi I, Farkas I and Vicsek T 2005 *Nature* **435**(7043) 814–8
[10] Spirin V and Mirny L A 2003 *Proc. Natl. Acad. Sci. USA* **100**(21) 12123–8
[11] Haraguchi M and Okubo Y 2006 in *Federation over the Web* vol 3847 of *Lecture Notes in Computer Science* (Springer)
[12] Zeng Z, Wang J, Zhou L and Karypis G 2006 in *KDD '06: Proc. of the 12th ACM SIGKDD Int. Conf. on Knowledge Discovery in Data Mining* (New York, NY, USA: ACM) pp 797–802 ISBN 1-59593-339-5
[13] Hanisch D, Zien A, Zimmer R and Lengauer T 2002 *Bioinformatics* **18**(suppl_1) S145–54
[14] Tanay A, Sharan R, Kupiec M and Shamir R 2004 *Proc. Natl. Acad. Sci. USA* **101**(9) 2981–6
[15] Segal E, Wang H and Koller D 2003 *Bioinformatics* **19**(suppl_1) i264–71
[16] Pei J, Jiang D and Zhang A 2005 in *ICDE '05: Proc. of the 21st Int. Conf. on Data Engineering (ICDE'05)* (Washington, DC, USA: IEEE Computer Society) pp 353–4 ISBN 0-7695-2285-8
[17] Ideker T, Ozier O, Schwikowski B and Siegel A F 2002 *Bioinformatics* **18**(suppl_1) S233–40
[18] Huang Y, Li H, Hu H, Yan X, Waterman M S, Huang H and Zhou X J 2007 *Bioinformatics* **23**(13) i222–9
[19] Yan X, Mehan M R, Huang Y, Waterman M S, Yu P S and Zhou X J 2007 *Bioinformatics* **23**(13) i577–86
[20] Avis D and Fukuda K 1996 *Discrete Appl. Math.* **65** 21–46
[21] Georgii E, Dietmann S, Uno T, Pagel P and Tsuda K 2009 *Bioinformatics* **25** 933–40
[22] Georgii E, Tsuda K and Schölkopf B 2009 in *Proc. of the 2009 Workshop on Data Mining using Matrices and Tensors (DMMT'09)* pp 32–41

[23] Han J and Kamber M 2006 *Data Mining: Concepts and Techniques* 2nd ed The Morgan Kaufmann Series in Data Management Systems (San Francisc, CA: Morgan Kaufmann Publishers)
[24] Koyuturk M, Szpankowski W and Grama A 2007 *J. Comput. Biol.* **14**(6) 747–64
[25] Farkas I J, Abel D, Palla G and Vicsek T 2007 *New J. Phys.* **9**(180)
[26] Guldener U, Munsterkotter M, Kastenmuller G, Strack N, van Helden J, Lemer C, Richelles J, Wodak S J, Garcia-Martinez J, Perez-Ortin J E, Michael H, Kaps A, Talla E, Dujon B, Andre B, Souciet J L, De Montigny J, Bon E, Gaillardin C and Mewes H W 2005 *Nucl. Acids Res.* **33**(suppl_1) D364–8
[27] Shamir R, Maron-Katz A, Tanay A, Linhart C, Steinfeld I, Sharan R, Shiloh Y and Elkon R 2005 *BMC Bioinformatics* **6**(1) 232 ISSN 1471-2105 URL `http://www.biomedcentral.com/1471-2105/6/232`
[28] Su A I, Wiltshire T, Batalov S, Lapp H, Ching K A, Block D, Zhang J, Soden R, Hayakawa M, Kreiman G, Cooke M P, Walker J R and Hogenesch J B 2004 *Proc. Natl. Acad. Sci. USA* **101**(16) 6062–7
[29] Ruepp A, Brauner B, Dunger-Kaltenbach I, Frishman G, Montrone C, Stransky M, Waegele B, Schmidt T, Doudieu O N, Stumpflen V and Mewes H W 2008 *Nucl. Acids Res.* **36**(suppl_1) D646–50
[30] Lei M and Tye B K 2001 *J. Cell Sci.* **114**(Pt 8) 1447–54
[31] Orphanides G, Wu W H, Lane W S, Hampsey M and Reinberg D 1999 *Nature* **400**(6741) 284–8
[32] Zheng J, Yang X, Harrell J M, Ryzhikov S, Shim E H, Lykke-Andersen K, Wei N, Sun H, Kobayashi R and Zhang H 2002 *Mol. Cell* **10**(6) 1519–1526
[33] Cenciarelli C, Chiaur D, Guardavaccaro D, Parks W, Vidal M and Pagano M 1999 *Curr. Biol.* **9** 1177–9
[34] Kipreos E and Pagano M 2000 *Genome Biology* **1**(5) reviews3002.1–reviews3002.7 ISSN 1465-6906
[35] Koepp D M, Schaefer L K, Ye X, Keyomarsi K, Chu C, Harper J W and Elledge S J 2001 *Science* **294**(5540) 173–7
[36] Janeway C, Travers P, Walport M and Shlomchik M 2005 *Immunobiology: Immune System in Health and Disease* (Garland Publishing)
[37] Zhao L and Zaki M J 2005 in *SIGMOD '05: Proc. of the 2005 ACM SIGMOD Int. Conf. on Management of Data* (New York, NY, USA: ACM) pp 694–705 ISBN 1-59593-060-4
[38] Chuang H Y, Lee E, Liu Y T, Lee D and Ideker T 2007 *Mol. Syst. Biol.* **3** 140

International Workshop on Statistical-Mechanical Informatics 2009 (IW-SMI 2009) IOP Publishing
Journal of Physics: Conference Series **197** (2009) 012013 doi:10.1088/1742-6596/197/1/012013

On the presence of high-order interactions among somatosensory neurons and their effect on information transmission

Robin A A Ince[1], Fernando Montani[2], Ehsan Arabzadeh[3], Mathew E Diamond[4] and Stefano Panzeri[2]

[1] Faculty of Life Science, University of Manchester, 3.431 Stopford Building, Oxford Road, Manchester M13 9PL, UK
[2] Robotics, Brain, and Cognitive Sciences Department, Italian Institute of Technology, Via Morego 30, 16163 Genova, Italy
[3] School of Psychology, University of New South Wales, Sydney, New South Wales, Australia
[4] Cognitive Neuroscience Sector, International School for Advanced Studies, Trieste, Italy, and the SISSA Unit, Italian Institute of Technology, Trieste, Italy

E-mail: stefano.panzeri@iit.it

Abstract.
In order to understand how populations of neurons encode information about external correlates, it is important to develop minimal models of the probability of neural population responses which capture all the salient changes of neural responses with stimuli. In this context, it is particularly useful to determine whether interactions among neurons responding to stimuli can be described by a pairwise interaction model, or whether a higher order interaction model is needed. To address this question, we compared real neural population activity obtained from the rat somatosensory cortex to maximum-entropy models which take into account only interaction of up any given order. By performing these comparisons, we found that interactions of order two were sufficient to explain a large amount of observed stimulus-response distributions, but not all of them. Triple-wise interactions were necessary to fully explain the data. We then used Shannon information to compute the impact of high order correlations on the amount of somatosensory information transmitted by the neural population. We found that correlations of order two gave a good approximation of information carried by the neural population, within 4% of the true value. Third order correlations gave an even better approximation, within 2% of the true value. Taken together, these results suggest that higher order interactions exist and shape the dynamics of cortical networks, but play a quantitatively minor role in determining the information capacity of neural populations.

1. Introduction

The activity of sensory neurons provides the information about the external world based upon which animals take decisions. Determining how the activity of populations of neurons represents sensory stimuli is thus of obvious importance for the progress of neuroscience and is a prerequisite for understanding sensory function. Yet, characterizing the details of the neural representations of sensory stimuli has proven to be a daunting challenge so far. This difficulty is due to several reasons. The first reason is that the presentation of even the simplest sensory stimulus elicits the activity of large numbers of neurons, which at any time may either fire or not fire a spike in response to the stimulus. Thus, neural responses are intrinsically high dimensional [1]. The second problem is that neural responses are variable, and this implies that the stimulus-response relationship must be characterized probabilistically rather than with just a simple tuning curve or transfer function.

However, even despite the above problems, it would be still easy to construct a low dimensional representation of neural populations responses to a stimulus if the responses of each neurons were statistically independent from that of other neurons. In this case, one could measure the response probabilities of each neuron separately, and then construct a population response probability by simply taking the product of the probabilities of the individual neurons. Unfortunately, that's not the case: the response of a neuron does not depend only on the stimulus but also on the activity of the other neurons [2–5]. Theoretical and experimental studies have suggested that such interactions can profoundly affect the information transmitted by neural populations, by increasing the information content of neural populations [6–8], by tagging features to be bound together [9], by stabilizing the temporal relationships between cells against the detrimental effect of trial-to-trial variability [10, 11], by implementing strategies for error correction [12], or by severely limiting the representational capacity of neural populations [13, 14]. As a consequence, neural population responses cannot be described simply using models of independent non-interacting neurons neurons: any model of neural responses must be able to capture all the salient interactions among neurons.

Given that models of the neural responses must include all the salient interactions among neurons, the best hope to find a compact low-dimensional representation of neural responses is if the structure of interactions within neurons in a population could be described only in terms of interactions between pairs of neurons, rather than in terms of interactions between whole large assemblies. A pairwise representation of neural interaction would greatly simplify the number of parameters needed to describe neural representations of stimuli, potentially leading to the development of simple but effective models of decoding neural population activity [1].

The question of whether the structure of correlated activity can be satisfactorily described by considering only pairwise interactions is only recently beginning to be studied systematically both by theory and experiments [15–22] A number of groups [17–19] have addressed this problem demonstrating that in several types of experimental preparations pairwise correlations account for the majority ($> 90\%$) of decrease of response entropy due to all possible interactions (of any order) between neurons. Given that entropy is a measure of variability of responses, these results suggests that pairwise interactions are sufficient to describe reasonably well how interactions constrain neural population activity. However, these results do not clarify fully whether or not high order interactions exist and are statistically significant. Moreover, given that response entropy and information about stimuli are not equivalent concepts, these results do not fully clarify whether high order interactions are quantitatively important to describe how neurons encode information about external stimuli.

In this study we aim at contributing to this research by evaluating whether high order interactions among neurons are present in the somatosensory cortical network, and by measuring what is the impact of such high order interactions on the amount of Shannon information that the neural population conveys about somatosensory stimuli.

This paper is organized as follows. We first describe how to use maximum entropy models to quantify whether the probabilities of neural population responses to stimuli can be described with only interactions up to a given order. We then apply these techniques to real neural population activity recorded from the rat somatosensory cortex, and establish whether there are statistically significant interactions among neurons of order higher than two. We then use Shannon information to quantify what is the impact of correlations of a given order on the sensory representation of information.

Journal of Physics: Conference Series **197** (2009) 012013 doi:10.1088/1742-6596/197/1/012013

2. Investigating the order of interaction through the maximum entropy principle

We consider a population of N neurons whose activity is simultaneously observed during a specified short time window following the presentation of a sensory stimulus s taken from a set of S different stimuli. Since the time window is short, we can use a binary representation of the activity of each neuron (with zero or one to respectively denote the presence or absence of spikes from the neuron). Thus, the neuronal population activity is represented by a binary array $\mathbf{x} = \{x_1, ..., x_N\}$ in the space X of all binary arrays of length N, where $x_i = 0$ if neuron i is silent in some time window and $x_i = 1$ if it is firing one or more spikes.

The question that we seek to address is whether we can describe all interactions between the neurons in terms of interactions between up to two neurons only, or whether there are interactions among groups of more than two neurons which cannot be explained in terms of pairwise interactions. A rigorous way to investigate the effects of different orders of interaction is provided by the technique of *maximum entropy*, which was originally introduced in statistical physics [23], and is now beginning to be used in neuroscience [1, 17–19, 24, 25]. In general, the idea of the maximum entropy (ME) principle is to first fix some constraints that are of interest and then seek the simplest, or most random, distribution subject to those constraints. Using entropy as a measure of randomness, asking for the most random distribution corresponds to asking for the distribution with maximal entropy subject to the constraints. This removes all types of correlation or structure in the data that does not result from the constrained features.

The ME formalism can be naturally used to to address the problem of whether we can describe all interactions between neurons in terms of interactions between up to k neurons only, or whether there are higher interactions among more than k neurons which cannot be possibly explained in terms of interactions of order up to k. Measuring all interactions of up to k variables means measuring all the marginal response probabilities involving up to k variables. Therefore any probability matching the observed interactions of up to k elements must obey (apart from the usual non negativity and normalization constraints) the following linear constraints:

$$\begin{aligned} P(x_{i_1} = 1|s) &= \eta_{i_1} \\ P(x_{i_1} = 1, x_{i_2} = 1|s) &= \eta_{i_1,i_2} \\ &\cdots \\ P(x_{i_1} = 1, x_{i_2} = 1, \cdots, x_{i_k} = 1|s) &= \eta_{i_1,i_2,\cdots,i_k} \end{aligned} \tag{1}$$

where (in agreement with the well-established notation of Amari [26]), the set of variables denoted η represent the values of the marginal probabilities, and the subscript indices of η represent the variables involved in the marginal probabilities.

The probability distribution $P_{ME}^{(k)}(\mathbf{x}|s)$ with maximum entropy among those satisfying the above constraints is the one that does not impose the presence of any additional interaction of order higher than k. (The case $k = 1$ is trivial and corresponds to the case in which all neurons fire independently at fixed stimulus). Following [26, 27], it can be shown that there is a unique solution to the constrained maximum entropy problem, which can be written in the following exponential form:

$$P_{ME}^{(k)}(\mathbf{x}|s) = \exp\left\{\sum_i x_i\theta_i + \sum_{i_1<i_2} x_{i_1}x_{i_2}\theta_{i_1i_2} + \cdots + \sum_{i_1<\cdots<i_k} x_i\cdots x_k\theta_{i_1\cdots i_k} - \psi\right\} \tag{2}$$

To quantify whether interactions of up to k neurons in a population are sufficient to describe the probabilities of neural responses to stimuli, we can quantitatively compare the true distribution $P(\mathbf{x}|s)$of neural responses to the stimulus s consider to the distribution $P_{ME}^{(k)}(\mathbf{x}|s)$. By performing this comparison over a range of values of k, we can empirically determine the minimal k necessary to fit well the empirically measured response probability.

In order to compute the maximum entropy distribution $P_{ME}^{(k)}(\mathbf{x}|s)$ of Eq. (2) from real data, we just need to find the θ coefficients with up to K indices entering the above equation to construct the solution above. These θ coefficients can be determined from the experimentally measured marginal probabilities η of up to k elements through a set of algebraic equations which were derived in the work of Amari [26, 28]. To solve these equations numerically, we used the publicly available `pyentropy` numerical package that we recently developed and published [29]. We refer to Ref [29] for full details and for download of the code implementing the numerical solutions.

3. The role of high order correlations in shaping synchronous discharge in somatosensory cortex

After having described the ME techniques to study the presence of interactions of up to any given order, we apply them to a population of neurons recorded from the whisker representation in the somatosensory cortex of urethane anaesthetized rats. We first describe the dataset; we then compare the ME probabilities at fixed interaction order with the real data; and we finally evaluate the effect of the interaction order on the information about the stimuli carried by the neural responses.

3.1. Neurophysiological data

The data set (previously published in [30, 31]) consists of 24 simultaneously recorded neural clusters, each sampled with a different electrode with a minimal inter-electrode distance of 400 μm. Spike times from each electrode were determined by a voltage threshold set to a value 2.5 times the root mean square voltage. Since it was not possible to sort well-isolated units from each channel, spikes from the same recording channel were considered together as a single neural cluster. It has been estimated that, under these recoding conditions, each cluster captured the spikes of approximately two to five neurons located near the tip of the electrode (see [32]). Neural activity was recorded in response to stimulation (with a piezoelectric wafer controlled by a voltage generator) consisting of sinusoidal whisker vibrations, each defined by a different value of vibration velocity and delivered for 500 ms (see [31] for full details). Thirteen different values of vibration velocity were tested, ranging between $Af = 0.15$ mm/s and $Af = 47.7$ mm/s. Each value of vibration velocity was treated as a different stimulus s (there were 13 stimulus classes in total). The number of recorded repetitions for each stimulus (called "trials" in neurophysiology), from which the probability of response at fixed stimulus is determined, varied between a minimum of 200 and a maximum of 1400 across stimuli.

It was previously shown [31] that the majority of the information is transmitted very early post stimulus onset (typically between 5 and 30 ms). We therefore concentrated on data taken from these early highly-informative windows.

3.2. Fitting the neurophysiological stimulus-conditional response probabilities to maximum entropy models

We used these data to study the shape of the distribution of neural population responses at fixed stimulus, and the order of neural interactions needed to describe this distribution. We note that some previous seminal studies [17, 18] focused on the probability of response to many different stimuli. However, this has the potential problem that the resulting correlation may arise both from correlations in the stimulus and from actual neural interactions, and it is difficult to separate them [1]. Thus, we considered distributions at fixed stimuli to ensure that we did not investigate neural interactions originating at least in part from correlations in the stimuli.

We decided to consider response distributions of $N = 8$ channels simultaneously recorded (out of the 24 available). The reason was that this population size was big enough to begin observing some effects of high order interactions, while was small enough to be sampled with the available data and be tested with both quantitative goodness-of-fit tests and the information analysis of the next subsection.

We first considered (figure 1) the distribution of neural responses to one particular stimulus (velocity = 4.3 mm/s; 1200 trials available). Neural responses evolve over time, and thus the distribution of neural responses depends on the particular post-stimulus window considered. Since it has been previously shown [31] that in this system the majority of the information is transmitted very early after stimulus onset (typically between 5 and 30 ms), and since the integration time constant of neurons post-synaptic to the ones analyzed here is likely to be in the range of 10-20 ms, we concentrated on responses taken from short, early, highly informative windows. In particular, we selected the 5-15 and the 5-25 ms post-stimulus windows. The distributions of neural responses in these windows are shown in figure 1 (5-15 ms post-stimulus in figure 1A; and 5-25 ms post-stimulus in figure 1B). In figure 1, the responses were arbitrarily ordered in the x axis to give a monotonically decreasing probability. We compared the real observed distribution with the ME distribution of a given order k ($k = 1, \cdots 3$) obtained with the numerical procedure outlined in the previous section. It is apparent than, for both time windows, the $k = 1$ distribution (which of course corresponds to the case in which all neurons fire independently at fixed stimulus) did not provide a good fit to the data, whereas ME probabilities with $k > 1$ provided a good qualitative agreement. The $k = 3$ distribution provided a small but visible improvement in the goodness of fit with respect to the $k = 2$ distribution.

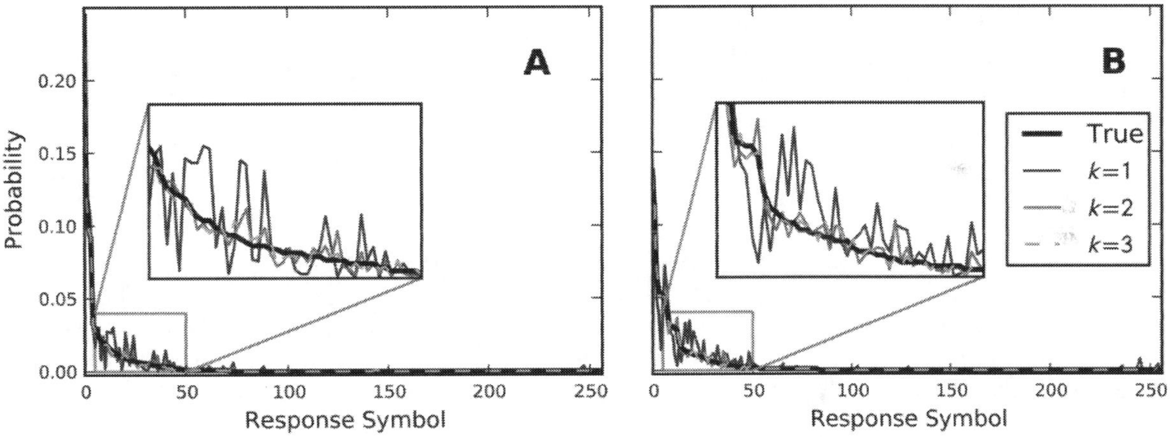

Figure 1. *Maximum entropy models at fixed stimulus.* Here the true distribution (solid black line) and maximum entropy distributions of orders $k = 1, 2, 3$ (solid blue, solid purple, dashed green respectively) are plotted for binary responses to a particular randomly selected group of 8 channels to a 4.3 mm/s stimulus (1200 trials) in the window 5-15 ms (panel **A**) and 5-25 ms (panel **B**) (see text). The response symbols are ordered by decreasing probability value.

To better quantify the goodness of fit of the maximum entropy models across all stimuli and all observed populations of $N = 8$ channels, we repeated the ME calculation of the probability models for each stimulus condition and for 1000 populations made of random combinations of 8 channels from the 24 available. We then computed the probability of rejection of the null hypothesis that the observed distribution comes from the considered ME model using a log-likelihood test (the g-test) [33]. This test produces a chi-square distributed test statistic, similar to the Pearson chi-squared test. Indeed the Pearson chi-squared test was originally developed as an approximation of the log-likelihood ratio test due to the difficulty of calculating logarithms before the advent of computers. The percentage of rejections of the null hypothesis at the

$p = 0.05$ level for the $k = 1$ and $k = 2$ model across all dataset are reported in figure 2 (for the 5-15 ms post-stimulus window) and figure 3 (for the 5-25 ms post-stimulus window). It is apparent that the $k = 1$ model had to be rejected most times, whereas for most stimulus conditions the number of rejections of the $k = 2$ model was below the 5% random level. The $k = 2$ model however failed above chance for longer windows and for the stimulus conditions with more trials available (velocity equal to 1.61, 2.66 and 4.3 mm/s). In contrast, the $k = 3$ model and higher had zero rejections at $p = 0.05$ for both windows and all stimuli.

In sum, the $k = 1$ order model does not explain the response probabiltiies at fixed stimulus in any satisfactory way. The $k = 2$ model explains most (but not all) probability distributions, whereas the $k = 3$ model was sufficient to fully explain the observed response distributions.

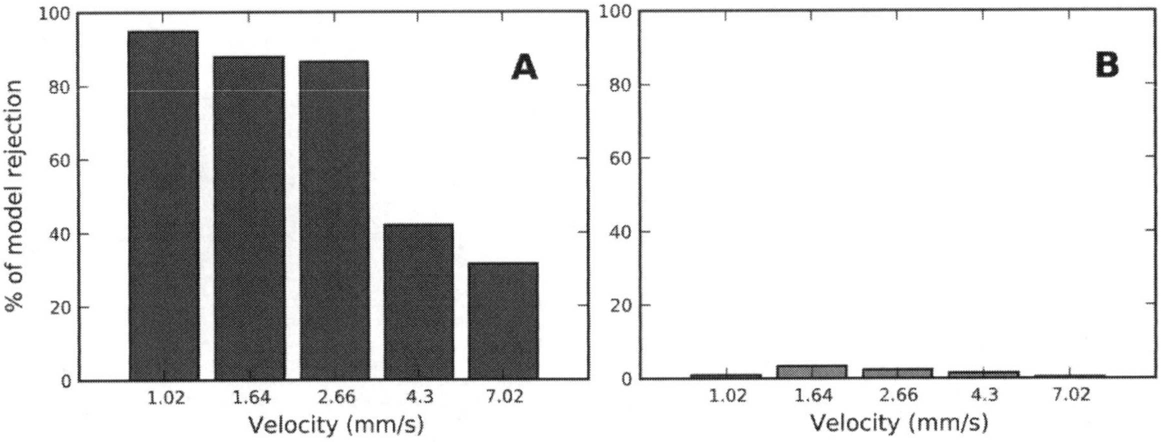

Figure 2. *Goodness of fit measures of maximum entropy distributions for 5-15 ms response window.* 1000 random combinations of 8 channels were chosen from the 24 available and the maximum entropy solutions of different orders were obtained. These were compared to the measured distribution using the log-likelihood g-test (see text) and the percentage of these 1000 trials for which the first order model (panel **A**) and second order model (panel **B**) was rejected (p=0.05) for 5 different fixed stimuli are shown. For the third order model (not shown) no trials were rejected for any stimuli.

3.3. Effect of Interactions on somatosensory information encoding

The above finding that there are significant high order interactions suggests that they should not be neglected in models of information transmission, but it does not tell how much these correlations are important. To quantify this, we next computed the information between the stimulus and the population activity, and we compared it to that derived from the maximum entropy models.

The mutual information between the stimuli and the neural population activity is defined as follows:

$$I(S;X) = H(X) - H(X|S) \tag{3}$$

where $H(X)$ and $H(X|S)$ are the response entropy and noise entropy respectively:

$$H(X) = -\sum_{\mathbf{x}} P(\mathbf{x}) \log_2 P(\mathbf{x}) \tag{4}$$

$$H(X|S) = -\sum_{\mathbf{x},s} P(s)P(\mathbf{x}|s) \log_2 P(\mathbf{x}|s) \tag{5}$$

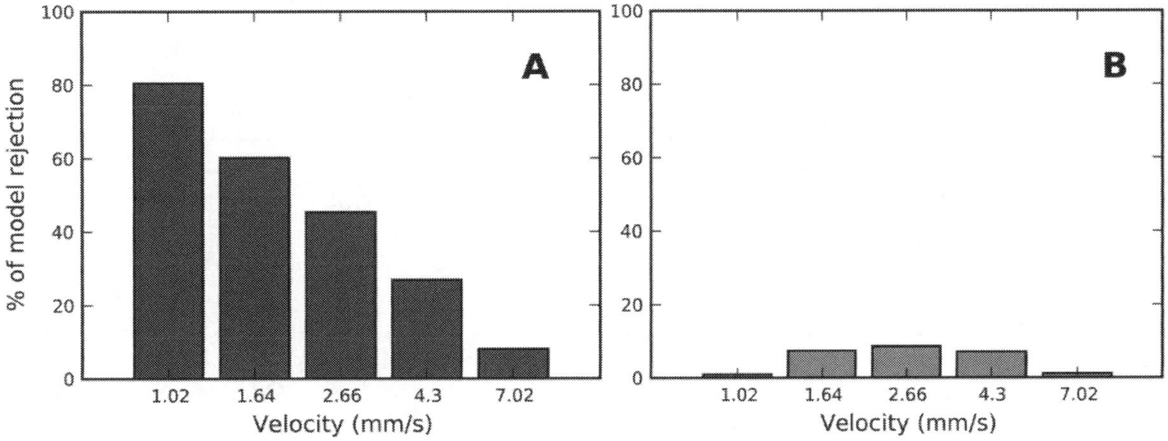

Figure 3. *Goodness of fit measures of maximum entropy distributions for 5-25 ms response window.* 1000 random combinations of 8 channels were chosen from the 24 available and the maximum entropy solutions of different orders were obtained. These were compared to the measured distribution using the log-likelihood g-test (see text) and the percentage of these 1000 trials for which the first order model $k = 1$ (panel **A**) and second order model $k = 2$ (panel **B**) was rejected (p=0.05) for 5 different fixed stimuli are shown. For the third order model (not shown) no trials were rejected for any stimuli.

where in the above $P(\mathbf{x}) = \sum_s P(\mathbf{x}|s)P(s)$. We computed $I(S;X)$ from the the population activity as follows. First, to improve the sampling properties of this data set we pooled the two lowest velocity stimuli (0.15 mm/s and 2.3 mm/s) and the two highest velocity stimuli (29.5 mm/s and 47.7 mm/s) resulting in 11 stimulus classes with a minimum of 600 trials and a maximum of 1400 trials. Then, we empirically measured $P(\mathbf{x}|s)$, $P(s)$ and $P(\mathbf{x})$ and plugged them into the above equations. The limited sampling bias [34] was corrected for using a combination of the shuffled estimator I_{sh} [25, 34] with the Panzeri-Treves analytic correction [35] from the Pyentropy library [29].

We investigated the impact of interactions at a given order k by calculating the mutual information that would result from a system exhibiting the probability distributions obtained from the maximum entropy solution, as follows:

$$I^{(k)}(S;X) = H^{(k)}(X) - H^{(k)}(X|S) \tag{6}$$

where $H^{(k)}(X)$ and $H^{(k)}(X|S)$ are the response and noise entropies respectively of the k-th order maximum entropy model. These entropies are obtained by replacing $P(\mathbf{x}|s)$ and $P(\mathbf{x})$ with $P^{(k)}_{ME}(\mathbf{x}|s)$ and $P^{(k)}_{ME}(\mathbf{x})$ in Eqs. (4,5), where $P^{(k)}_{ME}(\mathbf{x}) = \sum_s P^{(k)}_{ME}(\mathbf{x}|s)P(s)$. Then

$$I^{(k)}(S;X) = \sum_{\mathbf{x},s} P(s)P^{(k)}_{ME}(\mathbf{x}|s)\log_2 \frac{P^{(k)}_{ME}(\mathbf{x}|s)}{P^{(k)}_{ME}(\mathbf{x})} \tag{7}$$

$I_k(S;X)$ was computed from the data as follows. First, and as for $I(S;X)$, we pooled the two lowest velocity stimuli (0.15 mm/s and 2.3 mm/s) and the two highest velocity stimuli (29.5 mm/s and 47.7 mm/s) resulting in 11 stimulus classes. Then, for each stimulus class, we obtained the maximum entropy solution, $P^{(k)}_{ME}(\mathbf{x}|s)$, for each order of interest and for each stimulus conditional response. Then, from each of these stimulus-conditional maximum entropy solutions, we simulated data with the same number of trials as available in the experimental data set (this was different for each stimulus). These trials were generated using inverse transform sampling. This is done to ensure a fair comparison between the measured data and the generated data; any bias effects should affect both equally. The values obtained were averaged over 1000 repetitions to remove any trial to trial variation from the inverse transform sampling step.

Figure 4 shows the effect of including higher order interactions on information. The first order maximum entropy models conveyed significantly higher information than the true probabilities. This means that interactions in this system have mostly a limiting effect on information. The second and third order information was significantly lower than the first order one and was very close to to the true one. We found that correlations of order two gave a good approximation of information carried by the neural population, as they were within 3% and 4% of the true value for windows of 5-15 ms (figure 4A) and 5-25 ms (figure 4B) post-stimulus respectively. Third order correlations gave an even better approximation, within 2% of the true value for both post-stimulus windows (figure 4). This suggests that interactions of order two or higher are sufficient to describe well how neural interactions limit the rate of information transmission.

It is interesting that the mutual information of the system was already well approximated by models containing interactions of up to order 2 or 3. This is significant, since it greatly reduces the parameters required to describe the system. While it is still challenging to sample up to third order marginals reliably in an experiment, it is a much more tractable problem than the case where all orders of interaction must be accurately determined.

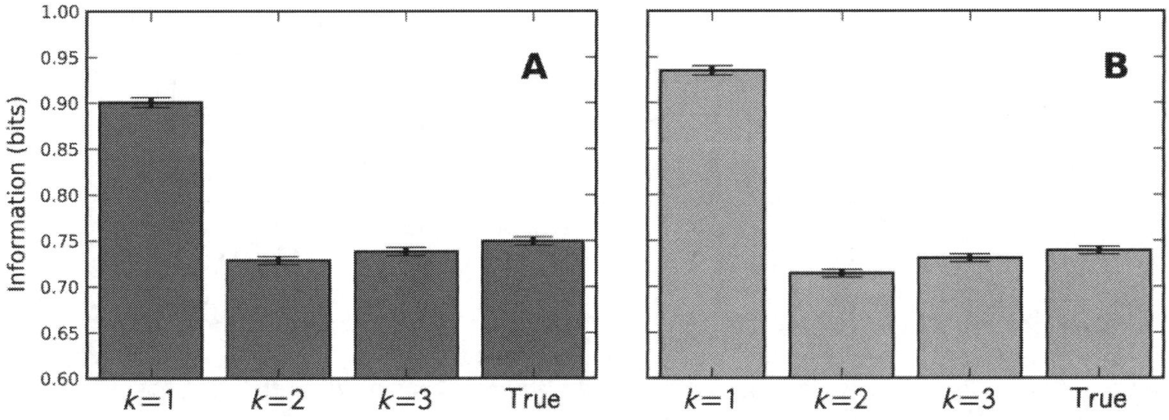

Figure 4. *Effect of interactions on mutual information.* 1000 random combinations of 8 channels were chosen from the 24 available and the maximum entropy solutions of different orders were obtained. These were used to compute the mutual information in a system with only interactions of up to order $k = 1, 2, 3$. The means of these values over the different channel combinations are compared to the mean of the true information for response windows of 5-15 ms (panel **A**) and 5-25 ms (panel **B**). Errorbars show ± 1 s.e.m.

4. Discussion

In this paper, we used maximum-entropy techniques to investigate the presence of interactions of order higher than two in the rat somatosensory cortex, and to evaluate their effect on the information that somatosensory cortical populations carry about the velocity of whisker stimulation. Considering population responses made of spiking activity collected from up to 8 simultaneously recorded locations, we found that interactions of order two were sufficient to explain a large amount of observed stimulus-response distributions, but not all of them. Triple-wise interactions were necessary to fully explain the data. We then used Shannon information to compute the impact of high order correlations on the amount of somatosensory information transmitted by the neural population. We found that correlations of order two gave a good approximation of information carried by the neural population, within 4% of the true value. Third order correlations did even better, proving an approximation within 2% of the true value of information carried by neural activity. In the following, we briefly discuss the implication of our results, as well as differences and similarities with respect to previous work.

4.1. Interactions of order higher than two are present in cortical populations

A first result is that pairwise interactions were not enough to fully explain the observed stimulus-response distributions, thereby confirming that higher order interactions between groups of neurons are indeed a part of cortical dynamics, as suggested by earlier studies demonstrating the existence of multineuronal firing patterns in a variety of experimental preparations [4, 24, 36–40]. An important and still open question is what may be the potential function of high order interactions. Modeling work suggests that higher order interactions may have a role in regularizing the overall rate of discharge of the population, for example by avoiding concentration of firing on the maximally firing state [41]. It would be interesting for future research to understand whether high order interaction serve this purpose in real local cortical networks.

4.2. Different measures of the impact of interactions at a given order

A series of recent studies [17–19] quantified the quality of pairwise models by using a measure called the fraction of full network information. The fraction of full network information that is captured by a second order ME model is defined [18, 42] as the the ratio between the reduction of total entropy accounted for by all interactions of order > 2 (computed as the difference between the pairwise maximum entropy and the the total entropy taking into account all correlations) and the reduction of entropy due to all interactions of any order (computed as the difference between the entropy of the independent probability model and the the total entropy taking into account all correlations). This entropy-based metric has many interesting properties and advantages, discussed in [42]. In particular, given that entropy is a measure of variability of responses, this metric is useful to investigate whether pairwise interactions are enough to describe how interactions shape the variability of neural responses. A novelty of our work, with respect to several previous studies, is that we introduced and used some new metrics to assess the quality of pairwise models in describing the response probabilities. These metrics are potentially useful because they allow addressing some questions which are in principle different from the ones that can be addressed using the fraction of network information, as we elaborate next.

The fraction of full network information cannot be used to clarify whether or not high order interactions exist and are statistically significant. This is because, as discussed in [22], this metric can be relatively small even the pairwise and the true probabilities have very different shapes. To overcome this potential problem, here we introduced a different procedure to assess the goodness of the pairwise model. This simple procedure, based on fitting ME distributions to the data numerically using Amari's formalism [26] and by assessing the difference between ME and true distribution using a g-test statistic, revealed that interactions of order higher than two are significant in the analyzed dataset.

Moreover, in this study we were interested in whether pairwise interactions were sufficient to describe how the population of neurons encodes information about the stimulus. This question cannot possibly be addressed by quantifying the fraction of network information. This is because of two reasons. The first is that the fraction of network information measures the reduction of network variability specifically attributable to correlations up to a given order, and the variability of the population response is not equivalent to the information about the stimuli carried by the population. The second is that typically mutual information is smaller than both the response and the noise entropy: an impact that may be proportionally small for entropy may be proportionally much larger when considering information. Thus, we had to introduce a new metric based on comparing the amount of Shannon information about the stimuli which is obtained when considering either the full response model or the one taking into account only correlations up to a given order k. The novel finding obtained using this metric was that interaction of order two gave a a very good (but not perfect) approximation of information carried by the neural population. These results extended a previous attempt from our group [20] to measure Shannon information from pairwise models which made the simplifying assumptions that all neurons and interactions are perfectly homogenous. This simplifying assumption was no longer used or necessary in the present work.

Another potentially interesting metric is to consider the information lost by a downstream decoder if it was to ignore higher order correlation when decoding. This metric has recently been investigated by Oizumi and colleagues [43].

4.3. Scaling of the effect of higher order with the population size

In this work, we decided to analysis populations responses made of spiking activity collected from up to 8 simultaneously recorded locations. Fixing the population size to 8 was somewhat arbitrary, but it was empirically the largest population size that allowed us to compute reliably, with the present methodology, both the goodness of fit of ME models and the information carried by the neuronal population. Analyzing 8 simultaneous recording sites was a significant improvement with respect to previous analyses based on Shannon information (which were usually limited to two or three cells), but is of course still far from enough to understand the behavior of the cortical network. A crucial question for the interpretation of these results is how the significance of interaction of order higher than two, and the information they carry, scales when increasing the population size. In fact, it would be unwise to assume that the behavior of these measures extrapolates smoothly from small to large population sizes [22]. Measuring information from large populations will require to further improve the techniques for estimating information from limited sampling [34], which is a topic of current research for neurostatisticians.

Acknowledgements

We are grateful to R Senatore, R Petersen and MA Montemurro for many useful discussions and interactions. This research was conducted under the framework of the BMI Project at the Department of Robotics, Brain and Cognitive Sciences of the Italian Institute of Technology. RAAI was supported by the EPSRC CARMEN grant.

References

[1] Nirenberg S H and Victor J D 2007 *Curr. Opin. Neurobiol.* **17** 397–400

[2] Li C L 1959 *Science* **129** 783–4

[3] Mastronarde D 1983 *J. Neurophysiol.* **49** 303–24

[4] Abeles M, Bergman H, Margalit E and Vaadia E 1993 *J. Neurophysiol.* **70** 1629–38

[5] Luczak A, Bartho P and Harris K D 2009 *Neuron* **62** 413–25

[6] Abbott L and Dayan P 1999 *Neural Comput.* **11** 91–101

[7] Oram M W, Földiák P, Perrett D I, Oram M W and Sengpiel F 1998 *Trends Neurosci.* **21** 259–65

[8] Pola G, Thiele A, Hoffmann K P and Panzeri S 2003 *Network: Comp. Neural.* **14** 35–60

[9] von der Malsburg C 1999 *Neuron* **24** 95–104

[10] Chase S and Young E 2007 *P. Natl. Acad. Sci. USA* **104** 5175–80

[11] Gollisch T and Meister M 2008 *Science* **319** 1108–11

[12] Schneidman E, Bialek W and Berry M J 2003 *J. Neurosci.* **23** 11539–53

[13] Zohary E, Shadlen M N and Newsome W T 1994 *Nature* **370** 140–43

[14] Mazurek M and Shadlen M 2002 *Nat. Neurosci.* **5** 463–71

[15] Bohte S, Spekreijse H and Roelfsema P 2000 *Neural Comput.* **12** 153–79

[16] Nakahara H and Amari S I 2002 *Neural Comput.* **14** 2269–316

[17] Shlens J, Field G D, Gauthier J L, Grivich M I, Petrusca D, Sher A, Litke A M and Chichilnisky E J 2006 *J. Neurosci.* **26** 8254–66

[18] Schneidman E, Berry II M J, Segev R and Bialek W 2006 *Nature* **440** 1007–12

[19] Tang A, Jackson D, Hobbs J, Chen W, Smith J L, Patel H, Prieto A, Petrusca D, Grivich M I, Sher A, Hottowy P, Dabrowski W, Litke A M and Beggs J M 2008 *J. Neurosci.* **28** 505–18

[20] Montani F, Ince R A A, Senatore R, Arabzadeh E, Diamond M and Panzeri S 2009 *Phil. Trans. R. Soc.* A **367** 3297–310

[21] Roudi Y, Tyrcha J and Hertz J 2009 *Phys. Rev.* E **79** 051915

[22] Roudi Y, Nirenberg S and Latham P E 2009 *PLoS Comput. Biol.* **15** e1000380

[23] Jaynes E 1957 *Phys. Rev.* **106** 620–30

[24] Martignon L, Deco G, Laskey K, Diamond M, Freiwald W and Vaadia E 2000 *Neural Comput.* **12** 2621–53

[25] Montemurro M A, Senatore R and Panzeri S 2007 *Neural Comput.* **19** 2913–57

[26] Amari S I 2001 *IEEE Trans. Inform. Theory* **47** 1701–11

[27] Cover T M and Thomas J A 2006 *Elements of Information Theory, 2nd Ed.* (John Wiley & sons)

[28] Amari S I and Nagaoka H 2000 *Methods of Information Geometry* (Oxford University Press)

[29] Ince R A A, Petersen R S, Swan D C and Panzeri S 2009 *Front. Neuroinformatics* **3** 4

[30] Arabzadeh E, Petersen R S and Diamond M E 2003 *J. Neurosci.* **23** 9146–54

[31] Arabzadeh E, Panzeri S and Diamond M E 2004 *J. Neurosci.* **24** 6011–20

[32] Petersen R S and Diamond M E 2000 *J. Neurosci.* **20** 6135–43

[33] Woolf B 1972 *Ann. Hum. Genet.* **35** 397–409

[34] Panzeri S, Senatore R, Montemurro M A and Petersen R S 2007 *J. Neurophysiol.* **98** 1064–72

[35] Panzeri S, Schultz S R, Treves A and Rolls E T 1999 *Proc. R. Soc.* B **266** 1001–12

[36] Schnitzer M J and Meister M 2003 *Neuron* **37** 499–511

[37] Schrader S, Gruen S, Diesmann M and Gerstein G 2008 *J. Neurophysiol.* **100** 2165–76

[38] Harris K D 2005 *Nat. Rev. Neurosci.* **6** 399–407

[39] Riehle A, Gruen S, Diesmann M and Aertsen A 1997 *Science* **278** 1950–53

[40] Gruen S, Diesmann M and Aertsen A 2002 *Neural Comput.* **14** 43–80

[41] Amari S I, Nakahara H, Wu S and Sakai Y 2003 *Neural Comput.* **15** 127–42

[42] Schneidman E, Still S, Berry M J and Bialek W 2003 *Phys. Rev. Lett.* **91** 238701

[43] Oizumi M, Ishii T, Ishibashi T, Hosoya T and Okada M 2009 *Advances in Neural Information Processing Systems* vol 21 ed Koller D, Schuurmans D, Bengio Y and Bottou L (Cambridge: MIT press) pp 1225–32

Capacity of a single spiking neuron

Shiro Ikeda[1] and Jonathan H Manton[2]

[1] The Institute of Statistical Mathematics, Tokyo, 106-8569, Japan
[2] The University of Melbourne, Victoria 3010, Australia

E-mail: `shiro@ism.ac.jp`,

Abstract. It is widely believed the neurons transmit information in the form of spikes. Since the spike patterns are known to be noisy, the neuron information channel is noisy. We have investigated the channel capacity of this "Spiking neuron channel" for both of the "temporal coding" and the "rate coding," which are two main coding considered in the neuroscience [1, 2]. As the result, we've proved that the distribution of inputs, which achieves the channel capacity, is a discrete distribution with finite mass points for temporal and rate coding under a reasonable assumption. In this draft, we show the details of the proof.

1. Introduction

Neurons transfer information to other neurons in the form of spike trains. Although precise control of spike timing is important for reliable information transfer, a lot of studies revealed that spike patterns are noisy. When a communication channel is noisy, the rate at which information can be transmitted reliably through the channel is limited. The upper bound on the rate is the channel capacity [3]. We have studied the capacity of a single neuron [1, 2] for two types of coding, temporal and rate coding. The temporal coding uses the inter-spike intervals (ISIs) to code information while the rate coding uses the number of spikes in a fixed interval.

The channel model is deeply related to the noise of ISIs. Many works reported [4–6] that the statistical properties of ISIs are similar to the gamma distribution. We employ this model, and ISIs are modelled with a gamma distribution. The capacity is defined as the supremum of mutual information over possible input distributions. We pose a natural assumption on the input distributions, and under the assumption, we proved the capacity of each coding is achieved by a discrete distribution which has only finite mass points [1, 2]. Although the proof for each coding shares the steps with other studies of information theory [7–10], the neuron channel is special and we have to prove each step. In this draft, we provide the details of the proof. Our result shows that the information is maximally transmitted through a single neuron when the inputs to the neuron have only a fixed number of modes.

The problem is formulated mathematically in section 2 and the discreteness for each coding is proved in section 3. Section 4 concludes the paper.

2. Single Neuron Channel

2.1. Communication Channel and Capacity

Let X be the input to a noisy channel and Y be the output. In the following, we assume $X \in \mathcal{X} \subseteq \mathbb{R}$ is a one-dimensional stochastic variable and let $F(\cdot)$ be a cumulative distribution function of X. Communication channel is defined as a stochastic model described as $p(y|x)$ and the mutual information is defined as

$$I(X;Y) = \int_{x\in\mathcal{X}}\int_{y\in\mathcal{Y}} p(y|x)\log\frac{p(y|x)}{p(y)}d\mu(y)dF(x), \quad \text{where} \quad p(y) = \int_{x\in\mathcal{X}} p(y|x)dF(x). \tag{1}$$

Here, $\mu(y)$ denotes the measure of $y \in \mathcal{Y}$. Since the channel is defined as $p(y|x)$, $I(X;Y)$ is a functional of $F(\cdot)$ and we denote it as $I(F)$.

Let $\mathcal{F}$ be the set of cumulative distribution functions of X. The channel capacity is defined as

$$C = \sup_{F\in\mathcal{F}} I(F). \tag{2}$$

For a noisy channel, one interesting fundamental problem is to compute the capacity C. Another interesting problem is to obtain the distribution, if it exists, which achieves the capacity.

2.2. Single Neuron: Channel and Coding

It has been reported that a gamma distribution is a suitable model to describe the stochastic nature of ISIs [4, 6]. The gamma distribution has two positive real parameters which are the shape parameter κ and the scale parameter θ. From some studies [11], κ of individual neuron appears to be constant (the value of κ may depends on the type of neuron), while θ changes dynamically over time. Let T denote an ISI, which is a stochastic variable following a gamma distribution. We assume κ of each neuron is fixed and known and the distribution of each ISI is independent.

Under these assumption, the scale parameter θ is the only variable parameter which plays the role of input, that is, X in §.2.1. The density function of t is

$$p(t|\theta;\kappa) = \left(\frac{t^{\kappa-1}}{\theta^\kappa}\right)\frac{\exp[-t/\theta]}{\Gamma(\kappa)}, \quad \kappa,\theta > 0,\ t \geq 0.$$

where we denote it as $p(t|\theta;\kappa)$ to show θ is a stochastic variable and κ is a parameter and $\Gamma(\kappa)$ is a gamma function defined as

$$\Gamma(z) = \int_0^\infty t^{z-1}e^{-t}dt.$$

The gamma distribution is an exponential family,

$$p(t|\theta;\kappa) = \exp\left[-\frac{1}{\theta}t + (\kappa-1)\log t - \log\Gamma(\kappa) - \kappa\log\theta\right]. \tag{3}$$

Next, let us consider the family of all the possible distributions of input θ. Noting that ISI is positive and is not infinite if the neuron is active, it is natural to assume that the average ISI, which depends on θ and κ, is limited between a_0 and b_0, that is,

$$a_0 \leq \overline{T} = \kappa\theta \leq b_0, \quad \text{where} \quad 0 < a_0 < b_0 < \infty.$$

Thus, θ is bounded in $\Theta(\kappa) = \{\theta \mid a(\kappa) \leq \theta \leq b(\kappa)\}$, where $a(\kappa)$ and $b(\kappa)$ are defined as

$$a(\kappa) = a_0/\kappa, \quad b(\kappa) = b_0/\kappa.$$

Let us define $F(\theta)$ as the cumulative distribution function of θ and $\mathcal{F}$ as the set of all possible $F(\theta)$, that is

$$\mathcal{F} = \{F : \mathbb{R} \to [0,1] \mid F(\theta) = 0, (\forall\theta < a),\ F(\theta) = 1, (\forall\theta > b)\}. \tag{4}$$

Next, let us consider what is Y, that is, "the output of the channel" of a neuron communication channel. There are mainly two different ideas in neuroscience. One is that Y is ISI, T, itself. This is called "temporal coding." The other is that Y is the rate, which is the number of spikes in fixed time intervals. This is called "rate coding"[1]. How to encode the input θ to the neuron channel depends on which coding is used. For the temporal coding, θ is fixed during the interval t while θ is fixed during Δ for the rate coding.

Temporal coding In temporal coding, received information is T. For a $F \in \mathcal{F}$, we define the marginal distribution as

$$p(t; F, \kappa) = \int_a^b p(t|\theta;\kappa)dF(\theta) \tag{5}$$

where $p(t|\theta;\kappa)$ is defined in eq.(3). The mutual information of T and θ is defined as

$$I_T(F) = \int_a^b i_T(\theta; F)dF(\theta), \quad \text{where} \quad i_T(\theta;F) = \int_0^\infty p(t|\theta;\kappa)\log\frac{p(t|\theta;\kappa)}{p(t;F,\kappa)}dt. \tag{6}$$

Let us define $g(t; F, \kappa)$ and rewrite $p(t; F, \kappa)$ as follows

$$g(t;F,\kappa) = \int_a^b \frac{\exp[-t/\theta]}{\theta^\kappa}dF(\theta), \quad p(t;F,\kappa) = \frac{t^{\kappa-1}}{\Gamma(\kappa)}g(t;F,\kappa). \tag{7}$$

The mutual information $I_T(F)$ is rewritten as follows

$$I_T(F) = h_T(F;\kappa) - \kappa\, h_{T|\theta}(F;\kappa) - \kappa, \quad \text{where}$$
$$h_T(F;\kappa) = -\int_0^\infty p(t;F,\kappa)\log g(t;F,\kappa)dt, \quad h_{T|\theta}(F;\kappa) = \int_a^b \log\theta dF(\theta).$$

Hence, the capacity per channel use or equivalently per spike is defined as

$$C_T = \sup_{F\in\mathcal{F}} I_T(F) = \sup_{F\in\mathcal{F}}\langle h_T(F;\kappa) - \kappa\, h_{T|\theta}(F;\kappa)\rangle - \kappa.$$

The capacity C_T and the distribution which achieves C_T will be studied in the next section.

Rate coding In rate coding, a time window is set and the spikes in the interval is counted. Let us denote the interval and the rate as Δ and R, respectively, and define the distribution of R as $p(r|\theta;\kappa,\Delta)$. The form of the distribution of R is shown in the following lemma.

Lemma 1. *The distribution* $p(r|\theta;\kappa,\Delta)$ [2] *has the following form*

$$p(r|\theta;\kappa,\Delta) = P(r\kappa, \Delta/\theta) - P((r+1)\kappa, \Delta/\theta), \quad r \in \mathbb{Z}^* \ \textit{(nonnegative integers)}, \tag{8}$$

where $P(\alpha, x)$ *is the regularized incomplete gamma function*

$$P(0,x) = 1, \quad P(\alpha,x) = \frac{1}{\Gamma(\alpha)}\int_0^x t^{\alpha-1}e^{-t}dt, \quad \text{for} \quad \alpha, x > 0.$$

[1] It seems the term "modulation" is more suitable to the above definition. However, we follow the standard usage of the neuroscience community

[2] The same distribution is discussed in [12].

For an $F \in \mathcal{F}$, let us define the following marginal distribution $p(r; F, \kappa, \Delta)$

$$p(r; F, \kappa, \Delta) = \int_a^b p(r|\theta; \kappa, \Delta) dF(\theta).$$

The mutual information of R and θ is defined as

$$I_R(F) = \int_a^b i_R(\theta, F) dF(\theta), \quad \text{where} \quad i_R(\theta, F) = \sum_{r=0}^{\infty} p(r|\theta; \kappa, \Delta) \log \frac{p(r|\theta; \kappa, \Delta)}{p(r; F, \kappa, \Delta)}. \tag{9}$$

Hence, the capacity per channel use or equivalently per Δ is defined as

$$C_R = \sup_{F \in \mathcal{F}} I_R(F).$$

The capacity C_R and the distribution which achieves C_R will be studied in the next section.

3. Theoretical Studies

The cumulative distribution $F \in \mathcal{F}$ is a right-continuous non-decreasing function on a interval Θ. Thus, θ can be a discrete or continuous random variable over Θ. In this section, the capacity achieving input distribution of a single neuron channel is proved to be a discrete distribution with finite mass points, for both temporal and rate coding.

For some channels, the capacity achieving input distributions have been shown to be discrete under some conditions [7–9, 13, 14]. The neuron channel with temporal coding is different from those and the proof must be provided independently.

3.1. Steps to Prove the Discreteness of the Capacity Achieving Distribution

We first show the common steps of the proof for the discreteness of the capacity achieving input distributions. In the following, results of optimization theory and probability theory will be used. Suppose X is a normed linear space. In optimization theory, the space of all bounded linear functionals of X is called the normed dual of X and is denoted X^*. The weak* convergence is defined as follows.

Definition 1. *A sequence $\{x_n^*\}$ in X^* is said to converge weak* to the element x^* if for every $x \in X$, $x_n^*(x) \to x^*(x)$. In this case we write $x_n^*(x) \xrightarrow{w^*} x^*(x)$ ([15], 5.10).*

If X is the real normed linear space of all bounded continuous functions on $\mathbb{R}$, X^* includes the set of all probability measures, and it is clear that "weak convergence" of probability measures is "weak* convergence" on X^*. The following theorem is used to prove the existence and the uniqueness of the capacity achieving input distribution.

Theorem 1. *Let J be a weak* continuous real-valued functional on a weak* compact subset S of X^*. Then J is bounded on S and achieves its maximum on S. If S is convex and J is strictly concave, then the following maximum is achieved by a unique x^* in S.*

$$C = \max_{x^* \in S} J(x^*).$$

Proof. See [15], 5.10, [9] and [13]. □

From the above discussion, $\mathcal{F}$ in eq.(4) is a subset of X^*. It is clear that $\mathcal{F}$ is convex. Thus, if $\mathcal{F}$ is weak* compact and $I_T(F)$ (or $I_R(F)$) is a weak* continuous function on $F \in \mathcal{F}$ and strictly concave in $\mathcal{F}$, the capacity is achieved by a unique distribution F_0 in $\mathcal{F}$. This is the first step of the proof. The following proposition states $\mathcal{F}$ is compact.

Proposition 1. *$\mathcal{F}$ in eq.(4) is compact in the Lévy metric topology.*

Proof. For the proof of compactness, see [7] (proof of proposition 1), the proof is a direct application of the Helly's compactness theorem ([16], section X). □

The Kuhn-Tucker (K-T) condition on the mutual information is used for the next step of the proof. Before showing the condition, let us define the weak differentiability.

Definition 2. *Let J be a function on a convex set $\mathcal{F}$. Let F_0 be a fixed element of $\mathcal{F}$, and $\eta \in [0,1]$. Suppose there exists a map $J'_{F_0} : \mathcal{F} \to \mathbb{R}$ such that*

$$J'_{F_0}(F) = \lim_{\eta \downarrow 0} \frac{J((1-\eta)F_0 + \eta F) - J(F_0)}{\eta}, \qquad F \in \mathcal{F}.$$

Then J is said to be weakly differentiable in $\mathcal{F}$ at F_0 and $J'_{F_0}(F)$ is the weak derivative in $\mathcal{F}$ at F_0. If J is weakly differentiable in $\mathcal{F}$ at F_0 for all $F \in \mathcal{F}$, J is weakly differentiable in $\mathcal{F}$.

And the K-T condition is described as follows,

Proposition 2. *Assume J is a weakly differentiable, concave functional on a convex set $\mathcal{F}$. If J achieves its maximum on $\mathcal{F}$ at F_0, then a necessary and sufficient condition for F_0 to attain the maximum is to satisfy the following inequality for all $F \in \mathcal{F}$*

$$J'_{F_0}(F) \leq 0.$$

Proof. See Proposition 1 in [7]. □

If $I_T(F)$ (or $I_R(F)$) is weakly differentiable, the K-T condition is derived with the theorem. Finally, the discreteness is proved by deriving a contradiction based on the K-T condition and the assumption that F_0 has infinite mass points as its support. In order to show the discreteness of the capacity achieving input distribution for temporal and rate codings, the following properties must be shown.

(i) $I_T(F)$ and $I_R(F)$ are weak* continuous on $\mathcal{F}$ and strictly concave.

(ii) $I_T(F)$ and $I_R(F)$ are weakly differentiable.

Then, the K-T condition is derived and the discreteness will be checked.

3.2. Discreteness of the Capacity Achieving Distribution for Temporal Coding

In this subsection, the capacity achieving input distribution for temporal coding is shown to be a discrete distribution with a finite number of points. Let us start with the following lemma.

Lemma 2. *$I_T(F)$ in eq.(6) is a weak* continuous function on $F \in \mathcal{F}$ and strictly concave in $\mathcal{F}$.*

Proof. $I_T(F)$ is weak* continuous if the following relation holds,

$$F_n \overset{w^*}{\to} F \implies I_T(F_n) \to I_T(F), \tag{10}$$

since $I_T(F) = h_T(F;\kappa) - \kappa\, h_{T|\theta}(F;\kappa) - \kappa$, more precisely,

$$F_n \overset{w^*}{\to} F \implies h_T(F_n;\kappa) \to h_T(F;\kappa) \ \text{ and } \ h_{T|\theta}(F_n;\kappa) \to h_{T|\theta}(F;\kappa).$$

$h_{T|\theta}(F_n;\kappa) \to h_{T|\theta}(F;\kappa)$ holds since $h_{T|\theta}(F_n;\kappa) = \int_a^b \log\theta dF_n(\theta)$ and $\log\theta$ is a bounded continuous function for $\theta \in \Theta$.

Next, we show the following equalities

$$\lim_n h_T(F_n;\kappa) = -\lim_n \int_0^\infty p(t;F_n,\kappa)\log g(t;F_n,\kappa)dt = -\int_0^\infty \lim_n p(t;F_n,\kappa)\log g(t;F_n,\kappa)dt \quad (11)$$

$$= -\int_0^\infty p(t;F,\kappa)\log g(t;F,\kappa)dt = h_T(F;\kappa) \quad (12)$$

The interchange of integral and limit in eq.(11) is justified as follows. From eqs.(5) and (7), $p(t;F,\kappa)$ and $g(t;F,\kappa)$ are bounded as follows.

$$\frac{t^{\kappa-1}}{b^\kappa}\frac{e^{-t/a}}{\Gamma(\kappa)} < p(t;F_n,\kappa) < \frac{t^{\kappa-1}}{a^\kappa}\frac{e^{-t/b}}{\Gamma(\kappa)}, \qquad -\frac{t}{a} - \kappa\log b < \log g(t;F_n,\kappa) < -\frac{t}{b} - \kappa\log a, \quad (13)$$

$p(t;F_n,\kappa)\log g(t;F_n,\kappa)$ is bounded for all F_n with finite A_1 and A_2 as follows

$$\left|p(t;F_n,\kappa)\log g(t;F_n,\kappa)\right| < A_1 t^{\kappa-1}e^{-t/b} + A_2 t^\kappa e^{-t/b}. \quad (14)$$

RHS is integrable as

$$\int_0^\infty \left[A_1 t^{\kappa-1}e^{-t/b} + A_2 t^\kappa e^{-t/b}\right]dt = \Gamma(\kappa)b^\kappa(A_1 + \kappa b A_2).$$

Since eq.(14) is bounded from above with an integrable function, eq.(11) is justified by the Lebesgue dominated convergence theorem. Since $p(t|\theta;\kappa)$ and $\exp[-t/\theta]/\theta^\kappa$ are continuous bounded functions of $\theta \in \Theta$, $p(t;F,\kappa)$ and $g(t;F,\kappa)$ are continuous function on F, hence $p(t;F_n,\kappa)\log g(t;F_n,\kappa)$ is also continuous for every $F_n \in \mathcal{F}$. These arguments justify eq.(12) and eq.(10) is justified.

$I_T(F)$ is also strictly concave following the proof of lemma 2 in [9]. □

Lemma 2 and theorem 1 imply the capacity for temporal coding is achieved by a unique distribution in $\mathcal{F}$. In order to show it is a discrete distribution, the following lemma and corollary are used.

Lemma 3. *$I_T(F)$ in eq.(6) is weakly differentiable in $\mathcal{F}$. The weak derivative at $F_0 \in \mathcal{F}$ has the form*

$$I'_{T,F_0}(F) = \int_a^b i_T(\theta;F_0)dF - I_T(F_0). \quad (15)$$

Proof. Let us define F_η and rewrite $i_T(\theta;F)$ in eq.(6) as follows

$$F_\eta = (1-\eta)F_0 + \eta F, \quad i_T(\theta;F) = -\kappa\log\theta - \int_0^\infty p(t|\theta;\kappa)\log g(t;F,\kappa)dt.$$

Then

$$I_T(F_\eta) - I_T(F_0) = \int_a^b i_T(\theta;F_\eta)dF_\eta - \int_a^b i_T(\theta;F_0)dF_0$$

$$= \eta\left[\int_a^b i_T(\theta;F_\eta)dF - \int_a^b i_T(\theta;F_\eta)dF_0\right] \quad (16)$$

$$+ \int_a^b \left[i_T(\theta;F_\eta) - i_T(\theta;F_0)\right]dF_0. \quad (17)$$

The weak derivative of $I_T(F)$ at F_0 is defined as $I'_{T,F_0}(F) = \lim_{\eta\downarrow 0}\left(I_T(F_\eta) - I_T(F_0)\right)/\eta$. By dividing the term in eq.(16) with η and by taking $\eta \downarrow 0$, it becomes

$$\int_a^b i_T(\theta; F_0)dF - \int_a^b i_T(\theta; F_0)dF_0 = \int_a^b i_T(\theta; F_0)dF - I_T(F_0).$$

By noting $g(t; F_\eta, \kappa) = (1-\eta)\, g(t; F_0, \kappa) + \eta\, g(t; F, \kappa)$, the term in eq.(17) becomes 0. This proves $I_T(F)$ is weakly differentiable. □

Corollary 1. *Let E_0 denote the points of increase of F_0 on $\theta \in \Theta$. F_0 is optimal if and only if*

$$\begin{aligned} i_T(\theta; F_0) &\le I_T(F_0), \quad \forall\theta \in \Theta \\ i_T(\theta; F_0) &= I_T(F_0), \quad \forall\theta \in E_0. \end{aligned} \tag{18}$$

Proof. This is proved following the same steps in [7] (Corollary 1) with eq.(15). □

The main result of this subsection is summarized in the following theorem.

Theorem 2. *Under the constraint $\theta \in \Theta$, the channel capacity of a single neuron channel with temporal coding is achieved by a discrete distribution with a finite number of mass points.*

Proof. The extension of $i_T(\theta; F_0)$ to the complex plain z is analytic for $\mathrm{Re}\, z > 0$

$$i_T(z; F_0) = -\kappa \log z - \int_0^\infty p(t|z; \kappa) g(t; F_0, \kappa)dt.$$

If E_0 in corollary 1 has infinite points, since Θ is bounded and closed, E_0 has a limit point. Hence, from corollary 1, the identity theorem implies $i_T(z; F_0) = I_T(F_0)$ for the region $\mathrm{Re}\, z > 0$. This region includes positive real line $\mathbb{R}^+$ and

$$-\int_0^\infty p(t|\theta; \kappa) \log g(t; F_0, \kappa)dt = \kappa \log\theta + I_T(F_0), \quad \theta \in \mathbb{R}^+ \tag{19}$$

is implied. The LHS is bound as follows (eq.(13)).

$$\frac{1}{b}\int_0^\infty t\, p(t|\theta; \kappa)dt + \kappa\log a \le -\int_0^\infty p(t|\theta; \kappa)\log g(t; F_0, \kappa)dt \le \frac{1}{a}\int_0^\infty t\, p(t|\theta;\kappa)dt + \kappa \log b. \tag{20}$$

Since the expectation of T w.r.t $p(t|\theta; \kappa)$ is $\kappa\theta$, eq.(20) shows the LHS of eq.(19) grows linearly with θ. Since the RHS increases only with $\log\theta$, eq.(19) cannot hold for all $\theta \in \mathbb{R}^+$. This is the contradiction and the optimal distribution has a finite number of mass points. □

3.3. Discreteness of the Capacity Achieving Distribution for Rate Coding

The capacity achieving input distribution for rate coding is shown to be a discrete distribution with a finite number of points. In [8], it has been proved that the capacity achieving input distribution of a Poisson channel under peak and average power constraints is a discrete distribution with a finite point of supports. Since $\theta \in \Theta$ is a peak constraint, this directly proves the case $\kappa = 1$. For $\kappa \ne 1$ further study is needed. First we the following proposition

Proposition 3. *The expectation of R with respect to $p(r|\theta; \kappa, \Delta)$ is finite.*

Proof of proposition 3. The expectation of R is

$$\overline{R}_{\kappa,\Delta_\theta} = \sum_{r=0}^{\infty} r\ p(r|\theta;\kappa,\Delta) = \sum_{r=1}^{\infty} P(r\kappa,\Delta_\theta) = e^{-\Delta_\theta}\sum_{r=1}^{\infty}\sum_{i=0}^{\infty}\frac{\Delta_\theta^{r\kappa+i}}{\Gamma(r\kappa+i+1)},$$

where $\Delta_\theta = \Delta/\theta$. Since $P(\alpha,x)$ is a strictly decreasing function of α for $\alpha>0$, $x>0$, if $\kappa\geq 1$

$$P(r\kappa,\Delta_\theta)\leq P(r\lfloor\kappa\rfloor,\Delta_\theta),\quad r\in\mathbb{Z}^+,$$

Thus, the upper bound is given as

$$\overline{R}_{\kappa,\Delta_\theta}\leq e^{-\Delta_\theta}\sum_{r=1}^{\infty}\sum_{i=0}^{\infty}\frac{\Delta_\theta^{r\lfloor\kappa\rfloor+i}}{\Gamma(r\lfloor\kappa\rfloor+i+1)}\leq e^{-\Delta_\theta}\sum_{r=1}^{\infty}\sum_{i=0}^{\infty}\frac{\Delta_\theta^{r+i}}{\Gamma(r+i+1)}=\overline{R}_{1,\Delta_\theta}=\Delta_\theta,$$

where $R_{1,\Delta_\theta}=\Delta_\theta$ holds from the fact that $p(r|\theta;1,\Delta_\theta)$ is a Poisson distribution. For $\kappa<1$, $P(r\kappa,\Delta_\theta)\leq P(\lfloor r\kappa\rfloor,\Delta_\theta)$ holds and $\overline{R}_{\kappa,\Delta_\theta}$ is bounded as follows

$$\overline{R}_{\kappa,\Delta_\theta}<e^{-\Delta_\theta}\sum_{r=1}^{\infty}\sum_{i=0}^{\infty}\frac{\Delta_\theta^{\lfloor r\kappa\rfloor+i}}{\Gamma(\lfloor r\kappa\rfloor+i+1)}\leq\lceil\frac{1}{\kappa}\rceil e^{-\Delta_\theta}\sum_{r=1}^{\infty}\sum_{i=0}^{\infty}\frac{\Delta_\theta^{r+i}}{\Gamma(r+i+1)}+\lfloor\frac{1}{\kappa}\rfloor\leq\lceil\frac{1}{\kappa}\rceil(\Delta_\theta+1).$$

□

Next, we show the following lemma.

Lemma 4. *$I_R(F)$ is a weak* continuous function on $F\in\mathcal{F}$ and strictly concave in $\mathcal{F}$.*

Proof. $I_R(F)$ is weak* continuous if the following relation holds,

$$F_n\overset{w^*}{\to}F\implies I_R(F_n)\to I_R(F),\tag{21}$$

From the definitions of $I_R(F)$ and $i_R(\theta,F)$ in eq.(9),

$$I_R(F)=\int_a^b i_R(\theta,F)dF(\theta).$$

Since $i_R(\theta,F)$ is a positive continuous function of θ, if it is bounded from above, this is justified from the Helly-Bray theorem. It will be shown separately for $\kappa\geq 1$ and $\kappa<1$.

For $\kappa\geq 1$: Since $P(\alpha,\Delta_\theta)$ is a decreasing function of a, the following inequality holds.

$$\begin{gathered}P(r\kappa,\Delta_\theta)-P(r\kappa+\lfloor\kappa\rfloor,\Delta_\theta)\leq p(r|\theta;\kappa,\Delta)\leq P(r\kappa,\Delta_\theta)-P(r\kappa+\lceil\kappa\rceil,\Delta_\theta)\\ e^{-\Delta_\theta}\sum_{i=0}^{\lfloor\kappa\rfloor-1}\frac{\Delta_\theta^{r\kappa+i}}{\Gamma(r\kappa+i+1)}\leq p(r|\theta;\kappa,\Delta)\leq e^{-\Delta_\theta}\sum_{i=0}^{\lceil\kappa\rceil-1}\frac{\Delta_\theta^{r\kappa+i}}{\Gamma(r\kappa+i+1)}.\end{gathered}\tag{22}$$

With the above equation, $p(r;F,\kappa,\Delta)$ is bounded from below as follows

$$p(r;F,\kappa,\Delta)=\int_a^b p(r|\theta;\kappa,\Delta)dF(\theta)>e^{-\Delta_M}\sum_{i=0}^{\lfloor\kappa\rfloor-1}\frac{\Delta_m^{r\kappa+i}}{\Gamma(r\kappa+i+1)}$$

where $\Delta_m = \Delta/b$ and $\Delta_M = \Delta/a$ are the minimum and the maximum of Δ_θ, respectively. Thus,

$$\frac{p(r|\theta;\kappa,\Delta)}{p(r;F,\kappa,\Delta)} < e^{\Delta_M-\Delta_\theta}\frac{\sum_{i=0}^{\lceil\kappa\rceil-1}\frac{\Delta_\theta^{r\kappa+i}}{\Gamma(r\kappa+i+1)}}{\sum_{i=0}^{\lfloor\kappa\rfloor-1}\frac{\Delta_m^{r\kappa+i}}{\Gamma(r\kappa+i+1)}} < B\,e^{\Delta_M-\Delta_\theta}\Big(\frac{\Delta_\theta}{\Delta_m}\Big)^{r\kappa},$$

where B is the following upper bound

$$\frac{\sum_{i=0}^{\lceil\kappa\rceil-1}\frac{\Delta_\theta^i}{\Gamma(r\kappa+i+1)}}{\sum_{i=0}^{\lfloor\kappa\rfloor-1}\frac{\Delta_m^i}{\Gamma(r\kappa+i+1)}} = \frac{1+\sum_{i=1}^{\lceil\kappa\rceil-1}\Delta_\theta^i\frac{\Gamma(r\kappa+1)}{\Gamma(r\kappa+i+1)}}{1+\sum_{i=1}^{\lfloor\kappa\rfloor-1}\Delta_m^i\frac{\Gamma(r\kappa+1)}{\Gamma(r\kappa+i+1)}} < 1+\sum_{i=1}^{\lceil\kappa\rceil-1}\Delta_M^i = B.$$

With the result of proposition 3, $i_R(\theta,F)$ is bounded from above

$$i_R(\theta,F) = \sum_{r=0}^{\infty} p(r|\theta;\kappa,\Delta)\log\frac{p(r|\theta;\kappa,\Delta)}{p(r;F,\kappa,\Delta)} < \kappa\Delta_\theta\log\frac{\Delta_\theta}{\Delta_m} - \Delta_\theta + \Delta_M + \log B.$$

For $\kappa < 1$: When $\kappa < 1$, the following relation holds.

$$p(r|\theta;\kappa,\Delta) = P(r\kappa,\Delta_\theta) - P((r+1)\kappa+1,\Delta_\theta) - \frac{e^{-\Delta_\theta}\Delta_\theta^{(r+1)\kappa}}{\Gamma((r+1)\kappa+1)}.$$

Since $P(\alpha,x)$ is a decreasing function, the following relation holds

$$P(r\kappa,\Delta_\theta) - P(r\kappa+1,\Delta_\theta) - \frac{e^{-\Delta_\theta}\Delta_\theta^{(r+1)\kappa}}{\Gamma((r+1)\kappa+1)} < p(r|\theta;\kappa,\Delta) < P(r\kappa,\Delta_\theta) - P(r\kappa+1,\Delta_\theta)$$

$$\frac{e^{-\Delta_\theta}\Delta_\theta^{r\kappa}}{\Gamma(r\kappa+1)}\Big(1-\Delta_\theta^\kappa\frac{\Gamma(r\kappa+1)}{\Gamma(r\kappa+\kappa+1)}\Big) < p(r|\theta;\kappa,\Delta) < \frac{e^{-\Delta_\theta}\Delta_\theta^{r\kappa}}{\Gamma(r\kappa+1)}. \quad (23)$$

Above equation gives the following bound of $p(r;F,\kappa,\Delta)$.

$$p(r;F,\kappa,\Delta) = \int_a^b p(r|\theta;\kappa,\Delta)dF(\theta) > \frac{e^{-\Delta_M}\Delta_m^{r\kappa}}{\Gamma(r\kappa+1)}\Big(1-\Delta_M^\kappa\frac{\Gamma(r\kappa+1)}{\Gamma(r\kappa+\kappa+1)}\Big).$$

From the property of the gamma-function, $\Gamma(r\kappa+1)/\Gamma(r\kappa+\kappa+1)$ decreases as r increases for $r > 1/\kappa$, and there exists a finite positive integer $r_0 \geq 1/\kappa$ such that, for all $r \geq r_0$, the following inequality holds for a positive real number C_1.

$$1-\Delta_M^\kappa\frac{\Gamma(r\kappa+1)}{\Gamma(r\kappa+\kappa+1)} > C_1.$$

Thus,

$$\frac{p(r|\theta;\kappa,\Delta)}{p(r;F,\kappa,\Delta)} < e^{\Delta_M-\Delta_\theta}\Big(\frac{\Delta_\theta}{\Delta_m}\Big)^{r\kappa}\frac{1}{1-\Delta_M^\kappa\frac{\Gamma(r\kappa+1)}{\Gamma(r\kappa+\kappa+1)}} < \frac{e^{\Delta_M-\Delta_\theta}}{C_1}\Big(\frac{\Delta_\theta}{\Delta_m}\Big)^{r\kappa},$$

With the result of proposition 3,

$$S_1 = \sum_{r=r_0}^{\infty} p(r|\theta;\kappa,\Delta)\log\frac{p(r|\theta;\kappa,\Delta)}{p(r;F,\kappa,\Delta)} < \kappa\lceil\frac{1}{\kappa}\rceil(\Delta_\theta+1)\log\frac{\Delta_\theta}{\Delta_m} - \Delta_\theta + \Delta_M - \log C_1,$$

where S_1 is finite. It can be shown that there exists a real number $C_2 > 0$, s.t., $p(r|\theta;\kappa,\Delta) > C_2$ for all $\theta \in \Theta$, $r \in \{0,\cdots,r_0-1\}$ and the following sum is finite,

$$S_2 = \sum_{r=0}^{r_0-1} p(r|\theta;\kappa,\Delta)\log\frac{p(r|\theta;\kappa,\Delta)}{p(r;F,\kappa,\Delta)}.$$

Thus $i_R(\theta,F) = S_1 + S_2$ is bounded from above.

The concavity of $I_R(F)$ can be proved as in [9]. The proof for the strict concavity follows the proof in §7.2 of [8], which is an application of Carleman's theorem [17]. □

Lemma 4 and theorem 1 imply the capacity for rate coding is achieved by a unique F in $\mathcal{F}$.

Lemma 5. *$I_R(F)$ in eq.(9) is weakly differentiable in $\mathcal{F}$. The weak derivative at $F_0 \in \mathcal{F}$ has the form*

$$I'_{R,F_0}(F) = \int_a^b i_R(\theta;F_0)dF - I_R(F_0). \tag{24}$$

Proof. The proof is identical to the proof of lemma 3. □

Corollary 2. *Let E_0 denote the points of increase of F_0 on $\theta \in \Theta$. F_0 is optimal if and only if*

$$\begin{aligned} i_R(\theta;F_0) &\le I_R(F_0), \quad \forall\theta \in \Theta \\ i_R(\theta;F_0) &= I_R(F_0), \quad \forall\theta \in E_0. \end{aligned} \tag{25}$$

Proof. This is proved following the same steps in [7] (Corollary 1) with eq.(24). □

Finally, we prove the capacity achieving input distribution is a discrete distribution with a finite number of mass points. We start with the following proposition and corollary.

Proposition 4. *As $x \to \infty$ ($x \in \mathbb{R}^+$), the following equation holds*

$$\lim_{x\to\infty}\frac{\sum_{r=1}^{\infty} P(rm,x)}{x} = \frac{1}{m}, \quad m \in \mathbb{Z}^+. \tag{26}$$

Proof of proposition 4. From proposition 3, $\sum_{r=1}^{\infty} P(rm,x)$ is bounded from above with a linear function of x. Let us define the sum as $S_m(x)$.

$$S_m(x) = \sum_{r=1}^{\infty} P(rm,x) = e^{-x}\sum_{r=1}^{\infty}\sum_{i=0}^{\infty}\frac{x^{rm+i}}{\Gamma(rm+i+1)}.$$

Here, following relation of $P(\alpha,x)$ and the beta function is used

$$P(\alpha,x) = e^{-x}\sum_{i=0}^{\infty}\frac{x^{\alpha+i}}{\Gamma(\alpha+i+1)}, \quad B(\beta,\gamma) = \int_0^1 t^{\beta-1}(1-t)^{\gamma-1}dt = \frac{\Gamma(\beta)\Gamma(\gamma)}{\Gamma(\beta+\gamma)}, \quad \alpha,\beta,\gamma,x > 0.$$

It is easily checked that

$$\left(\frac{d}{dx}+1\right)^k S_m(x) = e^{-x}\sum_{r=1}^{\infty}\sum_{i=0}^{\infty}\frac{x^{rm-k+i}}{\Gamma(rm-k+i+1)}, \quad k \in \{0,\cdots,m-1\}.$$

Thus, the following linear differential equation is derived.

$$\sum_{k=0}^{m-1}\left(\frac{d}{dx}+1\right)^k S_m(x) = e^{-x}\sum_{r=1}^{\infty}\sum_{i=0}^{\infty}\frac{x^{r+i}}{\Gamma(r+i+1)} = x.$$

Solving the differential equation, the general solution gives the following form of $S_m(x)$

$$S_m(x) = \frac{x}{m} + \sum_{k=1}^{m-1} C_k e^{(-1+\alpha_k)x} - \frac{m-1}{m^2}, \quad \alpha_k = \exp\left[\frac{2\pi k\sqrt{-1}}{m}\right]. \tag{27}$$

Since $|\mathrm{Re}\,\alpha_k| < 1$, $\mathrm{Re}(-1+\alpha_k) < 0$ holds for $k \in \{1, \cdots, m-1\}$, and $\lim_{x\to\infty} S_m(x)/x{=}1/m$. □

Corollary 3. *As* $\theta \downarrow 0$, *the expectation of* R *w.r.t.* $p(r|\theta;\kappa,\Delta)$ *grows proportional to* Δ_θ.

Proof of corollary 3. Since the expectation $\overline{R}_{\kappa,\Delta_\theta} = \sum_{r=0}^{\infty} r\, p(r|\theta;\kappa,\Delta)$ is bounded as follows

$$\sum_{r=1}^{\infty} P(r\lceil\kappa\rceil, \Delta_\theta) \le \overline{R}_{\kappa,\Delta_\theta} = \sum_{r=1}^{\infty} P(r\kappa, \Delta_\theta) \le \sum_{r=1}^{\infty} P(r\lfloor\kappa\rfloor, \Delta_\theta).$$

From proposition 4, $\sum_{r=1}^{\infty} P(r\lceil\kappa\rceil, \Delta_\theta)$ and $\sum_{r=1}^{\infty} P(r\lfloor\kappa\rfloor, \Delta_\theta)$ grows proportional to Δ_θ. □

Finally, the following theorem shows that the we discretenss of the capacity achieving input distribution for rate coding.

Theorem 3. *Under a bound constraint, the channel capacity of a single neuron channel with the rate coding is achieved by a discrete distribution with a finite number of mass points.*

Proof. The proof follows the same steps of theorem 2. The extension of $i_R(\theta; F)$ to the complex plain z is defined as

$$i_R(z;F) = \sum_{r=0}^{\infty} p(r|z;\kappa,\Delta)\log\frac{p(r|z;\kappa,\Delta)}{p(r;F,\kappa,\Delta)}, \quad p(r|z;\kappa,\Delta) = P(r\kappa, \Delta/z) - P((r+1)\kappa, \Delta/z).$$

Since $P(\alpha, z)$ and $\log z$ is analytic for $\mathrm{Re}\, z > 0$, $i_R(z; F_0)$ is analytic for $\mathrm{Re}\, z > 0$.

If E_0 in corollary 2 has infinite points, since Θ is bounded and closed, E_0 has a limit point and hence, from eq.(25), the identity theorem implies $i_R(z;F_0) = I_R(F_0)$ for the region $\mathrm{Re}\, z > 0$. This region includes positive real line $\mathbb{R}^+$ and

$$\sum_{r=0}^{\infty} p(r|\theta;\kappa,\Delta)\log\frac{p(r|\theta;\kappa,\Delta)}{p(r;F_0,\kappa,\Delta)} = I_R(F_0), \quad \theta\in\mathbb{R}^+ \tag{28}$$

is implied. The proof is completed by deriving a contradiction for eq.(28). The contradiction is derived for $\kappa \ge 1$ and $\kappa < 1$, separately.

For $\kappa \ge 1$**:** From eq.(22), $p(r;F,\kappa,\Delta)$ is bounded from above as follows

$$p(r;F,\kappa,\Delta) = \int_a^b p(r|\theta;\kappa,\Delta)dF(\theta) < e^{-\Delta_m}\sum_{i=0}^{\lceil\kappa\rceil-1}\frac{\Delta_M^{r\kappa+i}}{\Gamma(r\kappa+i+1)}$$

and

$$\frac{p(r|\theta;\kappa,\Delta)}{p(r;F,\kappa,\Delta)} > e^{\Delta_m-\Delta_\theta}\frac{\sum_{i=0}^{\lfloor\kappa\rfloor-1}\frac{\Delta_\theta^{r\kappa+i}}{\Gamma(r\kappa+i+1)}}{\sum_{i=0}^{\lceil\kappa\rceil-1}\frac{\Delta_M^{r\kappa+i}}{\Gamma(r\kappa+i+1)}} > De^{\Delta_m-\Delta_\theta}\left(\frac{\Delta_\theta}{\Delta_M}\right)^{r\kappa},$$

where D is the following lower bound.

$$\frac{\sum_{i=0}^{\lfloor\kappa\rfloor-1}\frac{\Delta_\theta^{i}}{\Gamma(r\kappa+i+1)}}{\sum_{i=0}^{\lceil\kappa\rceil-1}\frac{\Delta_M^{i}}{\Gamma(r\kappa+i+1)}} = \frac{\sum_{i=0}^{\lfloor\kappa\rfloor-1}\Delta_\theta^{i}\frac{\Gamma(r\kappa+1)}{\Gamma(r\kappa+i+1)}}{\sum_{i=0}^{\lceil\kappa\rceil-1}\Delta_M^{i}\frac{\Gamma(r\kappa+1)}{\Gamma(r\kappa+i+1)}} > \frac{1}{1+\sum_{i=1}^{\lceil\kappa\rceil-1}\Delta_M^{i}} = D.$$

This shows $i_R(\theta,F)$ is bounded from below as

$$i_R(\theta,F) > \kappa\,\overline{R}_{\kappa,\Delta_\theta}\log\frac{\Delta_\theta}{\Delta_M} - \Delta_\theta + \Delta_m + \log D.$$

Since $\overline{R}_{\kappa,\Delta_\theta}$ grows with Δ_θ as $\theta\downarrow 0$, the lower bound of $i_R(\theta,F)$ grows with $\Delta_\theta\log\Delta_\theta$. Thus, $i_R(\theta,F)$ cannot be finite and constant for $\forall\theta\in\mathbb{R}^+$, which brings the contradiction.

For $\kappa<1$: From eq.(23), $p(r;F,\kappa,\Delta)$ is bounded from above as follows

$$p(r;F,\kappa,\Delta) < \frac{e^{-\Delta_m}\Delta_M^{r\kappa}}{\Gamma(r\kappa+1)} \tag{29}$$

Let us denote r as

$$r = r'K+b, \quad \text{where} \quad K=\lceil\frac{1}{\kappa}\rceil,\ \ r'\in\mathbb{Z}^*,\ \ b\in\{0,\cdots,K-1\}.$$

r' and b can be considered as stochastic variables and the following relation holds,

$$p(r|\theta;\kappa,\Delta) = p(r'K+b|\theta;\kappa,\Delta) = q(r'|\theta;\kappa,\Delta)q(b|r',\theta;\kappa,\Delta),$$
$$q(r'|\theta;\kappa,\Delta) = \sum_{b=0}^{K-1}p(r'K+b|\theta;\kappa,\Delta),\quad q(b|r',\theta;\kappa,\Delta) = \frac{p(r'K+b|\theta;\kappa,\Delta)}{q(r'|\theta;\kappa,\Delta)}.$$

Let H_R and $H_{R'}$ be the entropy of R and R', respectively, and $H_{B|R'}$ be the conditional entropy of B given R'. The following relation holds

$$H_R = H_{R'} + H_{B|R'} \le H_{R'} + \log K,$$

which is justified from $0\le H_{B|R'}\le\log K$. With this result,

$$\sum_{r=0}^{\infty}p(r|\theta;\kappa,\Delta)\log p(r|\theta;\kappa,\Delta) \ge \sum_{r'=0}^{\infty}q(r'|\theta;\kappa,\Delta)\log q(r'|\theta;\kappa,\Delta) - \log K.$$

Since $\lfloor\kappa K\rfloor = 1$ holds, the probability $q(r'|\theta;\kappa,\Delta)$ is bounded as follows,

$$q(r'|\theta;\kappa,\Delta) = P(r'K\kappa,\Delta_\theta) - P((r'+1)K\kappa,\Delta_\theta) \ge e^{-\Delta_\theta}\frac{\Delta_\theta^{r'K\kappa}}{\Gamma(r'K\kappa+1)}. \tag{30}$$

With eqs.(29) and (30),

$$\frac{q(r'|\theta;\kappa,\Delta)}{p(r;F,\kappa,\Delta)} = \frac{q(r'|\theta;\kappa,\Delta)}{p(r'K+b;F,\kappa,\Delta)} > e^{\Delta_m-\Delta_\theta}\left(\frac{\Delta_\theta}{\Delta_M}\right)^{r'K\kappa}\Delta_M^{-b\kappa}\frac{\Gamma((r'K+b)\kappa+1)}{\Gamma(r'K\kappa+1)}$$
$$\geq E\, e^{\Delta_m-\Delta_\theta}\left(\frac{\Delta_\theta}{\Delta_M}\right)^{r'K\kappa},$$

where E is the following lower bound.

$$\Delta_M^{-b\kappa}\frac{\Gamma((r'K+b)\kappa+1)}{\Gamma(r'K\kappa+1)} \geq \min\left\{1,\ \Delta_M^{-\kappa}\Gamma(\kappa+1),\ \cdots,\ \Delta_M^{-(K-1)\kappa}\Gamma((K-1)\kappa+1)\right\} = E.$$

This shows $i_R(\theta,F)$ is bounded from below as

$$\begin{aligned} i_R(\theta,F) &= \sum_{r=0}^{\infty} p(r|\theta;\kappa,\Delta)\log\frac{p(r|\theta;\kappa,\Delta)}{p(r;F,\kappa,\Delta)} \\ &\geq \sum_{r'=0}^{\infty}\sum_{b=0}^{K-1} p(r'K+b|\theta;\kappa,\Delta)\log\frac{q(r'|\theta;\kappa,\Delta)}{p(r'K+b;F,\kappa,\Delta)} - \log K \\ &> \sum_{r'=0}^{\infty} q(r'|\theta;\kappa,\Delta)\log\left(E\, e^{\Delta_m-\Delta_\theta}\left(\frac{\Delta_\theta}{\Delta_M}\right)^{r'K\kappa}\right) - \log K \\ &= \left[\sum_{r'=0}^{\infty} r'\, q(r'|\theta;\kappa,\Delta)\right] K\kappa\log\left(\frac{\Delta_\theta}{\Delta_M}\right) - \Delta_\theta + \Delta_m + \log E - \log K. \end{aligned}$$

Since $\sum_{r'=0}^{\infty} r'\, q(r'|\theta;\kappa,\Delta)$ is equivalent to $\overline{R}_{K\kappa,\Delta_\theta}$, proposition 4 shows that it grows proportional to Δ_θ as $\theta \downarrow 0$. Thus, $i_R(\theta,F)$ is lower bounded with a term which grows with $\Delta_\theta\log\Delta_\theta$ and $i_R(\theta,F)$ cannot be finite and constant for $\forall\theta\in\mathbb{R}^+$, which brings the contradiction.

□

4. Discussion and Conclusion

We considered the channel capacity and the capacity achieving input distribution for a single neuron information channel. ISIs are modelled with a gamma distribution and two types of coding, temporal and rate, are considered. We have proved that the channel capacities of a single neuron with temporal and rate coding are achieved with discrete distributions[3].

However, this does not mean the capacity can be achieved biologically. Channel capacity is something similar to the maximum speed indicated in the speed meter of an automobile. Although you will never drive your vehicle with that speed, that speed tells us the potential of the automobile. Channel capacity not only provide the upper limit of the possible information transmission rate, but also describe how good the channel is.

Acknowledgements

This work was supported by the Grant-in-Aid for Scientific Research No. 18079013, MEXT, Japan.

[3] The numerical studies are shown in another paper [1].

References

[1] Ikeda S and Manton J H 2009 *Neural Comput.* **21** 1714–48
[2] Ikeda S and Manton J H 2009 Spiking neuron channel *2009 IEEE Int. Symp. on Information Theory* (Seoul, Korea) pp 1589–93
[3] Shannon C E 1948 *Bell System Tech. J.* **27** 379–423 and 623–56
[4] Baker S N and Lemon R N 2000 *J. Neurophysiol.* **84** 1770–80
[5] Stein R B 1967 *Biophys. J.* **7** 797–826
[6] Shinomoto S, Shima K and Tanji J 2003 *Neural Comput.* **15** 2823–42
[7] Smith J G 1971 *Information and Control* **18** 203–19
[8] Shamai (Shitz) S 1990 *IEE Proc. I Commun. Speech Vis.* **137** 424–30
[9] Abou-Faycal I C, Trott M D and Shamai (Shitz) S 2001 *IEEE Trans. Inform. Theory* **47** 1290–301
[10] Gursoy M C, Poor V and Verdú S 2005 *IEEE Trans. Wirel. Commun.* **4** 2193–206
[11] Shimokawa T and Shinomoto S 2009 *Neural Comput.* **21** 1931–51
[12] Pawlas Z, Klevanov L B and Prokop M 2008 *Neural Comput.* **20** 1325–43
[13] Gursoy M C, Poor H V and Verdú S 2002 The capacity of the noncoherent Rician fading channel *Princeton University Technical Report*
[14] Tchamkerten A 2004 *IEEE Trans. Inform. Theory* **50** 2773–8
[15] Luenberger D G 1969 *Optimization by Vector Space Method* (John Wiley & Sons, Inc.)
[16] Doob J 1994 *Measure Theory* (*Graduate texts in mathematics* vol 143) (Springer-Verlag)
[17] Akhiezer N 1965 *The Classical Moment Problem* (Oliver & Boyd) translated to English by N. Kemmer

The role of stochasticity in an information-optimal neural population code

N G Stocks[1], A P Nikitin[1], M D McDonnell[2] and R P Morse[3]

[1] School of Engineering, University of Warwick, Coventry CV4 7AL, UK
[2] Institute for Telecommunications Research, University of South Australia, SA 5095, Australia
[3] School of Life and Health Sciences, Aston University, Birmingham B4 7ET, UK

E-mail: n.g.stocks@warwick.ac.uk

Abstract. In this paper we consider the optimisation of Shannon mutual information (MI) in the context of two model neural systems. The first is a stochastic pooling network (population) of McCulloch-Pitts (MP) type neurons (logical threshold units) subject to stochastic forcing; the second is (in a rate coding paradigm) a population of neurons that each displays Poisson statistics (the so called 'Poisson neuron'). The mutual information is optimised as a function of a parameter that characterises the 'noise level'—in the MP array this parameter is the standard deviation of the noise; in the population of Poisson neurons it is the window length used to determine the spike count. In both systems we find that the emergent neural architecture and, hence, code that maximises the MI is strongly influenced by the noise level. Low noise levels leads to a heterogeneous distribution of neural parameters (diversity), whereas, medium to high noise levels result in the clustering of neural parameters into distinct groups that can be interpreted as subpopulations. In both cases the number of subpopulations increases with a decrease in noise level. Our results suggest that subpopulations are a generic feature of an information optimal neural population.

1. Introduction

It is well known in neuroscience that repeated presentation of identical stimuli to neurons (nerve cells) result in variable timings of resultant neural action potentials (voltage spikes). Thus neuronal responses are often referred to as 'noisy.' This variability arises due to noise sources that occur at all stages of neural processing e.g. from noise in the sensory receptors at the sensory periphery to 'network' and synaptic noise in the cortex [1]. Consequently, it is an unavoidable fact that neural coding and processing is dominated by stochasticity. However, what is still unclear is what impact the noise has had on the evolution of efficient neural coding and processing strategies.

In this study we investigate this issue in the context of neural coding. We consider the optimisation of Shannon Mutual information (MI) [2, 3] in neural populations as a function of the noise level. Two systems are studied; the first is an array of McCulloch-Pitts neurons and the second is a population of Poisson neurons. Both can be classified as *stochastic pooling networks* [4]. The objective of the study is to establish the neural parameters that maximise the MI. This, in turn, enables us to uncover the optimal neural architecture and hence gives an insight into the form an optimal neural code may take. It also enables us to investigate how the emergent architecture/codes depend on the level of stochasticity. As we will demonstrate, noise appears to be a significant organising influence behind an optimal neural structure.

The use of MI as an organising principle of neural coding was first proposed by Linsker in 1988 in the context of the visual system [5]. This is the so-called *infomax* principle. The idea is that neural systems organise themselves (i.e. their parameters) in such a way as to maximise the MI between their inputs and outputs subject to some constraints (such as energy, connectivity etc). The infomax principle is closely related to earlier work by Barlow [6] who proposed redundancy reduction as an organisational rule. More recently, the infomax principle has been extended by Bell and Sejnowski to independent components analysis [7]. The infomax principle has also recently been used to derive optimal codes for auditory coding that replicate closely the 'revcor' kernels observed in physiological experiments [8]. Consequently, there is growing evidence that the infomax principle is a useful paradigm for explaining neurophysiological structure and function. However, it is unlikely that such a principle can explain the structure of the brain as a whole because one would expect that the structure is related to function - and it is less than obvious that all brain function (e.g. detection, processing, computation, cognition, memory recall etc) can all be described by an optimisation of MI.

Nevertheless, where one might expect the infomax principle to be at work is at the sensory periphery. Information theory was originally developed to quantify the ability of systems to code and transmit information, which appears to be the same task as that faced by biological sensory systems. Furthermore, it is not unreasonable to assume that this information is coded in an efficient manner, especially with respect to metabolic energy. Inefficient coding is likely to result in a greater metabolic deficit and hence a survival disadvantage. Consequently, natural selection would be expected to deselect such strategies. These considerations lead us to postulate that information is coded using an infomax principle that is probably constrained by energy.

2. Optimization of neural systems

2.1. Population of McCulloch-Pitts neurons

We first consider the system shown in Fig. 1. Each neuron is connected to a common random

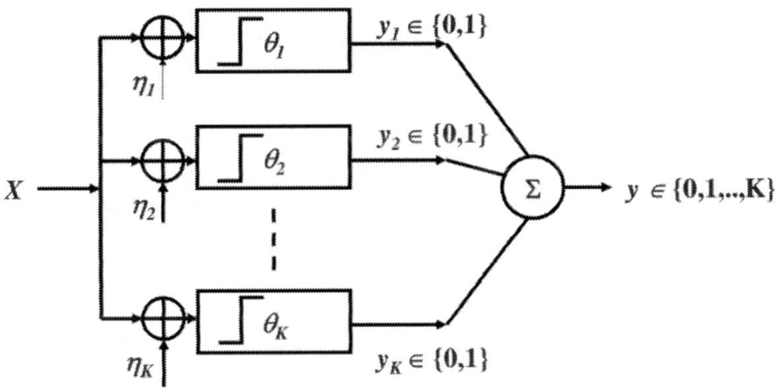

Figure 1. Array (population) of McCulloch-Pitts neuorns. Each of the K neurons, with threshold values θ_n, $n = 1, .., K$, is subject to independent additive noise, η_n, $n = 1, .., K$. The common input to each neuron is a random signal, X, which in the results presented here is assumed to be Gaussian. Each neuron thresholds its input to produce a binary signal, y_n. The overall output is the sum of each y_n.

signal, x, but has an independent noise source η_n. The response of each neuron is governed by

$$y_n = \Theta(x + \eta_n), \quad n = 1, .., K, \tag{1}$$

where $\Theta(.)$ is the Heaviside function and the response of the population, Y is

$$Y = \sum_{n=1}^{K} y_n. \tag{2}$$

If the input signal has probability density function $P_X(x)$, then the mutual information between the input and output is given by [9]

$$I(X;Y) = -\sum_{i=0}^{K} P_Y(i)\log_2 P_Y(i) - \left(-\int_{-\infty}^{\infty} P_X(x)\sum_{i=0}^{K} P_{Y|X}(i|x)\log_2 P_{Y|X}(i|x)dx\right), \tag{3}$$

where

$$P_Y(n) = \int_{-\infty}^{\infty} P_{Y|X}(n|x)P_X(x)dx.$$

Therefore for a given signal distribution, the mutual information depends entirely on the conditional probabilities, $P_{Y|X}(n|x)$. If the noise distribution is specified, the only free variables are the population size, K, and the threshold values. Let $\hat{P}_n$ be the probability of neuron n 'spiking' in response to signal value x. Then

$$\hat{P}_n = \int_{\theta_n - x}^{\infty} f_\eta(\eta)d\eta = 1 - F_\eta(\theta_n - x), \tag{4}$$

where F_η is the cumulative distribution function of the noise and $n = 1, .., K$.

Given a noise density and arbitrary threshold values, $\hat{P}_n$ can be calculated exactly for any value of x from (4). These can then by used in an algorithm specified in [10] for calculating the conditional probabilities, $P_{Y|X}(n|x)$.

We now aim to find the threshold settings that maximize the mutual information as the input SNR varies. This goal can be formulated as the following nonlinear optimization problem:

$$\begin{aligned} \text{Find:} \quad & \max_{\{\theta_n\}} \; I(X;Y) \\ \text{subject to:} \quad & \{\theta_n\} \in \mathbb{R}^K. \end{aligned} \tag{5}$$

It is necessary to solve (5) numerically using standard unconstrained non-linear optimization methods such as the conjugate gradient method [11]. However, note that the objective function is not convex, and there exist a number of local optima. This problem can be overcome by employing random search techniques such as simulated annealing. We present here in Figs. 2-5 results for the optimal thresholds obtained by solving Problem (5) for Gaussian signal and Gaussian noise, as a function of $\sigma := \frac{\sigma_\eta}{\sigma_x}$, where σ_η and σ_x are the noise and signal standard deviation respectively; we have set $\sigma_x = 1$ for all results. In Figs. 2, 3, 4, 5 we have $K = 3, 4, 5, 15$ respectively [12].

It can be observed that the dependence of the optimal threshold values on the noise level is similar for all results. For large σ it is optimal to set all the thresholds equal to the signal mean which is zero in this case. As the noise is reduced this optimal solution bifurcates, at a value σ_1, in favour of grouping all the thresholds at two distinct values. Further reduction in the noise level results in three distinct values becoming optimal. This pattern is repeated as the noise is

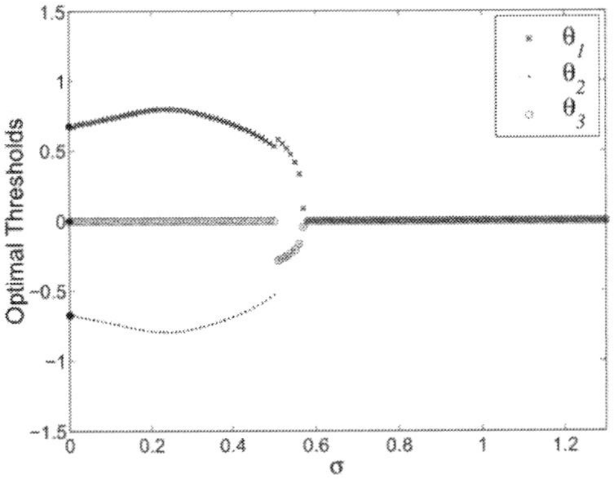

Figure 2. Optimal thresholds against noise intensity, σ, for $K = 3$. Reproduced from Ref. [12]

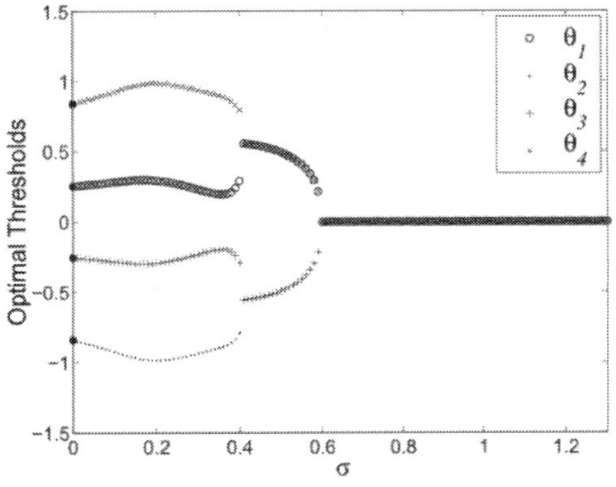

Figure 3. Optimal thresholds against noise intensity, σ, for $K = 4$. Reproduced from Ref. [12]

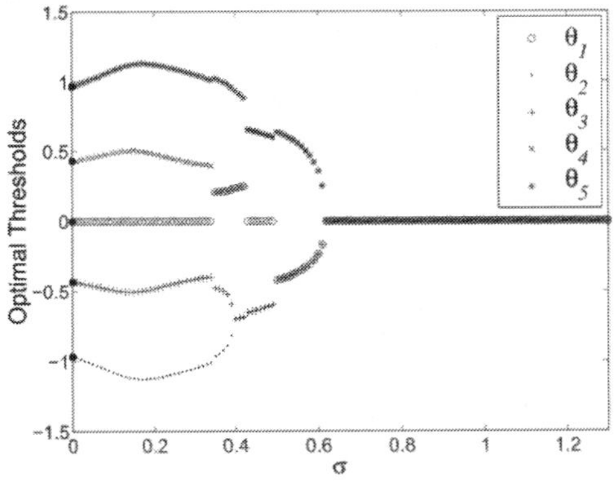

Figure 4. Optimal thresholds against noise intensity, σ, for $K = 5$. Reproduced from Ref. [12]

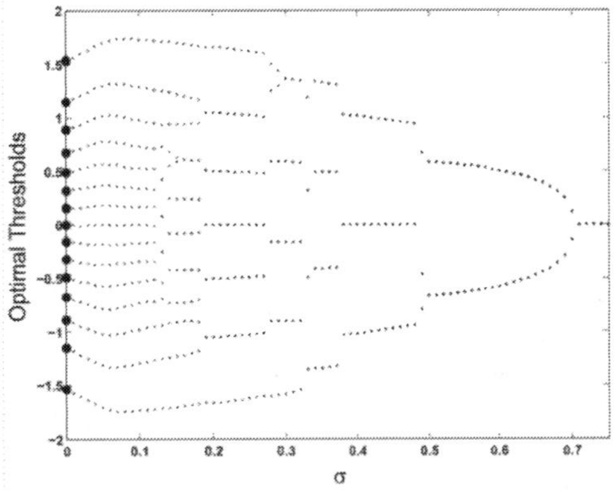

Figure 5. Optimal thresholds against noise intensity, σ, for $K = 15$. Reproduced from Ref. [12]

reduced further. In general it is observed that the number of distinct threshold values increases as $1 \to 2 \to 3 \to \ldots \to K$ as the noise level is reduced. It should be noted that the exact values of the noise at which the transitions occur (e.g. σ_1) depends on the size of the population size K. This problem is discussed in Chapter 8 of Ref. [12]. For the limit $K \to \infty$, it is observed that an asymptotic limit of σ_1 exists, but that its value is dependent on the distributions of the signal and noise. For example, for Gaussian signal and noise the limiting σ_1 is about 0.8, while for logistic signal and noise it is about 1.2. However, qualitatively, the pattern is always the same.

These results raise a number of interesting points. First, the structure of the optimal threshold distribution is discrete. Numerical experiments indicate that this bifurcational structure is not changed even in the limit $K \to \infty$. Consequently, the clustering of thresholds into distinct groups is not a consequence of the finite number of neurons but is rather a property of the system (channel) itself. Second, the existence of these distinct groups indicate an optimal coding strategy is to group neurons with similar thresholds together into *subpopulations*; the number of distinct subpopulations depends on the level of noise. Third, the optimal design solution is strongly dependent on the level of the noise. Indeed, this point is graphically highlighted in

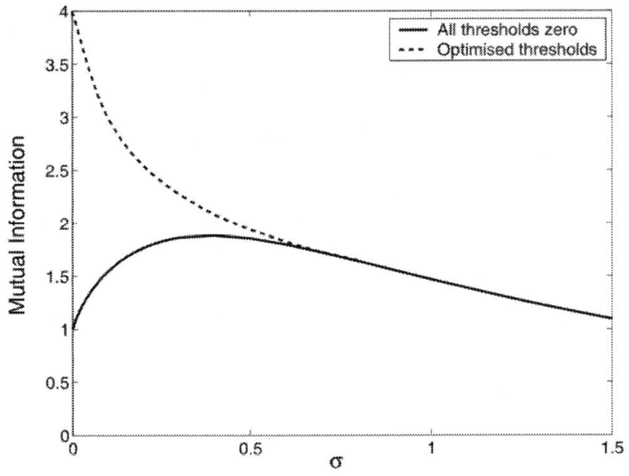

Figure 6. Plot of $I(X, Y)$ against noise intensity, σ, for $K = 15$, the optimal thresholds (dashed line) and $\theta_i = 0 \ \forall \ i$ (solid line). The solid line shows the suprathreshold stochastic resonance (SSR) effect [9, 10]. Reproduced from Ref. [12]

Fig. 6.

In Fig. 6 the MI is plotted as a function of the noise level for two different cases i) for the optimal threshold values (dashed line) shown in Fig. 5 and ii) for the optimal threshold solution for $\sigma \geqslant \sigma_1$ (solid line); i.e. the thresholds are held constant at $\theta_i = 0 \ \forall \ i$ as the noise level is varied. The two curves coalesce at σ_1. It can be observed that while the solution $\theta_i = 0 \ \forall \ i$ is optimal for $\sigma \geqslant \sigma_1$ this solution performs very badly in the limit $\sigma \to 0$. Indeed, it achieves a MI of only 1-bit compared to the optimal solution at $\sigma = 0$ which achieves 4-bits. Consequently, the correct choice of neural parameters (and hence the resulting neural code) is seen to strongly depend on the level of noise.

An interesting consequence of the sub-optimal nature of the solution $\theta_i = 0 \ \forall \ i$ at small σ is that it gives rise to a suprathreshold stochastic resonance [9, 10] i.e. the MI is maximised at a non-zero level of noise. Such counter-intuitive 'noise-benefits' rely on some aspect of a system being 'sub-optimal' [9, 12].

Some theoretical progress in calculating the optimal information can be made by utilising a connection between mutual information and Fisher information. Following [16, 18] we can write,

$$Y(x) = T(x) + \sqrt{V(x)}\xi \tag{6}$$

where $T(x)$ is the mean output for a give input x and $V(x)$ is the variance of the output conditioned on the input, and ξ is a random variable with zero mean and unit variance. For not too large $V(x)/N^2$ we can write

$$I(X;Y) \approx H(x) - \frac{1}{2}\int_x f_x(x) \log_2 \left(\frac{2\pi e}{J(x)}\right) dx, \tag{7}$$

where the Fisher information $J(x)$ is given by

$$J(x) \approx \frac{\left(\frac{dT(x)}{dx}\right)^2}{V(x)}. \tag{8}$$

Given that the input entropy $H(x)$ only depends on the statistics of the signal and not on the properties of the channel, maximisation of (7) can be seen to be equivalent to a minimisation of the function

$$F = -\langle \log_2 Z \rangle_x, \tag{9}$$

where $Z = J(x)$ and the average, $\langle . \rangle_x$ is taken over the input signal distribution. It is interesting to note that this looks similar to the minimisation of a Free Energy with a partition function Z.

These expressions can be simplified by noting that the responses of the neurons, $y_n(x)$, are (conditionally) independent. Consequently, the mean response and variance can be obtained by summing the individual means, $T_n(x)$, and variances,$\mathrm{Var}(y_n|x)$ [12]. Hence,

$$T(x) = \mathrm{E}(Y|x) = \sum_{n=1}^{N} T_n(x), \tag{10}$$

and

$$V(x) = \mathrm{Var}(Y|x) = \sum_{n=1}^{N} \mathrm{Var}(y_n|x) = \sum_{n=1}^{N} T_n(1 - T_n). \tag{11}$$

The final expression for the variance comes from the fact that $\hat{P}_n = T_n$ and the variance of each neuron is governed by Bernoulli statistics, which gives $\mathrm{Var}(y_n|x) = \hat{P}_n(1 - \hat{P}_n)$. Consequently, substituting these expressions into Eq. (8) yields

$$Z = J(x) = \frac{\left(\sum_{n=1}^{N} \frac{dT_n(x)}{dx}\right)^2}{\sum_{n=1}^{N} T_n(1 - T_n)}. \tag{12}$$

It is interesting to note that Eq. (12) only requires specification of the mean response $T_n(x)$, i.e. the tuning curve of each neuron. Furthermore, $T_n(x)$ is, itself, simply related to the cumulative noise distribution of each neuron as $T_n(x) = 1 - F_\eta(\theta_n - x)$. Consequently, it is largely the cumulative noise distribution that ultimately specifies the information flow in these systems.

A particularly simple expression for the mutual information can be derived under the assumption that the noise has a logistic distribution. In this case the cumulative distribution function of the noise is $F_\eta(\theta_n - x) = [1 + \exp(-\beta(\theta_n - x))]^{-1}$ where β is the inverse noise level. With this assumption it can be shown that T_n, via F_η, satisfies the differential equation associated with the logistic function, that is

$$\frac{dT_n}{dx} = \beta T_n(1 - T_n). \tag{13}$$

Noting that the expression for the Fisher information simplifies to $J(x) = \beta^2 \sum_{n=1}^{N} T_n(1 - T_n)$, Eq. (9) can be written

$$F(\{\theta_n\}, \beta) = -\left\langle \log_2 \sum \frac{\beta^2}{4} \left[\cosh^2\left(\frac{\beta}{2}(\theta_n - x)\right)\right]^{-1} \right\rangle_x. \tag{14}$$

To obtain the optimal information it is now necessary to find the set $\{\theta_n\}$ that minimises $F(\{\theta_n\}, \beta)$ for a given β. Although we have used a similar approach to find the location of the first bifurcation point for large N we have not been successful in describing the entire sequence of bifurcations. We speculate that further progress might be made by using a *replica* approach, but we have been unable to verify this.

2.2. Population of Poisson neurons

In this section we replace each MP neuron in Fig. 1 with a 'Poisson' neuron [13, 14]. Spike timing variability is often closely modelled by Poisson statistics and, hence, the Poisson neuron has become an archetypal model for neural rate coding. In this model the input signal x is coded in the mean firing rate $\nu = g(x)$ where $g(x)$ is known as the *tuning curve* (or the stimulus-response curve, gain function or rate-level function). Pooling (summing) the outputs of Poisson neurons results in a Poisson process, so without loss of generality we can treat the tuning function $g(x)$ as the tuning function for the whole population. While several definitions of *rate* exist [15], following related studies [16, 17], here we assume the *observable* output when the mean rate is ν is the number of spikes, k, that occur in a time window T. The input x is assumed to be a continuous variable, such as an external sensory stimulus.

To make connection with the studies presented in the preceding section we need to quantify the stochasticity in the system. Unlike the MP neuron, where we simply introduced additive noise and used the standard deviation of the noise as a measure of stochasticity, we cannot introduce noise in this fashion for a Poisson neuron. In a Poisson neuron the statistics of the action potentials (neural spikes) are defined through the Poisson distribution—see Eq. (16). Consequently, the level of stochasticity in some sense is fixed. However, in a rate coding paradigm the signal is coded in a spike count over a time window T, where it is further assumed that the rate is constant over this window. Consequently, the accuracy of the code will depend on the length of the window; the longer the window length the more accurate the estimate of the rate and hence, in principle, the more accurate our estimate of x.

We can quantify this accuracy in terms of a signal-to-noise ratio (SNR). The information received at the output about the input signal is based on a measurement of the output rate $z = k/T$, where z is a random variable. The 'signal' can be identified as E[z] and the uncertainty, or 'noise', is given by the variance, $\text{var}[z] = \text{E}[(\text{z} - \text{E}[\text{z}])^2]$. We can now define an SNR in the standard way as the ratio of the signal power to the noise power i.e. $\text{SNR} = \text{E}[z]^2/\text{var}[z]$. Noting that $\text{E}[z] = \text{E}[k]/T$ and $\text{var}[z] = \text{var}[k]/T^2$, and furthermore that for a Poisson process we have $\text{var}[k] = \text{E}[k]$, the SNR simplifies to $\text{SNR} = \text{E}[k]$. Given also we have $\text{E}[k] \sim T$ this implies $\text{SNR} \sim T$. Therefore, the SNR ratio improves proportionally with increasing T.

The $g(x)$ that maximizes MI has only previously been obtained in the limit $T \to \infty$ [16, 18]. Arguably, this limit is not relevant to a large number of biological sensory systems where it is well established that behavioural responses occur on timescales that imply short coding windows [14]. Here we present the optimal tuning curve for finite T and outline a proof that the tuning curve is discrete; full details of the proof can be found in [19].

To derive the optimal tuning curve, we make use of a known result from the photonics literature. Both the generation of neural action potentials, and the emission of photons can be modeled as Poisson point processes [20].

A classical problem in communication theory is that of finding the signal distribution that maximizes the mutual information for a channel. The resultant *optimal code* is said to achieve *channel capacity* [2]. The optimal input distribution for the direct detection photon channel has been proven to be discrete [21, 22, 23]. Indeed, the discreteness of optimal input distributions is the norm, regardless of whether the output distribution is discrete or continuous [24, 25] and is a property of the channel and not the output measure—see [25, 26].

The discreteness of the optimal signal for the optical Poisson channel implies that the optimal stimulus for a Poisson neural system is also discrete. However, the signal distribution is not something that a neural channel can control directly. However, instead, it is plausible that a neural system may have been optimized by evolution so that the *tuning curve* discretises its input to match the theoretical optimal source distribution and, indeed, our results suggest this is the case.

We now proceed to formulate the problem of optimising the MI. The MI between the (continuous) input random variable $x \in \mathcal{X} = [x_{\min}, x_{\max}]$, and the (discrete) output random variable, k, is

$$I(x;k) = \sum_{k=0}^{\infty} \int_{x \in \mathcal{X}} dx P_x(x) P[k|x] \log_2 \frac{P[k|x]}{P_k(k)}, \tag{15}$$

where $P_k(k) = \int_{x \in \mathcal{X}} dx P_x(x) P[k|x]$. Here $P_x(x)$, $P_k(k)$ and $P[k|x]$ are the distributions of the stimulus, the response and the conditional distribution respectively.

For Poisson statistics, the conditional distribution is

$$Q[k|\nu] = \frac{[T\nu]^k}{k!} \exp(-T\nu), \quad k = 0, .., \infty. \tag{16}$$

The mean firing rate is restricted to $\nu \in [\nu_{\min}, \nu_{\max}]$, where the upper bound $\nu_{\max}$ is due to physiological limits (metabolic and refractory), and we set $\nu_{\min} = 0$. Later, we use the notation $N = T\nu_{\max}$ to denote the *maximum* mean spike count.

The conversion of a signal follows the Markov chain $x \to \nu \to k$. We refer to $x \to \nu$ and $\nu \to k$ as separate 'subchannels.' To find the optimal channel, we maximize the MI by variation of the distribution $P_\nu(\nu)$ for given $P_x(x)$ and $Q[k|\nu]$. Since the distribution of ν is $P_\nu(\nu) = \int_{x \in \mathcal{X}} dx P_x(x) \delta(\nu - g(x))$, where $\delta(.)$ is the Dirac delta function, variation of $P_\nu(\nu)$ means variation of the tuning curve $g(x)$.

We have recently been able to prove the following theorem [19]: *The mutual information in the neural channel, $x \to \nu \to k$, is maximized when the distribution $P_\nu(\nu)$ is discrete.* The full proof can be found in [19] but, briefly, is as follows:

First we prove that $I(x;k) = I(\nu;k)$ using Theorem 5.2.8 of [27]. We then only need to consider the optimisation of the noisy neural subchannel $\nu \xrightarrow{Q[k|\nu]} k$. Here we make use of a result from [23] on the direct detection photon channel. The photon channel and our noisy neural *subchannel* are mathematically equivalent, by which we mean that they are both described by Eq. (2) except that for the photon channel k denotes the number of emitted photons rather than the number of action potentials. See [20, 28] for further discussion on the connection between these channels. Note that [23] assumes the rate is a continuous function of time, as is the case in our study, and yet Theorem 1 of [23] states that the optimal distribution of ν, $P_\nu(\nu)$, has a discrete form. This means that since $I(k;x) = I(k;\nu)$, the mutual information in the neural channel is maximized when the information in the noisy subchannel is maximized. By Theorem 1 of [23], this occurs when the distribution $P_\nu(\nu)$ is discrete.

Unfortunately, the theorem does not provide any means for finding a closed-form solution for the optimal discrete distribution, $P_\nu(\nu)$. However, its utility is that it allows a reduction in the set of functions we need to consider when optimizing $P_\nu(\nu)$ and/or the tuning curve $g(x)$.

Without loss of generality we can now introduce the following simplifying restriction for the function $g(x)$. Let $g(x)$ be a non-decreasing multi-step function

$$g(x) = \sum_{i=0}^{M-1} \gamma_i \sigma(x - \theta_i), \tag{17}$$

where M is the number of levels and $\sigma(.)$ is the Heaviside step function. Letting $\beta_i = \sum_{n=0}^{i} \gamma_n$ we have θ_{i+1} as the value of x at which $g(x)$ jumps from value β_i to β_{i+1}. Since we assume $x_{\min} = \theta_0 < \theta_1 < \theta_2 < ... < \theta_{M-1} < \theta_M = x_{\max}$, the optimal $g(x)$ is unique. This latter requirement means that we consider only the case of monotonically non-decreasing (sigmoidal) tuning curves. Without this restriction it is not possible to find a unique solution and hence this study does not generalize to non-monotonic tuning curves. This is not highly restrictive, since sigmoidal tuning curves are widely observed in many sensory modalities [29]. The mutual information of the neural channel can be written as

$$I(x;k) = \sum_{k=0}^{\infty} \sum_{i=0}^{M-1} \alpha_i Q[k|\beta_i] \log_2 \frac{Q[k|\beta_i]}{\sum_{n=0}^{M-1} \alpha_n Q[k|\beta_n]}, \tag{18}$$

where $\alpha_i = \int_{\theta_i}^{\theta_{i+1}} dx P_x(x)$. The optimal function $g(x)$ cannot be easily found in an analytical form using variational principles, because it leads to a set of transcendental equations. Therefore we use stochastic gradient descent methods to solve for the optimal $P_\nu(\nu)$.

Fig. 7 shows the results. The upper insets display the normalized optimal tuning curve, $f(x) \equiv g(x)/N$, for four different values of maximum mean spike count, N. Note that $I(x;k)$ in Eq. (18) is parameterized entirely by the set α_i, β_i , $i = 0, .., M-1$, and it is these parameters that are optimized. The set of θ_i–s required for the optimal $g(x)$ can be obtained for any given $P_x(x)$ from the α_i. Hence, in Fig 7, without loss of generality we have assumed that the stimulus is uniformly distributed on $[0,1]$. Similarly, the γ_i–s follow from β_i.

For small $N < 3$, only two firing rates are observed; for values of $x < \theta_1$, $f(x) = 0$ (the absence of firing) while for larger values of x, $f(x) = 1$ (firing at the maximum allowable spike rate). This form of optimal binary coding has been predicted previously for Poisson neurons using estimation theory [17]. It also agrees with the well known result that a binary source maximizes information through a Poisson channel when the input can switch instantaneously between states [21, 28].

As N is increased, the number of steps in the optimal tuning curve increases; e.g. for $N = 7$, two steps are observed giving rise to a ternary coding scheme, for $N = 15$ three steps are observed giving a 4-ary (quaternary) coding. In general, an M-ary code will be optimal with increasing N. As $N \to \infty$ we predict that the optimal tuning curve will converge to a continuous function [16, 18]. Fig. 7 shows how the partition boundaries, θ_i, vary as N is increased; new boundaries can be seen to emerge via phase transitions. These appear to be continuous and hence are akin to second order phase transitions of the optimal tuning curve.

We have shown that $T \sim$ SNR and hence, given $N \propto T$, we have $N \sim$ SNR; hence, increasing N can be interpreted as a reduction in noise. Consequently, the bifurcation diagram in Fig. 7 can be seen to be similar to those presented in the previous section and others previously reported in the context of information optimisation [30, 12]. This similarity is even more striking when one notes that the steps in the optimal tuning curve can be interpreted as thresholds at which the input signal is mapped to a different output rate and, hence, are akin to the bifurcation of the thresholds observed in Figs. 2–5.

However, the analogy between the bifurcation diagram displayed in Fig. 7 and those in the Figs. 2–5 can be made even closer. The steps in the optimal $f(x)$ can be viewed as partitioning the stimulus into regions associated with neural *subpopulations* where each neuron in a subpopulation can only fire with two rates. For example, suppose an overall population consists of K neurons and $M-1$ sub-populations, within which each neuron is identical, and binary with rates 0 and γ_i. Since the neurons are Poisson, the sum of the K individual normalized firing rates is equal to $f(x)$. For overall binary coding, the only way of achieving $f(x)$ would be a single sub-population, where each neuron is identical, and able to fire at two rates, $\phi_0/K = 0$ and $\phi_1/K = \frac{1}{K}$, where rate ϕ_1/K is activated when $x \geq \theta_1$. For the ternary case, there would be two subpopulations, of sizes J and $K-J$, with individual normalized firing rates ϕ_1/J and $(1-\phi_1)/(K-J)$, so that the overall population has 3 rates: 0, ϕ_1 and 1. The first subpopulation would only be activated when $x > \theta_1$ and the second when $x > \theta_2$.

We can estimate the sizes of the subpopulations in our example as follows. Since the sizes of the subpopulations are proportional to the integrated firing rates, the neurons for ternary coding are distributed with probabilities $P_1 = \phi_1 = \gamma_1/N$ and $P_2 = 1 - \phi_1$ respectively. The

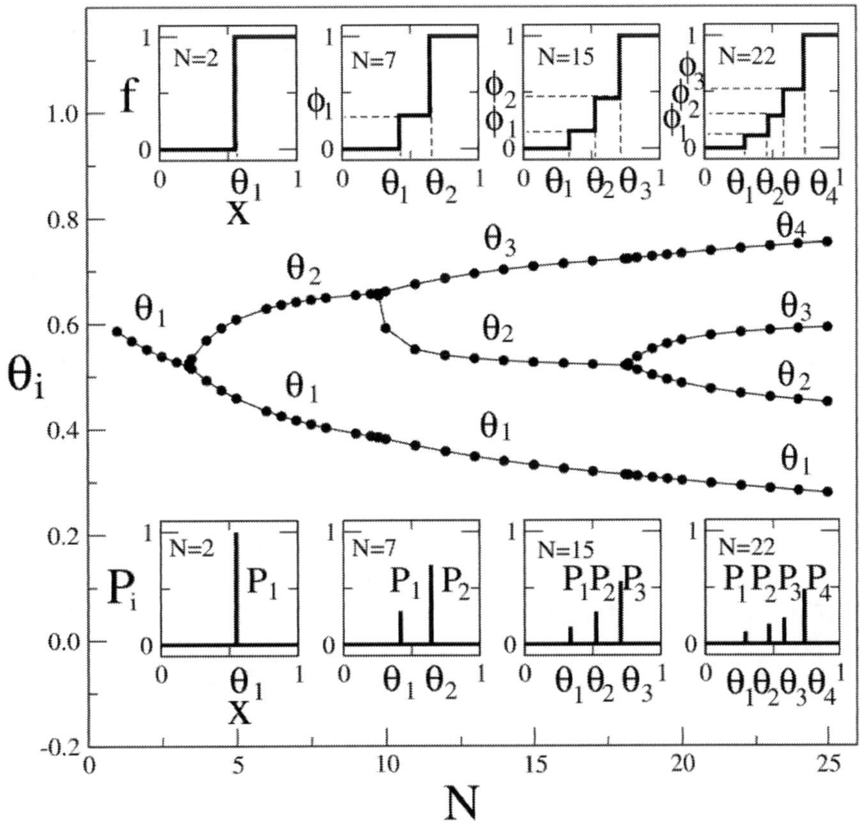

Figure 7. The partition boundaries θ_i against the maximum mean spike count, $N = T\nu_{\max}$. Also shown for $N = 2, 7, 15$ and 22, are the optimal $f(x)$ (top insets) and the population distributions (bottom insets). The parameters are $x_{\min} = 0$, $x_{\max} = 1$, $\nu_{\min} = 0$, and x uniformly distributed. Reproduced from Ref. [19]

quaternary coding scheme for $N = 15$ has three subpopulations with optimal individual firing rates proportional to γ_1, γ_2 and $1 - \gamma_1 - \gamma_2$, and overall rates 0, ϕ_1, ϕ_2 and 1. The sizes of the subpopulations are therefore $P_1 \propto \gamma_1 = N\phi_1$, $P_2 \propto \gamma_2 = N(\phi_2 - \phi_1)$ and $P_3 = 1 - P_1 - P_2$, as shown in the lower insets in Fig. 7.

3. Discussion and Conclusions

We have investigated the neural parameters that maximise the mutual information in two very different neural populations. In the first we used McCulloch-Pitts neurons and introduced noise by adding noise to each neuron's input. In the second system we considered each neuron to have Poisson statistics and quantified the amount of 'noise' in the system as the time window used to define the rate. In both these studies we came to identical conclusions. To optimise the mutual information it is necessary to choose the neural parameters so as to form subpopulations. More specifically, each subpopulation has identical neurons that have the same threshold. However, this threshold varies between subpopulations and the relationship between thresholds is chosen so as to maximise the information flow in the population as a whole. Furthermore, the number of subpopulations is strongly linked to the degree of stochasticity; the stronger the noise the fewer the number of subpopulations.

The fact that two seemingly very different systems maximise mutual information in the same manner suggests that our results are potentially fundamental and generic. The generic nature of the results appears to stem from the fact that the majority of channels known in engineering achieve channel capacity for discrete input signal distributions [21, 22, 23, 24, 25]. Whilst neural channels cannot control directly the signal distribution, it appears that they organise their thresholds to effect directly a quantization of the signal; thus approximating the capacity achieving input distribution. Consequently, if neural channels maximise Shannon mutual information we expect to observe subpopulations.

Furthermore, we have shown that the level of noise strongly influences the structure of the neural system. Small noise leads to a diversity in the selection of neural parameters whereas large noise results in a small number of subpopulations and hence a large degree of commonality in the neural parameters. This indicates that the level of noise intrinsic to the sensory system is a significant influence on the form that an optimal code will take.

Acknowledgments

We thank Steven Holmes, Riccardo Mannella and Jan A. Freund for fruitful discussions. This work was funded by the EPSRC (grant GR/R35650/01 and EP/D05/1894/1(P)). Mark D. McDonnell is funded by the Australian Research Council, Grant No. DP0770747, and Australia's ACoRN (ARC Communications Research Network).

References

[1] Faisal A A, Selen L P J and Wolpert D M 2008 *Nat. Rev. Neurosci.* **9** 292
[2] Shannon C E and Weaver W 1949 *The mathematical theory of communication* (Urbana: University of Illinois Press)
[3] Quiroga R Q and Panzeri S 2009 *Nat. Rev. Neurosci.* **10** 173
[4] McDonnell M D, Amblard P O and Stocks N G 2009 *J. Stat. Mech.* P01012
[5] Linsker R 1988 *IEEE Computer* **21** 105
[6] Barlow H 1961 *Possible principles underlying the transformations of sensory messages Sensory Communication* (Cambridge, MA: MIT press)
[7] Bell A J and Sejnowski T J 1995 *Neural Comput.* **7** 1129
[8] Smith E C and Lewicki M 2006 *Nature* **439** 978
[9] Stocks N G 2000 *Phys. Rev. Lett.* **84** 2310
[10] McDonnell M D, Abbott D and Pearce C E M 2002 *Microelectron. J.* **33** 1079
[11] Nocedal J and Wright S J 1999 *Numerical Optimization* (New York: Springer-Verlag)
[12] McDonnell M D, Stocks N G, Pearce C E M and Abbott D 2008 *Stochastic Resonance: From Suprathreshold Stochastic Resonance to Stochastic Signal Quantisation* (Cambridge: Cambridge University Press)
[13] Gerstner W and Kistler W 2002 *Spiking Neuron Models* (Cambridge: Cambridge University Press)
[14] Dayan P and Abbott L F 2001 *Theoretical Neuroscience* (Cambridge: MIT Press)
[15] Lansky P, Rodriguez R and Sacerdote L 2004 *Neural Comput.* **16** 477
[16] Brunel N and Nadal J 1998 *Neural Comput.* **10** 1731
[17] Bethge M, Rotermund D and Pawelzik K 2003 *Phys. Rev. Lett.* **90** 088104
[18] McDonnell M D and Stocks N G 2008 *Phys. Rev. Lett.* **101** 058103
[19] Nikitin A P, Stocks N G, Morse R P and McDonnell M D 2009 *Phys. Rev. Lett.* **103** 138101
[20] Teich M C and McGill W J 1976 *Phys. Rev. Lett.* **36** 754
[21] Kabanov Y M 1978 *Theory Probab. Appl.* **23** 143
[22] Davis M H A 1980 *IEEE Trans. Inform. Theory* **IT-26** 710
[23] Shamai S 1990 *IEE Proc. I Commun. Speech Vis.* **137** 424
[24] Smith J G 1971 *Information and Control* **18** 203
[25] Huang J and Meyn S P 2005 *IEEE Trans. Inform. Theory* **51** 2336
[26] Ikeda S and Manton J H 2009 *Neural Comput.* **21** 1714
[27] Blahut R E 1987 *Principles and practice of information theory* (Reading, Massachusetts: Addison-Wesley)
[28] Johnson D H and Goodman I N 2008 *Network: Computation in Neural Systems* **19** 13
[29] Salinas E 2006 *PLoS Biology* **4** 2383
[30] Rose K 1994 *IEEE Trans. Inform. Theory* **40** 1939

International Workshop on Statistical-Mechanical Informatics 2009 (IW-SMI 2009) IOP Publishing
Journal of Physics: Conference Series **197** (2009) 012016 doi:10.1088/1742-6596/197/1/012016

Is the Langevin phase equation an efficient model for oscillating neurons?

Keisuke Ota[1,2], Takamasa Tsunoda[1], Toshiaki Omori[3,4], Shigeo Watanabe[5], Hiroyoshi Miyakawa[5], Masato Okada[3,4] and Toru Aonishi[1,4]

[1] Interdisciplinary Graduate School of Science and Engineering, Tokyo Institute of Technology, Yokohama, Japan
[2] Research Fellow of the Japan Society for the Promotion of Science
[3] Graduate School of Frontier Sciences, The University of Tokyo, Kashiwa, Japan
[4] RIKEN Brain Science Institute, Wako, Japan
[5] School of Life Science, Tokyo University of Pharmacy and Life Science, Hachiohji, Japan

E-mail: keisuke@acs.dis.titech.ac.jp

Abstract. The Langevin phase model is an important canonical model for capturing coherent oscillations of neural populations. However, little attention has been given to verifying its applicability. In this paper, we demonstrate that the Langevin phase equation is an efficient model for neural oscillators by using the machine learning method in two steps: (a) Learning of the Langevin phase model. We estimated the parameters of the Langevin phase equation, i.e., a phase response curve and the intensity of white noise from physiological data measured in the hippocampal CA1 pyramidal neurons. (b) Test of the estimated model. We verified whether a Fokker-Planck equation derived from the Langevin phase equation with the estimated parameters could capture the stochastic oscillatory behavior of the same neurons disturbed by periodic perturbations. The estimated model could predict the neural behavior, so we can say that the Langevin phase equation is an efficient model for oscillating neurons.

1. Introduction

Coherent oscillation of neural populations has been reported in various regions of the brain, such as the olfactory bulb, hippocampus, and cortical layers 2/3. Phase reduction [1, 2] is an efficient way to gain an understanding of such collective behavior of neural populations. For a quarter century, many researchers have conducted phase-reduction-based theoretical analyses of these coherent oscillations, for instance, mean field analyses of globally coupled oscillatory neurons [3, 4] and synchronization of neural populations induced by a common noise [5–8]. Inspired by these theoretical studies, many research groups have sought to measure phase response curves (PRCs) from real neurons, e.g., from mitral cells of mouse olfactory bulb [9, 10], stellate cells of rat entorhinal cortex [11, 12], pyramidal cells of rat neocortex layer 5 [13, 14], pyramidal cells of rat hippocampal CA3 [15], interneurons of rat and mouse barrel cortex [16] and so on [17–21]. Several of them provide insight into the nature of computation [10, 15] as well as neural synchrony.

We should point out that the above theoretical studies implicitly or explicitly assume that the neural dynamics obey the Langevin phase equation defined in equation (2). The neural dynamics

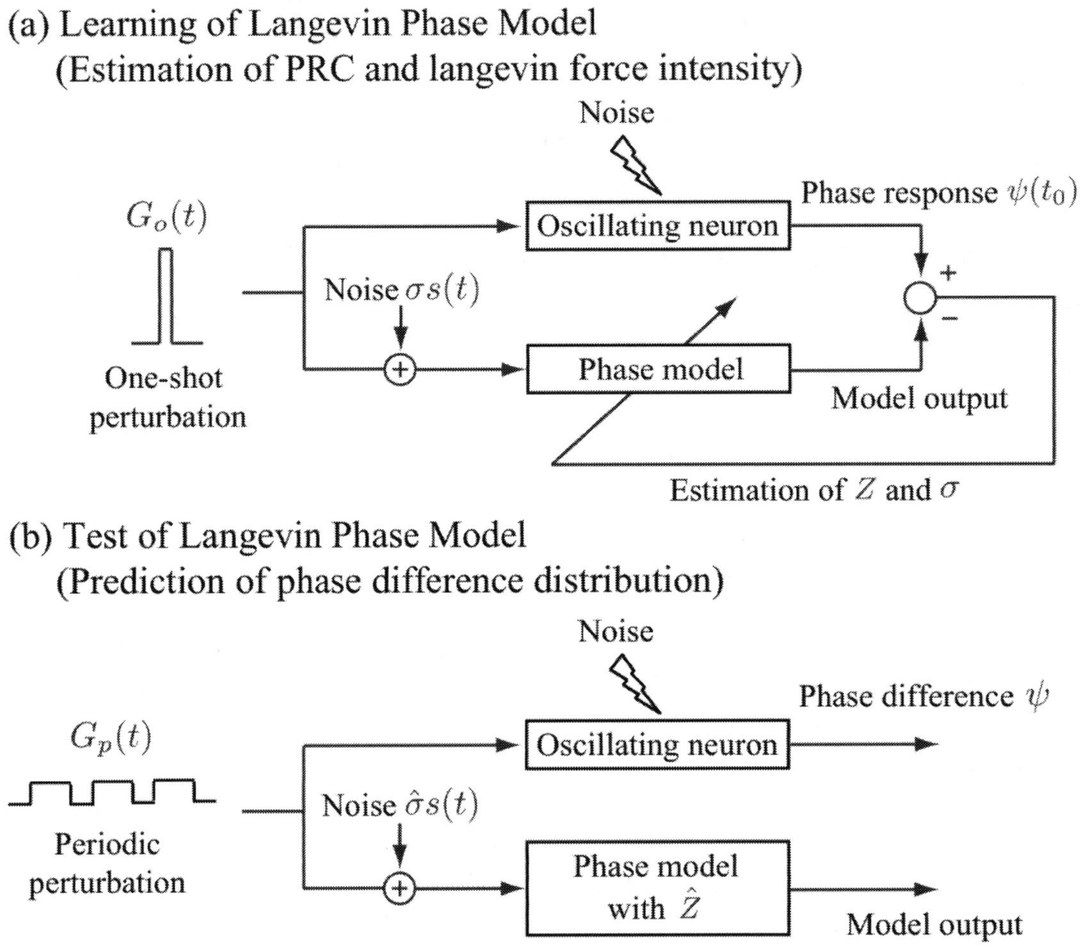

Figure 1. We will verify whether the Langevin phase equation (2) is an efficient model for neural oscillators through (a) learning of the Langevin phase model and (b) testing of the estimated model. (a) We observe the phase response $\psi(t_0)$ of a neural oscillator disturbed by a one-shot perturbation $G_o(t-t_0)$ and estimate the Z and σ that optimally reproduce the observed data set $\{G_o(t-t_0), \psi_0(t_0)\}$ (Section 3). (b) We predict the stochastic behavior of neural oscillators disturbed by a periodic perturbation $G_p(t)$ by using the Langevin phase model with the estimated parameters $\hat{Z}$ and $\hat{\sigma}$ and compare the model output with experimental data (phase difference ψ) (Section 4). If the estimated model can predict the neural dynamics, we can say that the Langevin phase equation is an efficient model for oscillating neurons.

are thus formulated in terms of the Langevin phase equation, which is a common starting point for these studies. However, little attention has been given to verifying the applicability of the equation. In previous experimental studies, various phase equations have been shown to be able to describe the oscillatory behavior of real neurons [11, 18]. However, the versatility and mathematical tractability of these models are lower than those of the Langevin phase equation. In this paper, we use machine learning technique in an attempt to demonstrate that the Langevin phase equation is an efficient model for oscillating neurons.

We conducted an electrophysiological experiment on pyramidal neurons of the rat

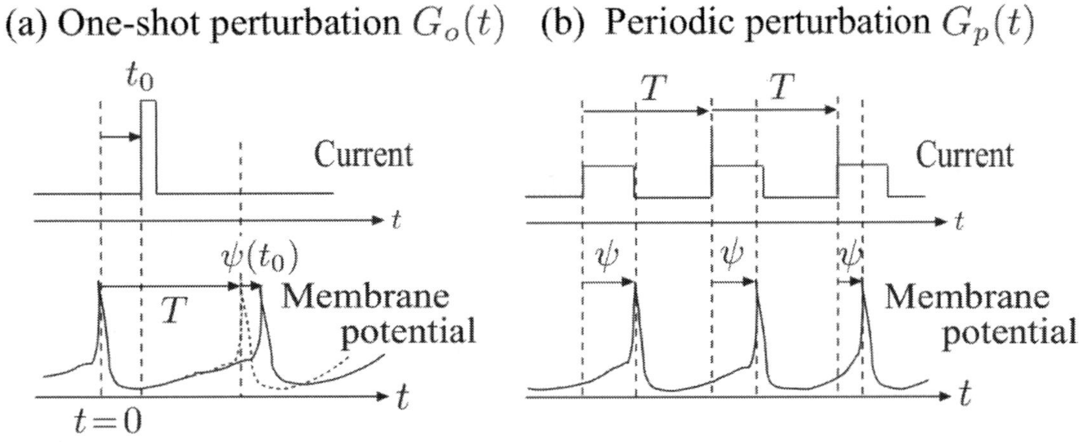

Figure 2. Schematic diagram of perturbation-response experiments for observing phase shifts. In figure. (a) and (b), regular firing induced by a depolarizing constant current is disturbed by a small perturbation current and inherent noise. Two different perturbations are used: (a) A one-shot perturbation $G_o(t - t_0)$ is superimposed on a constant current at various timings t_0 relative to the last spike, and a the phase shift $\psi(t_0)$ is measured to see how much the period of the oscillator increases or decreases. This experiment gives an estimate of PRC Z and the intensity of Langevin force σ. (b) A periodic perturbation $G_p(t)$ consisting of successive pulses is superimposed on a constant current, and the phase differences ψ between spikes and pulses are measured. This experiment gives the histogram of the phase difference, which can then be compared with the stationary distribution derived analytically.

hippocampal CA1 and verified whether the Langevin phase equation described their neural dynamics. First, by using the Bayesian approach from our previous study [22], we estimated the parameters of the Langevin phase model, i.e., the PRC and the intensity of noise in physiological data measured in perturbation-response experiments (Figure 1(a)). This procedure corresponds to the learning of the Langevin phase model. Next, we verified whether the Fokker-Planck equation (FPE) derived from the Langevin phase model with the estimated parameters could capture the stochastic oscillatory behavior of real neurons disturbed by periodic perturbations (Figure 1(b)). This procedure corresponds to the test of the Langevin phase model.

2. Theoretical Background

We consider the situation in figure 2 in which rhythmical firing with an average period T is disturbed by a sufficiently small perturbation current $\varepsilon G(t)$ and additive noise $\sqrt{\varepsilon}\sigma s(t)$:

$$\frac{dX}{dt} = F(X) + \varepsilon G(t) + \sqrt{\varepsilon}\sigma s(t), \tag{1}$$

where $G(t)$ is a representation including two perturbation types: one-shot perturbation $G_o(t)$ and periodic perturbation $G_p(t)$, $\sqrt{\varepsilon}\sigma$ is the intensity of the noise, ε is a scaling parameter of the perturbation response, and $s(t)$ is Gaussian white noise (Langevin force) with zero mean and a correlation function given by $\langle s(t)s(t')\rangle = 2\delta(t - t')$. We assume that the unperturbed neural oscillator $dX/dt = F(X)$ has a stable limit cycle solution $X_0(t)$. When ε is sufficiently small ($1 \gg \epsilon > 0$), the high-dimensional system (1) can be reduced to a one-dimensional system expressing the motion of the phase $\phi(t)$ defined along the limit cycle solution $X_0(t)$ as

$$\frac{d\phi}{dt} = 1 + Z(\phi)(\varepsilon G(t) + \sqrt{\varepsilon}\sigma s(t)). \tag{2}$$

This is the Langevin phase equation. Here, $Z(\phi)$ is called the phase-dependent sensitivity, or PRC, and it measures how sensitively the neural oscillator responds to an external perturbation [1, 2]. When we apply the variable transformation $\phi(t) = 1 + \psi(\varepsilon t)$ to equation (2), we can write the slow dynamics of the initial phase ψ as

$$\frac{d\psi}{dt} = Z(t + \psi(\varepsilon t))(\varepsilon G(t) + \sqrt{\varepsilon}\sigma s(t)). \tag{3}$$

By integrating both sides of equation (3) from 0 to T, we get

$$\psi(\varepsilon T) - \psi(0) = \varepsilon \int_0^T dt Z(t + \psi(\varepsilon t)) \cdot G(t) + \sqrt{\varepsilon}\sigma \int_0^T dt Z(t + \psi(\varepsilon t)) \cdot s(t) \tag{4}$$

Then, if $1 \gg \epsilon > 0$, we can expand the right-hand side of equation (4) [23],

$$\begin{aligned} \psi(\varepsilon T) - \psi(0) &= \sqrt{\varepsilon}\sigma \int_0^T dt Z(t + \psi(0)) \cdot s(t) + \varepsilon \int_0^T dt Z(t + \psi(0)) \cdot G(t) \\ &+ \varepsilon \int_0^T dt \int_0^t dt' Z'(t + \psi(0)) Z(t' + \psi(0)) \cdot s(t)s(t') + O(\varepsilon^{3/2}). \end{aligned} \tag{5}$$

The first moment about the variation of the initial phase ψ during εT is

$$\begin{aligned} \frac{\langle \psi(\varepsilon T) - \psi(0) \rangle}{\varepsilon T} &= \frac{1}{T} \int_0^T dt Z(t + \psi(0)) \cdot G(t) \\ &+ \frac{\sigma^2}{T} \int_0^T dt \int_0^t dt' Z'(t + \psi(0)) Z(t' + \psi(0)) \langle s(t)s(t') \rangle + O(\sqrt{\varepsilon}), \end{aligned} \tag{6}$$

where $\langle \; \rangle$ denotes ensemble averaging. Note that the second term on the right-hand side, which is called the noise-induced drift [24], vanishes because of the periodicity of the PRC. Similarly, the second moment is

$$\frac{\langle (\psi(\varepsilon T) - \psi(0))^2 \rangle}{\varepsilon T} = \frac{\sigma^2}{2!T} \int_0^T dt \int_0^T dt' Z(t + \psi(0)) Z(t' + \psi(0)) \langle s(t)s(t') \rangle + O(\sqrt{\varepsilon}). \tag{7}$$

From equations (6) and (7), we can obtain a Fokker-Planck equation (FPE) in the form [24]:

$$\frac{\partial P(\psi, \tau)}{\partial \tau} = -\frac{\partial}{\partial \psi} D^{(1)}(\psi) P(\psi, \tau) + D^{(2)} \frac{\partial^2}{\partial \psi^2} P(\psi, \tau), \tag{8}$$

$$D^{(1)}(\psi) = \frac{1}{T} \int_0^T dt Z(t + \psi) \cdot G(t), \; D^{(2)} = \frac{\sigma^2}{T} \int_0^T dt Z(t)^2,$$

where $\tau = \epsilon t$. τ denotes a slowly varying time. Therefore, the smallest time step in this FPE becomes ϵT. Note that $D^{(1)}(\psi)$ is the convolution between Z and G, and $D^{(2)}$ is the power of PRC times the variance of the Langevin force σ^2.

3. Estimation of Langevin phase model

We propose an algorithm for estimating the PRC and the noise intensity in the Langevin phase equation (2) from physiological data [22]. First, we formulate the PRC measurement experiment shown in figure 2(a) and analytically derive the probability density of the phase shift evoked by a one-shot perturbation. In the PRC observation process, we observe the neural dynamics in a transient state during $\tau = \varepsilon T$ after the previous spike is elicited. In this case, we can approximate $D^{(1)}(\psi)$ in equation (8) as a constant and obtain an analytical solution of equation (8) under the free boundary condition as

$$P(\psi(t_0), \varepsilon T|\, Z(t), \sigma; G_o(t-t_0)) = \frac{1}{\mathcal{Z}_{like}} \exp\left(- \frac{\left(\psi(t_0) - \varepsilon \int_0^T dt Z(t) G_o(t-t_0)\right)^2}{4\varepsilon\sigma^2 \int_0^T dt Z(t)^2} \right), \tag{9}$$

where the initial distribution is $P(\psi, 0) = \delta(\psi)$, G_o is the one-shot perturbation shown in figure 2(a), and $\mathcal{Z}_{like} = 2\sqrt{\pi\varepsilon\sigma^2 \int_0^T dt Z(t)^2}$ is the partition function for normalizing this probability density. This probability density represents the stochastic fluctuation of the spike timing disturbed by G_o, and it corresponds to the likelihood function of $Z(t)$. (A more detailed derivation of this observation process is given in [22].)

We introduce the following prior probability density to represent the smoothness of the PRC:

$$P(Z(t)|D) = \frac{1}{\mathcal{Z}_{pr}} \exp\left(-\frac{ND}{2} \int_0^T \left(\frac{\partial^n Z(t')}{\partial t'^n}\right)^2 dt' \right) \tag{10}$$

This prior probability density has a $2n$-th order differential smoothness constraint. (In this paper, we use $n = 1$.) N is the number of samples, and D is a hyper-parameter to control the smoothness of $Z(t)$, which is scaled to ND in order to obtain a consistent estimate. $\mathcal{Z}_{ps}$ is the partition function for normalizing this probability density. Because each trial for measuring the phase response $\psi_i(t_0)$ $(i = 1, \cdots, N)$ is independent, Bayes' theorem holds:

$$\begin{aligned} &P(Z(t)|\{\psi_i(\Delta tj)\}, \sigma, D; G_o(t-\Delta tj)) \\ &= \frac{\mathcal{Z}_{pr} \mathcal{Z}_{like}^{LN}}{\mathcal{Z}_{ps}} P(Z(t)|D) \prod_{i=1}^{N} \prod_{j=0}^{L-1} P(\{\psi_i(\Delta tj)\}|Z(t), \sigma; G_o(t-\Delta tj)), \end{aligned} \tag{11}$$

where $\mathcal{Z}_{ps}$ is a partition function for normalizing the posterior probability density, and ψ is sampled as a sequence of equidistant points $\Delta tj (j = 0, 1, \cdots, L-1)$ whose sampling period satisfies $\Delta t = T/L$. We obtain N samples $\{\psi_1(\Delta tj), \cdots, \psi_N(\Delta tj)\}$ at each sampling point $t_0 = \Delta tj$. Therefore, the total number of experimental data becomes LN. Here, we apply the maximum a posteriori (MAP) to equation (11) and estimate PRC as $\hat{Z}(t) = \underset{Z(t)}{\operatorname{argmax}}\, P(Z(t)|\{\psi_i(\Delta tj)\}, \sigma, D; G_o(t-\Delta tj))$.

Note that $\hat{Z}(t)$ depends on the hyper-parameters, i.e., the smoothness parameter D and the intensity of the Langevin force σ. Therefore, we should also estimate these parameters. For this, we can utilize the maximization of the following marginal likelihood, (i.e., type 2 maximum likelihood):

$$\mathcal{F}(D, \sigma|\{\psi_i(\Delta tj)\}) = \log \mathcal{Z}_{ps} - \log \mathcal{Z}_{like}^{LN} - \log \mathcal{Z}_{pr}, \tag{12}$$

which is referred to as the free energy in physics. Because the variance $2\varepsilon\sigma^2 \int_0^T dt Z(t)^2$ in equation (9) is proportional to the power of $Z(t)$, the posterior probability density is equivalent to the spherical spin model [25]. We can obtain an analytical solution of $\mathcal{F}$ and estimate the optimal hyper-parameters $\hat{D}$ and $\hat{\sigma}$ that maximize the marginal likelihood $\mathcal{F}$ [22].

We applied our algorithm to physiological data of rat hippocampal CA1 pyramidal neurons. Figure 3 shows the estimated PRC and hyper-parameters. The estimated PRC was fully positive, which indicates that hippocampal CA1 pyramidal neurons could be classified as Type I excitability neurons [26], while the estimated intensity of the Langevin force was 9.0. This result suggests that the noise inherent in the neuron is 9.0 pA.

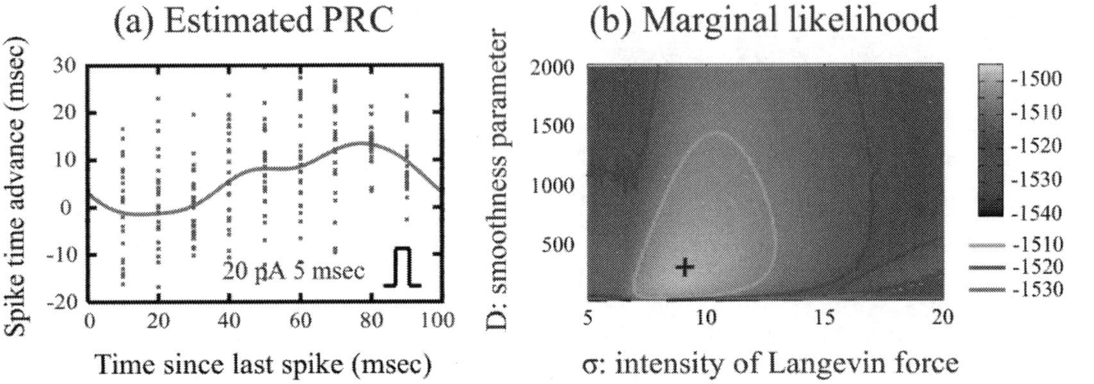

Figure 3. Estimated PRC and hyper-parameters of a pyramidal neuron in a rat hippocampal CA1. (a) The plotted crosses indicate the phase responses at various timings (x-axis) measured in the electrophysiological experiment (Figure 2(a)). A positive spike time advance (y-axis) means that the inter-spike interval (ISI) becomes shorter than the unperturbed period (100 msec), and a negative value means a longer ISI. The regularly firing with an average period 100 msec was disturbed by rectangular pulses with height of 20.0 pA and width of 5.0 msec displayed at the bottom right, and a total of 260 samples were obtained. The solid line denotes the MAP solution with the optimal hyper-parameters $(\hat{D}, \hat{\sigma})$ estimated by maximizing the marginal likelihood. (b) The landscape of the marginal likelihood indicates that the optimal values of the hyper-parameters, i.e., the smoothness parameter D, and the intensity of Langevin force σ, are $(\hat{D}, \hat{\sigma}) = (224.0, 9.0)$ (cross).

4. Test of Langevin phase model

Next, we validate the Langevin phase model with the parameters estimated in Section 3: $\hat{Z}$ and $\hat{\sigma}$. Here, we ascertain whether the stochastic behavior of oscillating neurons receiving the periodic perturbations $G_p(t)$ consisting of successive pulses as illustrated in figure 2(b) and satisfying $G_p(t) = G_p(t+T)$ could be predicted by the FPE (8) with $\hat{Z}$ and $\hat{\sigma}$. This FPE describes the evolution of the probability density of the phase difference ψ between spikes and pulses. Here, we focus on the stationary stochastic phase motion of the neurons whose spikes are entrained by periodic perturbations $G_p(t)$ and disturbed by a Langevin force $s(t)$. The stationary distribution of the phase difference is obtained by solving equation (8) under periodic boundary conditions:

$$P(\psi|\hat{Z}(t), \hat{\sigma}; G_p(t)) = \frac{1}{\mathcal{Z}_p}\left\{\exp\left(\int_0^{\psi} d\psi' \frac{D^{(1)}(\psi')}{D^{(2)}}\right) + C\int_0^{\psi} d\psi' \exp\left(\int_{\psi'}^{\psi} d\psi'' \frac{D^{(1)}(\psi'')}{D^{(2)}}\right)\right\}, \quad (13)$$

$$C = \frac{1 - \exp\left(\int_0^T d\psi' \frac{D^{(1)}(\psi')}{D^{(2)}}\right)}{\int_0^T d\psi' \exp\left(\int_{\psi'}^T d\psi'' \frac{D^{(1)}(\psi'')}{D^{(2)}}\right)},$$

$$D^{(1)}(\psi) = \frac{1}{T}\int_0^T dt \hat{Z}(t+\psi) \cdot G_p(t), \; D^{(2)} = \frac{\hat{\sigma}^2}{T}\int_0^T dt \hat{Z}(t)^2,$$

where $\hat{Z}$ is the estimated PRC, $\hat{\sigma}$ the estimated intensity of the Langevin force, and $\mathcal{Z}_p$ an integration constant. When $G_p(t)$ does not include the DC component, C is zero. In this case,

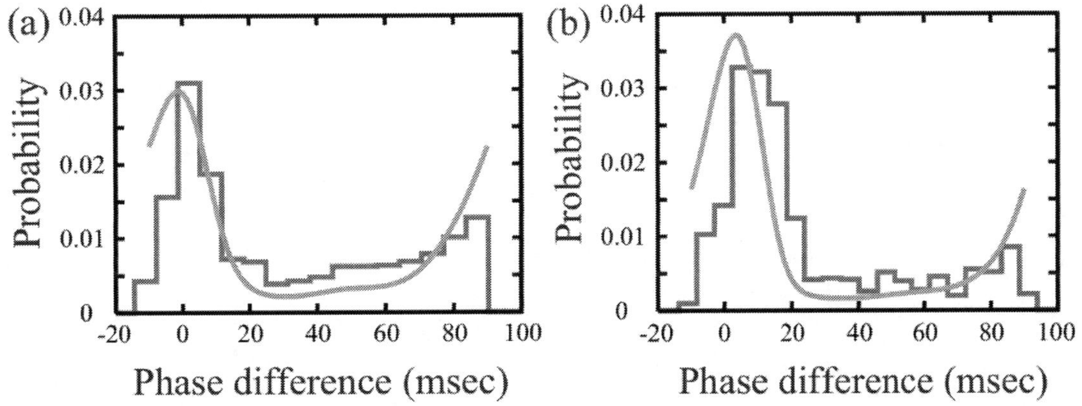

Figure 4. Comparison of the probability density of phase difference obtained from FPE (13) (green line) and the histogram of phase difference calculated from the physiological data in the perturbation-response experiment shown in figure 2(b) (red line), where the frequency of the periodic perturbation is the same as the average period of the neural spikes without the periodic perturbation, 100 msec. The x-axis indicates the phase difference between the spikes and pulses. A positive x-axis value means that the spike timing was delayed relative to the periodic perturbation, and a negative value represents advanced spike timing. The shape of each periodic rectangular perturbation was (a) height 20.0 pA, width 5.0 msec, interval 95.0 msec (the same value used in the PRC estimation) and (b) height 10.0 pA, width 20.0 msec, interval 80.0 msec (different value). Both perturbations include the DC component. The histograms were obtained from (a) 950 samples and (b) 1200 samples.

equation (13) is described by the just first term and we can explicitly define the potential as $\Phi(\psi) = -\int_0^{\psi} d\psi' D^{(1)}(\psi')/D^{(2)}$.

We measured the phase differences ψ between neural spikes and pulses in the same neuron whose PRC and noise intensity had been estimated. The frequency of the periodic perturbation was the same as the average period of the unperturbed neural oscillator. As figure 4 shows, the stationary distribution $P(\psi)$ obtained from equation (13) (green line) matches the histogram obtained from the experimental data (red line), even when the shape of the perturbation is not the same as the one used in the PRC measurement (Figure 4(b)). This result indicates that equation (13) can predict the stochastic phase motion of neurons driven by a periodic perturbation; that is, the Langevin phase equation has a great ability to describe real neural oscillators.

5. Conclusion

We verified that the Langevin phase equation actually describes neural dynamics. First, by using the Bayesian approach, we estimated the PRC and the Langevin force intensity from physiological data measured in rat hippocampal CA1 pyramidal neurons. Then, we demonstrated that the FPE derived from the estimated model could predict the stochastic oscillatory behavior of real neurons, as revealed in figure 4. Thus, we conclude that the Langevin phase equation can describe the oscillatory neurons. A future direction of this study will be to ascertain whether the Langevin phase equation can predict the stochastic behavior of the oscillating neurons given more realistic periodic perturbations, e.g., periodic excitatory postsynaptic current (EPSC) and inhibitory postsynaptic current (IPSC) or a sinusoidal current. If we can predict the neural dynamics not only in pyramidal cells, but also in other types of

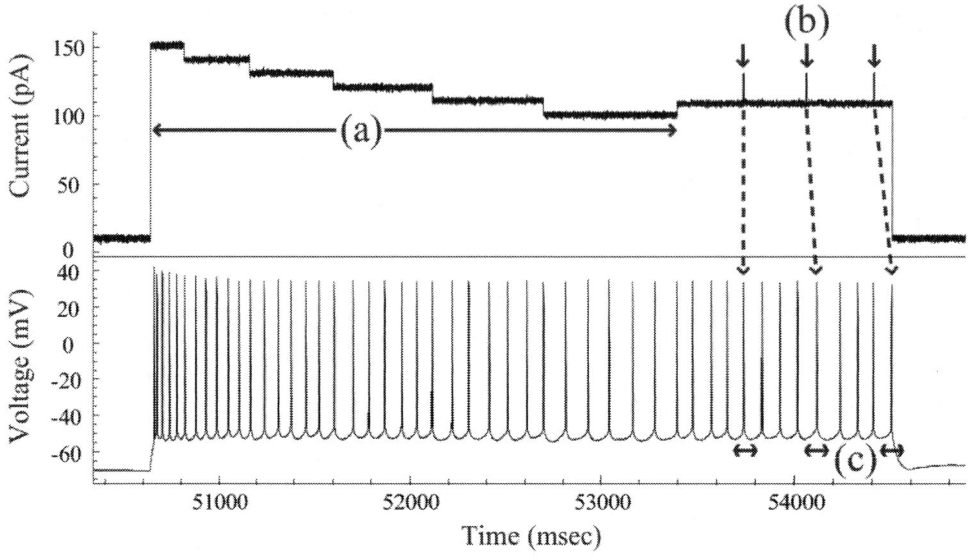

Figure 5. Protocol of the perturbation-response experiment using the dynamic clamp method for a one-shot perturbation G_o. Top and bottom traces are examples of the injected current and membrane potential of the rat hippocampal CA1 pyramidal neuron, respectively. In (a), the mean ISI is adjusted to 100 msec by tuning the DC depolarizing current. In (b), a one-shot perturbation G_o is evoked using various timings relative to the last spike, and in (c), we measure how the perturbation current disturbs the timing of the next spike firing. Here, the neuron was perturbed every five periods in order to reduce the effect of the last perturbation. The dynamic clamp automatically repeats this protocol for measuring over 200 samples.

neurons like interneurons, this would prove the efficacy of the Langevin phase equation more generally.

Slice preparation and recording

Whole-cell patch clamp recordings were obtained from the somas of hippocampal CA1 pyramidal neurons of wister rats (27, 28-day old). The membrane potentials of neurons were recorded in current clamp mode with an internal solution containing the following (in mM): 120 mM K gluconate, 20 mM KCl, 10 mM HEPES, 4.0 mM NaCl, 4.0 mM Mg_2-ATP, 0.2 mM EGTA, 4.0 mM Na_2-ATP, 0.3 mM Tris-GTP, 14 mM Phosphocreatine-disodium salt. During the recordings, the slices were perfused continuously with normal ACSF ($29-31$ °C) [which contained (in mM) 125 mM NaCl, 2.5 mM KCl, 1.25 mM NaH_2PO_4, 25 mM $NaHCO_3$, 2.5 mM $CaCl_2$, 1.5 mM $MgCl_2$, 10 mM dextrose], in which 100 μM PTX, 10 μM CNQX, 50 μM APV were included to block most intrinsic synaptic conductances. We measured 2 neurons from 2 slices taken from 2 rats.

According to the definition of PRC, the inter-spike interval (ISI) must be steady as long as the perturbation-response experiment for measuring the PRC lasts. However, it is difficult to keep the ISI constant for a long time because of variations in intracellular ionic concentrations. In much of the previous work, the ISI has been normalized for assigning phase values to spike response data of real neurons. In our experiment, we tried to measure mathematically ideal PRCs from real neurons by using the dynamic clamp [27, 28]. Figure 5 shows the protocol of the perturbation-response experiment for the PRC measurement. In the first step, the mean

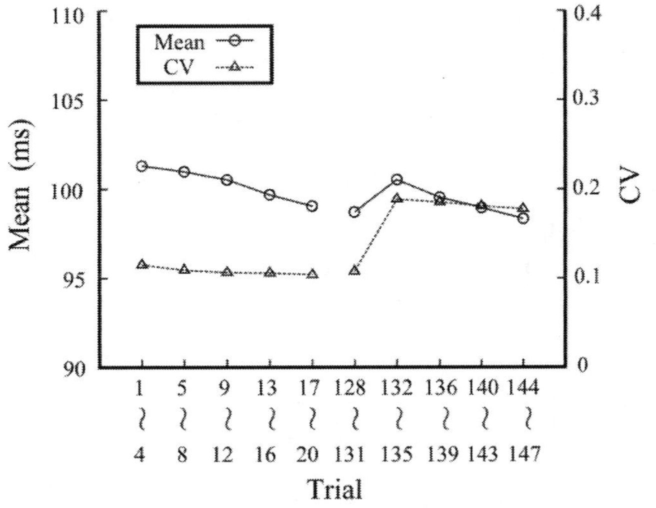

Figure 6. Mean and CV of unperturbed ISI in the first and last 20 trails of the PRC measurement.

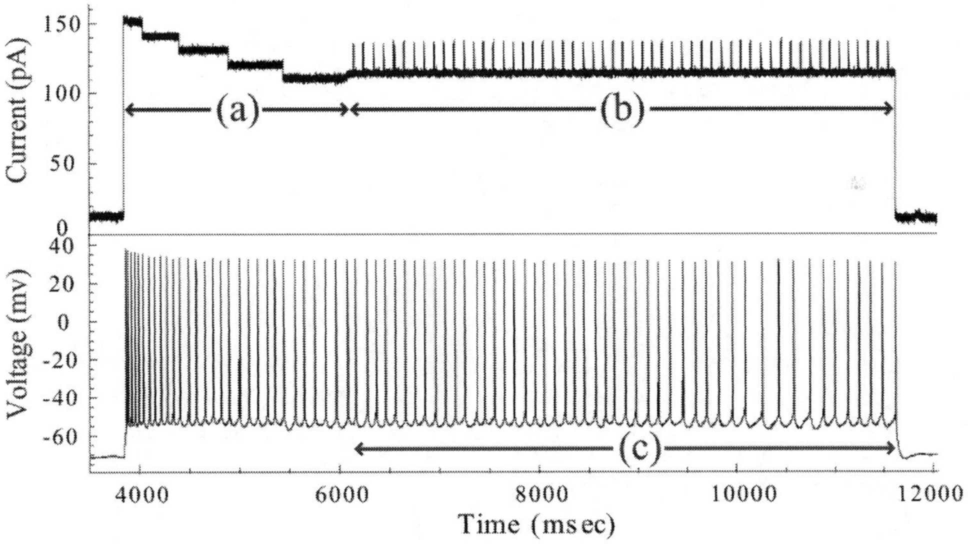

Figure 7. Protocol of the perturbation-response experiment using the dynamic clamp method for a periodic perturbation G_p. The traces are examples of the injected current and membrane potential of the same neuron used for the PRC measurement. In (a), the mean ISI is adjusted to 100 msec by tuning the DC depolarizing current as in figure 5. In (b), a periodic perturbation G_p with a period of 100 msec is superimposed on the DC depolarizing current. In (c), phase differences between the periodic perturbation G_p and spikes are measured. The dynamic clamp automatically repeats this protocol for measuring about 1000 samples.

ISI is adjusted to 100 msec by tuning the DC depolarizing current. In the next step, every five periods, one-shot perturbation G_o is evoked using various timings relative to the last spike, and we measure how the perturbation current disturbs the timing of the next spike firing. The dynamic clamp automatically repeats this protocol for measuring over 200 samples. As shown in figure 6, by using this protocol, the mean ISI could be maintained for almost 100 msec (as long as the perturbation experiment), and the CV (Coefficient of Variation) of ISI is from 10 to 20 % of the mean ISI. We also used the dynamic clamp to measure the phase differences between the periodic perturbation G_p and spikes. Figure 7 shows the protocol for the phase difference measurement. In the first step, the mean ISI is adjusted to 100 msec by tuning the DC depolarizing current. In the next step, and successive rectangle pulses are superimposed on the DC depolarizing current, and differences between the periodic perturbation G_p and spikes were measured. In all experiments, the spike time was defined as the time at which the membrane potential crossed 0 mV going upward.

Acknowledgments

This work was partially supported by Grants-in-Aid for Scientific Research (No. 20509001 for T. O., No.19300096 for H. M., Nos. 18079003, 20240020, 20650019 for M. O., No. 20500201 for T. A.) from MEXT and JSPS, Japan. K. O. was supported by JSPS Research Fellowship for Young Scientists.

References

[1] Winfree A 1967 *J. Theor. Biol.* **16** 15
[2] Kuramoto Y 1984 *Chemical Oscillations, Waves, and Turbulence* (Berlin: Springer-Verlag)
[3] Tsubo Y, Teramae J and Fukai T 2007 *Phys. Rev. Lett.* **99** 228101
[4] Neltner L, Hansel D, Mato G and Meunier C 2000 *Neural Comput.* **12** 1607
[5] Teramae J and Tanaka D 2004 *Phys. Rev. Lett.* **93** 204103
[6] Galán R F, Ermentrout G B and Urban N N 2007 *Phys. Rev.* E **76** 056110
[7] Nakao H, Arai K and Kawamura Y 2007 *Phys. Rev. Lett.* **98** 184101
[8] Marella S and Ermentrout G B 2008 *Phys. Rev.* E **77** 041918
[9] Galán R F, Ermentrout G B and Urban N N 2005 *Phys. Rev. Lett.* **94** 158101
[10] Ermentrout G B, Galán R F and Urban N N 2007 *Phys. Rev. Lett.* **99** 248103
[11] Netoff T I, Banks M I, Dorval A D, Acker C D, Haas J S, Kopell N and White J A 2005 *J. Neurophysiol.* **93** 1197
[12] Netoff T I, Acker C D and Bettencourt J C 2004 *J. Comput. Neurosci.* **18** 287
[13] Stoop S, Schindler K and Bunimovich L 2000 *Nonlinearity* **13** 1515
[14] Goldberg J A, Deister C A and Wilson C J 2007 *J. Neurophysiol.* **97** 208
[15] Lengyel M, Kwag J, Paulsen O and Dayan P 2005 *Nature Neurosci.* **8** 1677
[16] Mancilla J G, Lewis T J, Pinto D J, Rinzel J and Connors B W 2007 *J. Neurosci.* **27** 2058
[17] Preyer A J and Butera R J 2005 *Phys. Rev. Lett.* **95** 138103
[18] Oprisan S A, Prinz A A and Canavier C C 2004 *Biophys. J.* **87** 2283
[19] Tateno T and Robinson H P C 2007 *Biophys. J.* **92** 683
[20] Reyes A D and Fez E E 1993 *J. Neurophysiol.* **69** 1661
[21] Tsubo Y, Takada M, Reyes A D and Fukai T 2007 *Eur. J. Neurosci.* **25** 3429
[22] Ota K, Omori T and Aonishi T 2009 *J. Comput. Neurosci.* **26** 185
[23] Teramae J and Fukai T 2008 *Phys. Rev. Lett.* **101** 248105
[24] Risken H 1989 *The Fokker-Planck Equation: Methods of Solution and Applications* (Berlin: Springer)
[25] Berlin T H and Kac M 1952 *Phys. Rev.* **86** 821
[26] Rinzel J R and Ermentrout G B 1998 *Methods in Neuronal Modeling* ed Koch C and Segev I (Cambridge: MIT Press) p 251
[27] Robinson H and Kawai N 1993 *J. Neurosci. Methods* **49** 157
[28] Sharp A A, O'Neil M B, Abbott L F and Marder E 1993 *J. Neurophysiol.* **69** 992

Journal of Physics: Conference Series **197** (2009) 012017 doi:10.1088/1742-6596/197/1/012017

Tracking dynamics of two-dimensional continuous attractor neural networks

C C Alan Fung[1], K Y Michael Wong[1] and Si Wu[2]

[1]Department of Physics, The Hong Kong University of Science and Technology, Hong Kong, China

[2]Lab of Neural Information Processing, Institute of Neuroscience, Shanghai, China

E-mail: alanfung@ust.hk, phkywong@ust.hk, siwu@ion.ac.cn

Abstract. We introduce an analytically solvable model of two-dimensional continuous attractor neural networks (CANNs). The synaptic input and the neuronal response form Gaussian bumps in the absence of external stimuli, and enable the network to track external stimuli by its translational displacement in the two-dimensional space. Basis functions of the two-dimensional quantum harmonic oscillator in polar coordinates are introduced to describe the distortion modes of the Gaussian bump. The perturbative method is applied to analyze its dynamics. Testing the method by considering the network behavior when the external stimulus abruptly changes its position, we obtain results of the reaction time and the amplitudes of various distortion modes, with excellent agreement with simulation results.

1. Introduction

Continuous attractor neural networks (CANNs) are very useful models for describing the encoding of continuous stimuli in neural systems [1–5]. The encoded stimuli can either be some simple features of objects, such as their orientations [6], moving directions [7] and spatial locations [8], or some complicated rules that underly the categorization of objects [9]. Compared with other attractor models, CANNs have the distinctive feature that they hold a family of stationary states which can be translated into each other without the need to overcome energy barriers. In the continuum limit, these stationary states form a continuous manifold in which the system is neutrally stable, and the network state can translate easily when the external stimulus changes continuously. Beyond pure memory retrieval, this large-scale structure of the state space endows the neural system with a tracking capacity.

To construct a model for CANN, the key is that the neuronal interactions should be properly balanced in excitation and inhibition (e.g., of the Mexican-hat shape) and be translationally invariant. The former enables the network to have a local persistent bump solution and the latter ensures that the network has a continuous family of such solutions. Although mathematically it is possible to construct a CANN of dimensionality larger than two, the research on CANNs has so far been mainly focused on one or two dimensional case. This is because in the cortex, neurons are essentially distributed in a two-dimensional sheet. To maintain a CANN of dimensionality larger than two, it is difficult to wire neurons in a two-dimensional space without their signals interfering with each other. To encode continuous features of high dimensionality, the brain may employ layers of neurons to combine low-dimensional CANNs hierarchically.

The tracking dynamics of a CANN has been theoretically investigated by several authors in the literature [5, 6, 10–12]. These studies have demonstrated that a CANN has the capacity of tracking a moving stimulus continuously and that this tracking behavior can describe many brain functions well. Despite these successes, however, detailed rigorous analysis of tracking behaviors of a CANN is still lacking. In a recent work [13, 14], the authors have developed a perturbative approach to elucidate the tracking performance of a one-dimensional CANN clearly. Because of the bump shape of the stationary states of the network, we used the wave functions of the quantum harmonic oscillator as the basis to decompose the network dynamics into different motion modes. These modes have clear physical meanings, corresponding to distortions in the amplitude, position, width or skewness of the network state. Due to the neutral stability of network states, the dynamics of a CANN is typically dominated by a few motion modes, with their contributions determined by the corresponding eigenvalues. We therefore can project the network dynamics on these dominating modes and simplify the network dynamics significantly. In this study, we extend the perturbative approach to a two-dimensional CANN. The two-dimensional CANN has much richer dynamics and distortion patterns than the one-dimensional case [15–17]. To elucidate the effect of different distortion patterns on the network dynamics clearly, we develop the perturbative approach in both rectangular and polar coordinates. To test our method, we study the tracking performance of the network when the external stimulus position experiences an abrupt change. Simulation results confirm that our method works very well.

2. The Model

We consider a two-dimensional neural network coding the stimulus $\mathbf{x} = (x_1, x_2)$, with N neurons distributed over this space. For simplicity, the neurons are assumed to be uniformly distributed in the space. Considering the common case that the range of possible values of the stimulus being much larger than the range of neuronal interactions, we can effectively take $-\infty < x_1, x_2 < \infty$. The dynamics of the synaptic input $U(\mathbf{x}, t)$ and neuronal response $r(\mathbf{x}, t)$ is given by

$$\tau \frac{\partial U(\mathbf{x},t)}{\partial t} = I^{\rm ext}(\mathbf{x}) + \rho \int_{-\infty}^{\infty} \int_{-\infty}^{\infty} d\mathbf{x}' J(\mathbf{x}, \mathbf{x}') r(\mathbf{x}') - U(\mathbf{x}, t); \quad (1)$$

$$r(\mathbf{x}, t) = \frac{U(\mathbf{x}, t)^2}{1 + k\rho \int_{-\infty}^{\infty} \int_{-\infty}^{\infty} d\mathbf{x}' U(\mathbf{x}', t)^2}, \quad (2)$$

where $J(\mathbf{x}, \mathbf{x}')$ is the translationally invariant coupling function defined by

$$J(\mathbf{x}, \mathbf{x}') = \frac{A}{2\pi a^2} \exp\left[-\frac{|\mathbf{x} - \mathbf{x}'|^2}{2a^2}\right], \quad (3)$$

a is the tunning width of the neural network, k is the global inhibition, and ρ is the density of neurons over the space. When $I^{\rm ext} = 0$ and $0 < k < k_c \equiv A^2\rho/(32\pi a^2)$, we have the steady solutions, or *stationary states*, given by (see figure 1)

$$\tilde{U}(\mathbf{x}|\mathbf{z}) = U_0 \exp\left[-\frac{|\mathbf{x} - \mathbf{z}|^2}{4a^2}\right], \quad (4)$$

$$\tilde{r}(\mathbf{x}|\mathbf{z}) = r_0 \exp\left[-\frac{|\mathbf{x} - \mathbf{z}|^2}{2a^2}\right], \quad (5)$$

where $U_0 = [1 + (1 - k/k_c)^{1/2}]A/(8\pi a^2 k)$ and $r_0 = [1 + (1 - k/k_c)^{1/2}]/(4\pi a^2 k\rho)$. It is notable that Eqs. (4) and (5) are valid for any $\mathbf{z}$. For simplicity, we consider $I^{\rm ext} = \alpha U_0 \exp[-|\mathbf{x} - \mathbf{z}_0|^2/(4a^2)]$, where α is the strength of the stimulus. Thanks to the translational invariance of the coupling function, the stationary state solution can be peaked at any point in the space. In this paper, we consider the network response to a stimulus abruptly changed from $(x_1, x_2) = (0, 0)$ to (z_{01}, z_{02}) at $t = 0$. As shown in the simulation result in figure 2, the synaptic input can track the change.

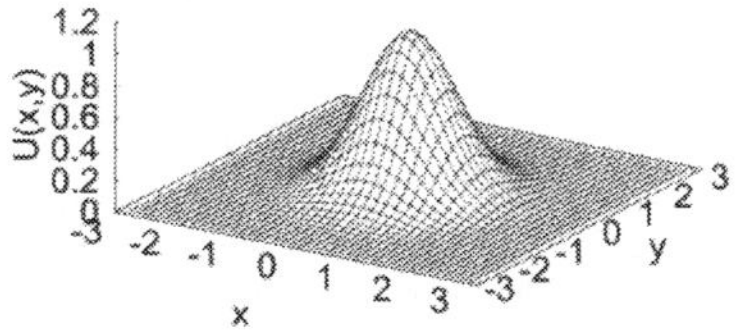

Figure 1. A stationary state $\tilde{U}(\mathbf{x}, 0)$. Parameters: $N = 40 \times 40$, $a = 0.5$, $k = 0.5$, $\tau = 1$ and $\rho = N/(2\pi)^2$.

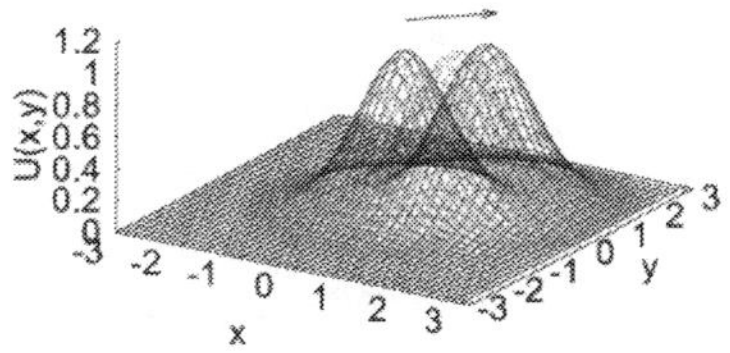

Figure 2. The synaptic input in the space due to the abruptly changed stimulus. Parameters: $\alpha = 0.05$, $\mathbf{z}_0(t < 0) = 0$, $\mathbf{z}_0(t \geq 0) = (1, 1)$ and the rest are the same as figure 1.

3. Solution to the Model

Under the driving of an external stimulus, the network state (i.e, the bump) moves from its initial position to the target one, with its shape distorted during the tracking process. Thus, to describe the tracking performance of a CANN, the key is to know the distortion patterns and their effects on the network dynamics. We denote the the state distortion to be $\delta U(\mathbf{x}, t) \equiv U(\mathbf{x}, t) - \tilde{U}(\mathbf{x}|\mathbf{z})$, whose dynamics is given by linearizing Eq.(1) at $\tilde{U}(\mathbf{x}|\mathbf{z})$ [13, 14],

$$\tau \frac{\partial}{\partial t} \delta U(\mathbf{x}, t) = \int_{-\infty}^{\infty} \int_{-\infty}^{\infty} d\mathbf{x}' F(\mathbf{x}, \mathbf{x}'|\mathbf{z}) \delta U(\mathbf{x}', t) - \delta U(\mathbf{x}, t), \tag{6}$$

where the interaction kernel $F(\mathbf{x}, \mathbf{x}'|\mathbf{z})$ is

$$F(\mathbf{x}, \mathbf{x}'|\mathbf{z}) = \frac{2\rho U(\mathbf{x}')}{B} \left[J(\mathbf{x}, \mathbf{x}') - k\rho \int_{-\infty}^{\infty} \int_{-\infty}^{\infty} d\mathbf{x}'' J(\mathbf{x}, \mathbf{x}'') r(\mathbf{x}'') \right], \tag{7}$$

where $B = 1 + k\rho \int_{-\infty}^{\infty} \int_{-\infty}^{\infty} \tilde{U}(\mathbf{x}'|\mathbf{z})^2 d\mathbf{x}'$ is a constant. The network dynamics is determined by the eigenfunctions and eigenvalues of the kernel $F(\mathbf{x}, \mathbf{x}'|\mathbf{z})$. To compute them, we choose the eigenfunctions of the quantum harmonic oscillator as the basis. While the results are more clearly presented using basis functions in polar coordinates, the analysis is more conveniently done in the rectangular coordinates. Hence we will describe the basis functions in both coordinates.

3.1. Using Basis Functions in Rectangular Coordinates

Under the rectangular coordinates, the basis functions are

$$u_{n,m}(\mathbf{x}|\mathbf{z}) = \sqrt{\frac{1}{2\pi a^2 m! n! 2^{n+m}}} H_n \left(\frac{x_1 - z_1}{\sqrt{2}a} \right) H_m \left(\frac{x_2 - z_2}{\sqrt{2}a} \right) \exp \left[-\frac{|\mathbf{x} - \mathbf{z}|^2}{4a^2} \right], \tag{8}$$

where $H_n(x)$ is the nth order Hermite Polynomial [18]. By the completeness of the basis functions, we have

$$U(\mathbf{x},t) = \tilde{U}(\mathbf{x}|\mathbf{z}(t)) + \sum_{n,m} a_{n,m}(t) u_{n,m}(\mathbf{x}|\mathbf{z}(t)). \tag{9}$$

Thanks to the orthonormality of the basis functions, we have derived [13, 14]

$$\begin{aligned}
\tau \frac{da_{n,m}}{dt} &= I_{n,m} + \sum_{n',m'} \mathcal{F}_{n,m,n',m'} a_{n',m'}(t) - a_{n,m} \\
&\quad - \frac{\tau}{2a}\frac{dz_1}{dt}\left[\sqrt{2\pi} a U_0 \delta_{n1,m0} + \sqrt{n} a_{n-1,m}(t) - \sqrt{n+1} a_{n+1,m}(t)\right] \\
&\quad - \frac{\tau}{2a}\frac{dz_2}{dt}\left[\sqrt{2\pi} a U_0 \delta_{n0,m1} + \sqrt{m} a_{n,m-1}(t) - \sqrt{m+1} a_{n,m+1}(t)\right],
\end{aligned} \tag{10}$$

where $I_{n,m}$ is the projection of I^{ext} onto $u_{n,m}$, given by

$$I_{n,m}(\mathbf{x}|\mathbf{z}_0) = \alpha U_0 \sqrt{\frac{2\pi a^2}{n!m!}} \left(\frac{z_{01}-z_1}{2a}\right)^n \left(\frac{z_{02}-z_2}{2a}\right)^m \exp\left[-\frac{|\mathbf{z}-\mathbf{z}_0|^2}{8a^2}\right], \tag{11}$$

and $\mathcal{F}_{n,m,n',m'}$, the interaction matrix, is given by

$$\mathcal{F}_{0,0,0,0} = 1 - \left(1 - \frac{k}{k_c}\right)^{\frac{1}{2}}, \tag{12}$$

$$\mathcal{F}_{n,m,n'm'} = \sqrt{\frac{m'!n'!}{m!n!}} \frac{2}{2^{m'+n'}} \frac{(-)^{\frac{m'+n'-m-n}{2}}}{2^{\frac{m'+n'-m-n}{2}} (\frac{m'-m}{2})!(\frac{n'-n}{2})!}, \quad \text{if } \tfrac{n'-n}{2} \text{ and } \tfrac{m'-m}{2} \text{ are positive integers} \tag{13}$$

$$\mathcal{F}_{n,m,n'm'} = 0, \quad \text{otherwise.} \tag{14}$$

The center of mass is given by the self-consistent condition $\mathbf{z}(t) = \iint d\mathbf{x} U(\mathbf{x},t)\mathbf{x} / \iint d\mathbf{x} U(\mathbf{x},t)$. If the external stimulus is symmetric with respect to the x-axis, then $dz_2/dt = 0$ and

$$\frac{dz_1}{dt} = \frac{2a}{\tau} \frac{\sum_{\text{odd } n, \text{ even } m} \sqrt{\frac{(m-1)!!}{m!!}\frac{n!!}{(n-1)!!}} \left(I_{n,m} + \sum_{n',m'} \mathcal{F}_{n,m,n',m'} a_{n',m'}\right)}{\sqrt{2\pi} a U_0 + \sum_{\text{even } n,\ m} \sqrt{\frac{(m-1)!!}{m!!}\frac{(n-1)!!}{n!!}} a_{n,m}}. \tag{15}$$

The eigenvalues of $\mathcal{F}$ are $\lambda_{0,0} = 1 - (1 - k/k_c)^{1/2}$, $\lambda_{n,0} = \lambda_n$ for $n \neq 0$, $\lambda_{0,m} = \lambda_m$ for $m \neq 0$, $\lambda_{n,m} = \lambda_n \lambda_m$ for $n \neq 0$ and $m \neq 0$, where $\lambda_n = 2/2^n$. From this result, one can conclude that, if the stimulus is absent, all modes of distortion will decay exponentially in time, except for the eigenfunctions $u_{1,0}$ and $u_{0,1}$, whose eigenvalues are 1. To prove this, we define $v^R_{n,m}(\mathbf{x}|\mathbf{z})$ to be the right eigenfunctions of $\mathcal{F}$. Then we may express $\delta U(\mathbf{x},t)$ in Eq. (6) as $\delta U(\mathbf{x},t) = \sum_{n,m} \delta U_{n,m}(\mathbf{z},t) v^R_{n,m}(\mathbf{x}|\mathbf{z})$. Using the orthonormality of the left and right eigenfunctions of $F(\mathbf{x},\mathbf{x}'|\mathbf{z})$, we have

$$\delta U_{n,m}(z,t) = \int_{-\infty}^{\infty}\int_{-\infty}^{\infty} d\mathbf{x} \delta U(\mathbf{x},t) v^L_{n,m}(\mathbf{x}|\mathbf{z}). \tag{16}$$

Assume that the motion of the bump is slow, so that $d\mathbf{z}/dt$ becomes negligible in Eq. (6); as we shall see, this assumption is valid as long as the external stimulus is sufficiently weak. Then, the projection of Eq. (6) on the eigenfunctions become

$$\tau \frac{d}{dt} \delta U_{n,m}(\mathbf{z},t) = (\lambda_{n,m} - 1)\delta U_{n,m}(z,t). \tag{17}$$

Hence,

$$\delta U_{n,m}(\mathbf{z},t) = \delta U_{n,m}(\mathbf{z},0)\exp\left[-\frac{(1-\lambda_{n,m})t}{\tau}\right], \tag{18}$$

where $\delta U_{n,m}(\mathbf{z},0)$ is the initial value of the projection. $u_{1,0}$ and $u_{0,1}$ correspond to the trackability of the synaptic input as well as the neuronal response, because $u_{1,0} \sim \partial\tilde{U}/\partial x_1$ and $u_{0,1} \sim \partial\tilde{U}/\partial x_2$. $u_{1,0}$ and $u_{0,1}$ are the modes of the position shift of the synaptic input. Thus, it guarantees the stability of the stationary solution and trackability of the synaptic input.

We are now ready to find the tracking solution to the stimulus abruptly changed from $(0,0)$ to $(z_{01},0)$ at $t=0$. Neglecting the depdendence on all $a_{n,m}$ terms in Eq. (15), we have

$$\frac{dz_1}{dt} = \frac{\alpha}{\tau}(z_{01}-z_1)\exp\left[-\frac{(z_{01}-z_1)^2}{8a^2}\right], \tag{19}$$

which is consistent with the result obtained from the one-dimensional case [13, 14]. This approximation is useful when $|\mathbf{z}-\mathbf{z}_0|$ is small and the stimulus is weak. It also shows that the tracking behavior is similar to the one-dimensional case.

3.2. Using Basis Functions in Polar Coordinates

Since the system is also rotationally invariant, the analysis can proceed by using the eigenfunctions with polar coordinates. The eigenfunctions of quantum harmonic oscillators are

$$\psi_{l,j}(r,\theta) = \sqrt{\frac{(\frac{l-j}{2})!(\frac{l+j}{2})!}{2\pi a}}\left[\sum_{t=0}^{\frac{l-|j|}{2}}\frac{(-1)^t(\frac{r}{\sqrt{2}a})}{(\frac{l-j}{2})!(\frac{l+j}{2})!t!}\right]e^{-\frac{r^2}{4a^2}+ij\theta}, \tag{20}$$

where $i=\sqrt{-1}$, and l, j are the radial and angular quantum numbers respectively. Decomposing the distortion term, we have $\delta U = \sum_{l,j} b_{l,j}(t)\psi_{l,j}$. The matrix elements of the transformation matrix are

$$[T]_{n,m,l,j} \equiv \langle u_{n,m}|\psi_{l,j}\rangle = i^m\sqrt{\frac{\left(\frac{l+j}{2}\right)!\left(\frac{l-j}{2}\right)!}{n!m!2^l}}\left[\sum_t(-1)^t\binom{m}{t}\binom{n}{\frac{l-j}{2}-t}\right], \tag{21}$$

Similar to Eq. (10), there is an interaction kernel $\mathcal{G}$ that represents the interaction between different $\psi_{l,j}$, which can be obtained by $\mathcal{G} = T^{-1}\mathcal{F}T$. The first few eigenfunctions of $\mathcal{G}$ are

$$\Psi_{00} = \psi_{00}, \tag{22}$$

$$\Psi_{1\pm1} = \psi_{1\pm1}, \tag{23}$$

$$\Psi_{20} = \frac{1}{\sqrt{1+(\lambda_{00}-1/2)^2}}\left[\psi_{00}+(\lambda_{00}-1/2)\psi_{20}\right], \tag{24}$$

$$\Psi_{2\pm2} = \psi_{2\pm2}, \tag{25}$$

$$\Psi_{3\pm1} = \frac{1}{\sqrt{3^2+2}}\left[\sqrt{2}\psi_{1\pm1}+3\psi_{3\pm1}\right], \text{ and} \tag{26}$$

$$\Psi_{3\pm3} = \psi_{3\pm3}, \tag{27}$$

where the indices l and j of $\Psi_{l,j}$ represent the highest basis function it contains. Their eigenvalues are λ_{00}, 1, 1/2, 1/2, 1/4, and 1/4 respectively.

As shown in Figs. 3 to 8, the eigenfunctions are symmetric with respect to the origin. The eigenfunctions correspond to different modes of the distortion of the synaptic input during the

Journal of Physics: Conference Series **197** (2009) 012017 doi:10.1088/1742-6596/197/1/012017

motion. Ψ_{00} corresponds to the change in height. It can describe, say, the reduction of the bump height during the process to catch up with the new position of the stimulus, as shown in figure 2. $\Psi_{1,\pm1}$ can describe the movement of the bump towards the preferred positions. $\Psi_{2,0}$ describes not only changes in the height of the bump, but also changes in the width of the bump during the motion. $\Psi_{3,\pm1}$ describes the skewing of the bump due to the stimulus and other modes. While the above distortion modes are apparently extensions of those in the one-dimension case, the modes $\Psi_{2,\pm2}$ and $\Psi_{3,\pm3}$ are unique to the two-dimensional case. The former corresponds to an elliptical distortion of the bump shape, and the latter to a three-fold distortion.

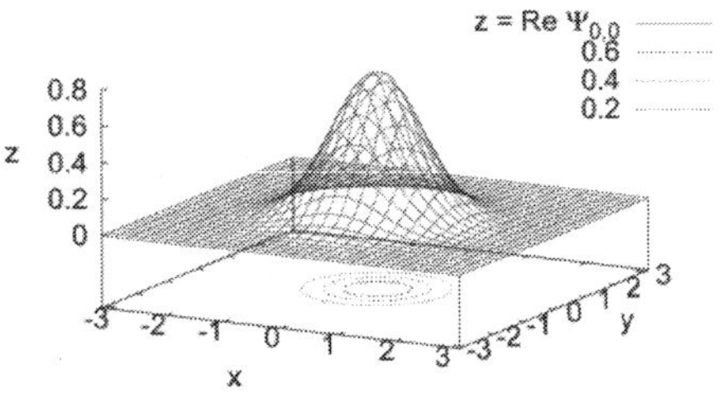

Figure 3. Real part of the eigenfunction $\Psi_{0,0}$.

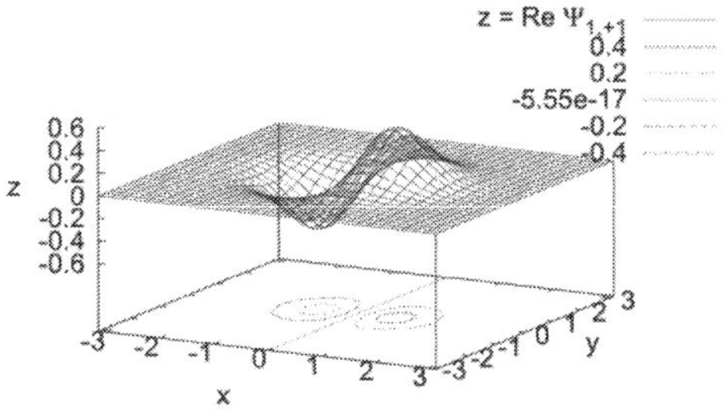

Figure 4. Real part of the eigenfunction $\Psi_{1,+1}$.

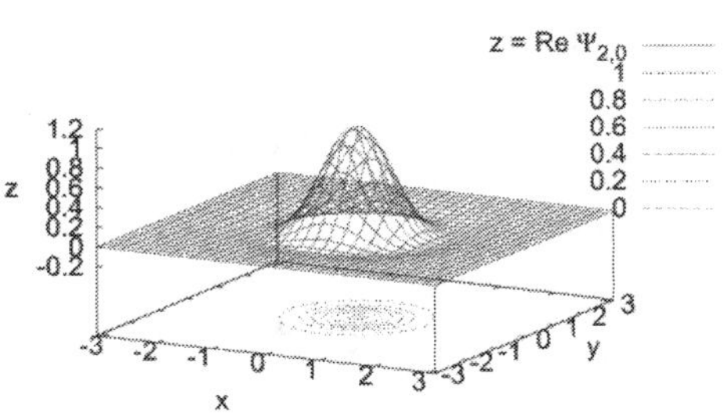

Figure 5. Real part of the eigenfunction $\Psi_{2,0}$.

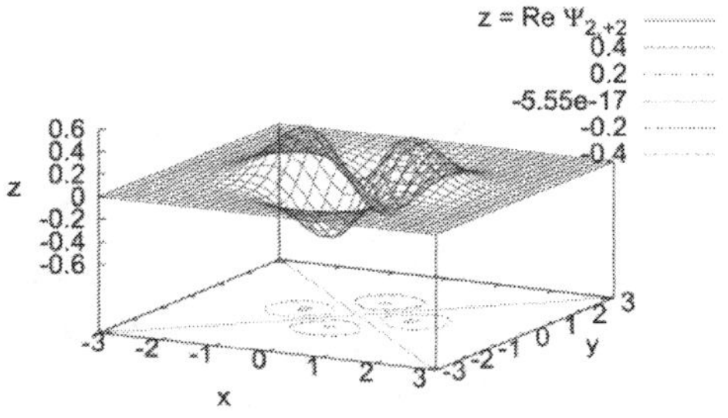

Figure 6. Real part of the eigenfunction $\Psi_{3,+1}$.

By using the transformation matrix in Eq. (21), Eq. (10) can be transformed from rectangular to polar coordinates up to arbitrary order. For the perturbation up to $l = 2$, and for external stimuli symmetric with respect to the x-axis, we have

$$\left(\tau\frac{d}{dt} + 1 - \lambda_{00}\right) b_{00} = I_{00} - \tau\frac{dR}{dt}\left[-\frac{1}{\sqrt{8}a}(b_{1-1} + b_{1+1})\right] - b_{20}, \tag{28}$$

$$\tau\frac{d}{dt}b_{1\pm1} = I_{1\pm1} - \tau\frac{dR}{dt}\left(\frac{\sqrt{2\pi}aU_0 + b_{00}}{2\sqrt{2}a} - \frac{1}{2\sqrt{2}a}b_{20} - \frac{1}{2a}b_{2\pm2}\right), \tag{29}$$

$$\left(\tau\frac{d}{dt} + \frac{1}{2}\right) b_{20} = I_{20} - \tau\frac{dR}{dt}\left(-\frac{1}{2\sqrt{2}a}(b_{1-1} + b_{1+1})\right), \tag{30}$$

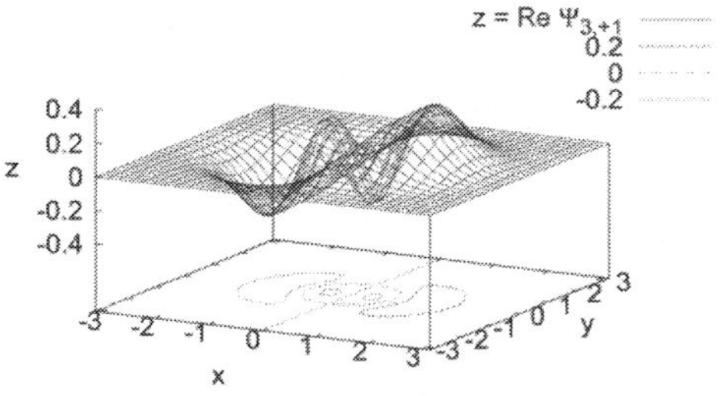

Figure 7. Real part of the eigenfunction $\Psi_{2,+2}$.

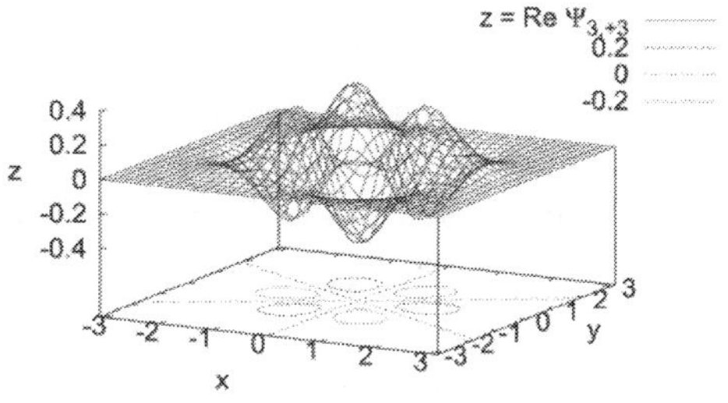

Figure 8. Real part of the eigenfunction $\Psi_{3,+3}$.

$$\left(\tau\frac{d}{dt}+\frac{1}{2}\right)b_{2\pm 2}=I_{2\pm 2}-\tau\frac{dR}{dt}\frac{1}{2a}b_{1\pm 1}, \tag{31}$$

$$\tau\frac{dR}{dt}=2\sqrt{2}a\frac{I_{11}}{\sqrt{2\pi}aU_0+b_{00}-\sqrt{2}b_{20}-2\sqrt{2}b_{22}}, \tag{32}$$

where R is the radial distance from the origin. Note that $I_{l,j}=I_{l,-j}$ and $b_{l,j}=b_{l,-j}$ due to the symmetry when the stimulus lies on the x-axis.

4. Simulation Experiments

In the simulation experiments, the number of neurons is $N=N_x\times N_y$, and the range of (x_1,x_2) is $-\pi\le x_1,x_2<\pi$. The boundary condition is periodic.

4.1. Reaction Time to an Abrupt Change of the Stimulus

In this experiment, the stimulus is centered at $(0,0)$ until the synaptic input $U(\mathbf{x},t)$ becomes steady. At $t=0$, the stimulus abruptly changes from $(0,0)$ to $\mathbf{z}_0\equiv(z_{01},0)$, and we observe the dependence of the reaction time on the distance z_{01}. Then, the bump will track the stimulus, as shown in figure 2. The reaction time is defined by the time needed to have $|\mathbf{z}(t)-\mathbf{z}_0|<\Theta$, where Θ is the threshold. This threshold is necessary in this experiment, because the motion of $\mathbf{z}(t)$ will become very slow when it approaches the stimulus, as implied by Eq. (19). Also, the assumption is reasonable because in real biological systems, we do not need to have $\mathbf{z}(t)=\mathbf{z}_0(t)$ to make decisions.

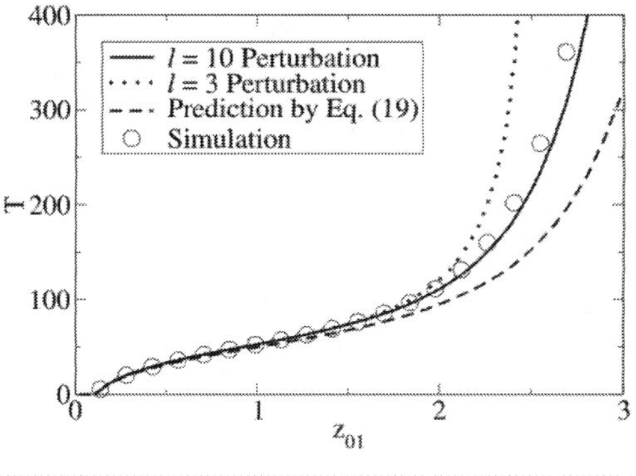

Figure 9. The reaction time T for the syanptic input to catch up the stimulus position change from 0 to z_{01}. Parameters: $N=N_x\times N_y=40\times 40$, $\Theta=\pi\sqrt{2/N}$, $\alpha=0.05$ and the rest are the same as figure 1.

As shown in figure 9, the prediction given by Eq. (19) works well only when the change in the position of the stimulus is small, while the $l = 3$ perturbation works well up to $|\mathbf{z}_0| = 2$. The prediction of the $l = 10$ perturbation is the best among the three. From this result, one can state that, when the position change of the stimulus is small, only $\psi_{l,j}$ with small l will be activated. However, when the change in stimulus position is larger, higher order distortions are activated. This is reasonable, because, for smaller l, the distortions are concentrated around $\mathbf{z}(t)$, but if the stimulus is far away from $\mathbf{z}(t)$, the tail part of the bump will be distorted first, leading to higher order distortions.

4.2. Amplitudes of the Basis Function Distortion Modes

From Eqs. (10) and (21), $b_{l,j}$ can be predicted by the projections of δU onto the basis functions $\psi_{l,j}$. The experimental settings are the same as above, but $\mathbf{z}_0$ was fixed to be $\mathbf{z}_0 = (2, 0)$ in polar coordinates.

As shown in figure 10, the predicted $b_{l,j}$'s agree with the simulation results well. It confirms that the perturbative method can also predict the motion of the synaptic input in detail. $b_{0,0}$ indicates that the height drops from its initial value after the stimulus is shifted. The $\psi_{0,0}$ component of the distortion is reduced by the inhibition. It approaches 0 roughly, as if there were no external stimulus, and there is even a slight overshoot. Afterwards, the distortion relaxes smoothly to the equilibrium value when it approaches the shifted position of the stimulus. Similarly, the amplitude of the $\psi_{1,\pm 1}$ components falls abruptly to a negative value initially. This is due to the tail of the bump being pulled by the newly positioned stimulus, causing the peak to lag behind the center of mass. Afterwards, it relaxes smoothly to 0. The initial change in $b_{2,0}$ indicates an increase in width along the direction of the stimulus, and that of $b_{2,\pm 2}$ signals a cigar-shaped distortion when the bump is being pulled by the stimulus. The amplitude $b_{3,\pm 1}$ describes the skewness of the bump, and $b_{3,\pm 3}$ describes the bump being distorted when its tip is pulled by the stimulus, with the posterior part lagging in motion, causing a triangular-shaped distortion.

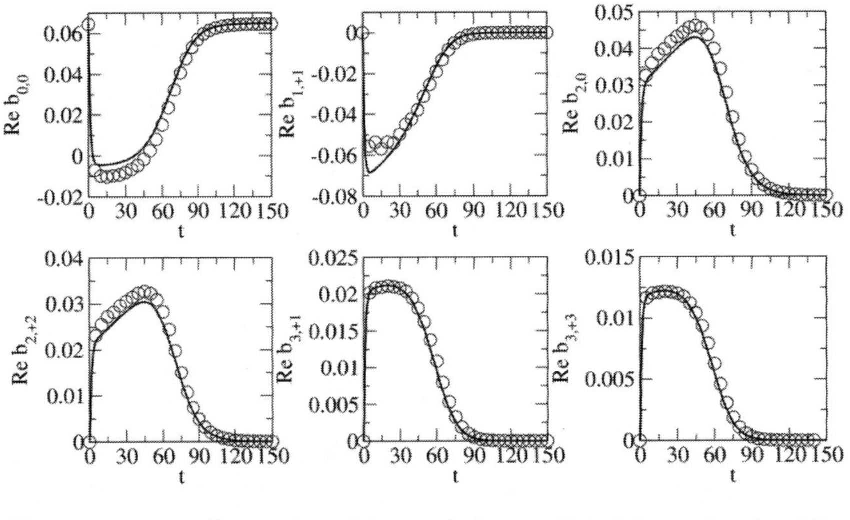

Figure 10. The experimentally projected $b_{l,j}$ and the predicted $b_{l,j}$ using $l = 10$ perturbation. Solid lines: corresponding predictions. Circles: simulation results. Parameters: same as figure 9.

5. Conclusion and Discussion

In this paper, a perturbative method to deal with continuous attractor neural networks is presented in two-dimensional space. We have introduced a simple solvable model to demonstrate how to use perturbative method to analyze the dynamics of distortions and synaptic input. Since the coupling factors are defined to be translational invariant, a family of stationary states can be sustained anywhere in the preferred stimulus space. Furthermore, the synaptic input is able to track the stimulus in the space. By studying the dynamics, one can deduce the tracking time and the distortions of the synaptic input.

For one-dimensional continuous attractor neural networks, the tracking speed can be roughly approximated by

$$\frac{dz}{dt} = \frac{\alpha}{\tau}(z_0 - z)\exp\left[-\frac{|z_0 - z|^2}{8a^2}\right], \tag{33}$$

which is similar to Eq. (19). Although they are rough approximations, the basic properties of tracking in one and two dimensions are similar. The key predictions in the one-dimensional case, the maximum trackable speed [11, 13] and the lag behind a continuously moving stimulus, are also applicable in two dimensions. These similarities arise from the rotational invariance of the interaction kernel and the bump in the limit of weak stimulus, since the description of the dynamics by the translational mode in sufficient. Note, However, for the stronger stimuli, the dynamics is richer in the two-dimensional case, since distortion modes unique to the two-dimensional case need to be considered; examples of cigar-shaped and triangular-shaped distortions are shown in figure 10.

For this particular model, the eigenfunctions of the interaction matrix in polar coordinates are also studied. By Eq. (21), the interaction matrix $\mathcal{G}$ can be obtained. However, due to complications in the calculation, we can only calculate it term by term. Since the matrix is upper triangular, the eigenvalues are the diagonal entries, $\lambda_{n,m}$. As the eigenvalues are at most 1, one can show that the synaptic input has a stable Gaussian form. The eigenfunctions corresponding to eigenvalue 1 represent the positional shift. As studied above, different modes of distortion correspond to different kinds of distortion. For example, the component of $\psi_{0,0}$ corresponds to the change in height, while $\psi_{2,0}$ represents the change in width.

Concerning the robustness of the method, we remark that it can be applied to other types of networks with tracking behavior. A common example is the continuous attractor neural network with the Mexican hat interaction. Using the basis functions of the two-dimensional quantum harmonic oscillator, we can obtain the matrix elements of the interaction kernel numerically, although elegant expressions such as those obtained here may not be available. Perturbation dynamics can then be worked out analogously. This proposed extension can be able to address a recent issue of interest, namely, the instability of bumps and rings in a two-dimensional neural field of Amari type [19]. Meanwhile, we note in passing that the present model with a quadratic response and a global inhibition does not suffer from the stability problem.

Acknowledgement

This work is partially supported by the Research Grants Council of Hong Kong (Grant Nos. HKUST 603607 and HKUST 604008).

References

[1] Amari S 1977 *Biol. Cybern.* **27** 77–87
[2] Ermentrout B 1998 *Rep. Prog. Phys.* **61** 353–430
[3] Seung H S 1996 *Proc. Natl. Acad. Sci. USA* **93** 13339–44
[4] Wu S, Hamaguchi K and Amari S 2008 *Neural Comput.* **20** 994–1025
[5] Zhang K C 1996 *J. Neurosci.* **16** 2112–26
[6] Ben-Yishai R, Lev Bar-Or R and Sompolinsky H 1995 *Proc. Natl. Acad. Sci. USA* **92** 3844–8

[7] Georgopoulos A P, Taira M and Lukashin A 1993 *Science* **260** 47–52
[8] Samsonovich A and McNaughton B L 1997 *J. Neurosci.* **7** 5900–20
[9] Jastorff J, Kourtzi Z and Giese M 2006 *J. Vision* **6** 791
[10] Folias E and Bressloff P 2004 *SIAM J. Appl. Dyn. Syst.* **3** 378–407
[11] Hansel D and Sompolinsky H 1998 *Methods in Neuronal Modeling: From Ions to Networks* ed Koch C and Segev I (Cambridge: MIT Press)
[12] Wu S and Amari S 2005 *Neural Comput.* **17** 2215–39
[13] Fung C C A, Wong K Y M and Wu S 2008 *Europhys. Lett.* **84** 18002
[14] Fung C C A, Wong K Y M and Wu S *Neural Comput. (in press)*
[15] Coombes S and Owen M R 2005 *Phys. Rev. Lett.* **94** 148102
[16] Taylor J G 1999 *Biol. Cybern.* **80** 393–409
[17] Werner H and Richter T 2001 *Biol. Cybern.* **85** 211–7
[18] Griffiths D 2004 *Introduction to Quantum Mechanics* (Prentice Hall)
[19] Owen M R, Liang C R and Coombes S 2007 *New J. Phys.* **9** 378

Statistical mechanics of attractor neural network models with synaptic depression

Yasuhiko Igarashi[1], Masafumi Oizumi[1,2], Yosuke Otsubo[1], Kenji Nagata[1] and Masato Okada[1,3]

[1] Graduate School of Frontier Sciences, The University of Tokyo, Kashiwa, Japan
[2] Research Fellow of the Japan Society for the Promotion of Science
[3] RIKEN Brain Science Institute, Wako, Japan

E-mail: okada@k.u-tokyo.ac.jp

Abstract. Synaptic depression is known to control gain for presynaptic inputs. Since cortical neurons receive thousands of presynaptic inputs, and their outputs are fed into thousands of other neurons, the synaptic depression should influence macroscopic properties of neural networks. We employ simple neural network models to explore the macroscopic effects of synaptic depression. Systems with the synaptic depression cannot be analyzed due to asymmetry of connections with the conventional equilibrium statistical-mechanical approach. Thus, we first propose a microscopic dynamical mean field theory. Next, we derive macroscopic steady state equations and discuss the stabilities of steady states for various types of neural network models.

1. Introduction

Neurophysiological experiments have shown that high-frequency inputs induce the decrease of synaptic weights [1]. Synaptic depression is known to control gain for presynaptic inputs [2, 3]. The synaptic depression not only influences the activity of a single neuron but also the overall activity in neural networks, i.e., their macroscopic properties.

Many authors have attempted to study the storage capacity of associative memory models with the synaptic depression [4, 5, 6, 7]. Matsumoto *et al.* proposed a mean field theory in a case of $T = 0$, and obtained steady state order parameter equations for an associative memory model [6]. They found that the synaptic depression does not affect the steady state of the network if the threshold is set at an appropriate value. Dynamic properties, as well as the static properties, have been analyzed by several researchers [6, 8]. Matsumoto *et al.* also reported that the synaptic depression enlarges the basins of memory retrieval.

The present paper first proposes a microscopic dynamical mean field theory for a stochastic neural network model with the synaptic depression. The derived microscopic equation coincides with that of an analog neural network model with the synaptic depression [9]. Based on this microscopic equation, we derive macroscopic steady state equations, and analyze the stabilities of macroscopic states for various types of attractor neural network models.

2. Formulation

2.1. Network and synapse dynamics

We discuss an attractor neural network model with N binary neurons, and consider the thermodynamic limit of $N \to \infty$. $s_i(t)$ represents a state of the i-th neuron at time t, whose state takes resting state $s_i(t) = 0$ or firing state $s_i(t) = 1$, i.e., we employ the McCulloch-Pitts model. $J_{ij}(t)$ is a synaptic connection from the j-th neuron to the i-th neuron at time t. The synaptic connection $J_{ij}(t)$ dynamically changes in accordance with a presynaptic neuron state, $s_j(t)$, as follows. The synaptic weight $J_{ij}(t)$ incorporating the synaptic depression is obtained by multiplying a fixed synaptic weight J_{ij}, and a dynamic amplitude factor $x_j(t)$ of the presynaptic neuron $s_j(t)$. The dynamic amplitude factor at time t is determined by the corresponding neuron state and itself at the preceding time $t-1$,

$$J_{ij}(t) = J_{ij}x_j(t), \tag{1}$$

$$x_j(t) = x_j(t-1) + \frac{1 - x_j(t-1)}{\tau} - Ux_j(t-1)s_j(t-1). \tag{2}$$

The phenomenological model of synaptic depression described by equations (1) and (2) was proposed by several researchers [2, 3] The synapses exhaust neurotransmitters when they transmit signals. The dynamical amplitude factor $x_j(t)$ decreases by a certain fraction $Ux_j(t-1)$ after the firing of the presynaptic neuron $s_j(t-1) = 1$, and recovers with a time constant τ as shown in equation (2). The post-synaptic potential or local field $h_i(t)$ of the i-th neuron is defined as

$$h_i(t+1) = \sum_{j \neq i}^{N} J_{ij}(2x_j(t)s_j(t) - 1). \tag{3}$$

In a case without the synaptic depression, i.e., $x_j(t) = 1$, the system in equation (3) is reduced to the well known Ising spin system with $\boldsymbol{\sigma}(\mathrm{t}) = (\sigma_1(t), \sigma_2(t), \cdots, \sigma_N(t))$, $\sigma_i(t) = \pm 1$,

$$h_i(t+1) = \sum_{j \neq i}^{N} J_{ij}\sigma_j(t). \tag{4}$$

The system is simultaneously updated, i.e., the synchronous rule, and the state of the i-th neuron stochastically changes by using the following conditional probability given the local field $h_i(t+1)$,

$$\mathrm{Prob}[s_i(t+1) = 1|h_i(t+1)] = 1 - \mathrm{Prob}[s_i(t+1) = 0|h_i(t+1)] = g_\beta(h_i(t+1)),$$

$$g_\beta(h) = \frac{1}{2}(1 + \tanh \beta h), \tag{5}$$

where β is the inverse temperature ($\beta = 1/T$).

2.2. Microscopic mean field theory

Since the synaptic weight $J_{ij}(t) = J_{ij}x_j(t)$ is asymmetric due to the dynamic amplitude factor $x_j(t)$ even if J_{ij} is symmetric, the present system described from equations (1) to (5) does not have a Lyapnov function, and therefore cannot be analyzed with the conventional equilibrium statistical-mechanical approach. Thus, we propose a dynamical mean field theory assuming that the synaptic weight $J_{ij}(t)$ is on the order of $1/N$ with respect to the number of neurons N.

First, we consider a case of $T = 0$. Matsumoto *et al.* proposed a mean field theory of $T = 0$ limit and obtained steady state order parameter equations for an associative memory model [6]. Here, we briefly introduce their work. They assumed the steady state condition,

$$s_j(\infty) = s_j, \quad x_j(\infty) = x_j, \quad h_i(\infty) = h_i. \tag{6}$$

For $T = 0$, the equilibrium state of the i-th neuron s_i is given by using equations (3), (5) and (6),

$$s_i = \Theta(h_i) = \Theta(\sum_{j\neq i}^{N} J_{ij}(2x_j s_j - 1)). \tag{7}$$

The value of dynamical amplitude factor $x_j(t)$ in the steady state is given by

$$x_j = \frac{1}{1+\gamma s_j}, \quad \gamma = \tau U. \tag{8}$$

Since the neuron state s_j takes 0 or 1, the value of $x_j s_j$ in equation (7) can be written as

$$x_j s_j = \frac{s_j}{1+\gamma s_j} = \frac{s_j}{1+\gamma}. \tag{9}$$

Finally, the microscopic steady state equation for $T = 0$ is,

$$s_i = \Theta(h_i) = \Theta(\sum_{j\neq i}^{N} J_{ij}(\frac{2}{1+\gamma} s_j - 1)). \tag{10}$$

This equation means that only the synaptic interaction J_{ij} is rescaled by a factor $1/(1+\gamma)$ in the case of $T = 0$. Matsumoto *et al.* calculated the storage capacity of the associative memory model based on this steady state equation in equation (10) using the self-consistent signal to noise analysis (SCSNA) proposed by Shiino and Fukai [6, 10]. Mejias and Torres recently extended Matsumoto's work to treat a low temperature case $T \simeq 0$ $(1 \ll \beta < \infty)$ [7]. They assumed the following microscopic steady state equation for the low temperature case based on equation (9),

$$\langle x_j s_j \rangle = \langle \frac{s_j}{1+\gamma s_j} \rangle = \frac{\langle s_j \rangle}{1+\gamma}, \tag{11}$$

where $\langle\cdot\rangle$ denotes the thermal average with respect to equation (5).

We propose a dynamical mean field theory for the finite temperature inspired by their work [7]. First, we consider the thermal average of the i-th neuron at time t, i.e., $\langle s_i(t)\rangle = \langle g_\beta(h_i(t))\rangle$. Since we have assumed $J_{ij} \sim O(1/N)$, we obtain

$$\langle s_i(t)\rangle = g_\beta(\langle h_i(t)\rangle), \quad \langle h_i(t)\rangle = \sum_{j\neq i}^{N} J_{ij}\big(2\langle x_j(t-1)s_j(t-1)\rangle - 1\big). \tag{12}$$

We take the thermal average of equation (2) for dynamical amplitude factor $x_j(t)$,

$$\langle x_j(t)\rangle = \langle x_j(t-1)\rangle + \frac{1-\langle x_j(t-1)\rangle}{\tau} - U\langle x_j(t-1)s_j(t-1)\rangle. \tag{13}$$

Although $x_j(t)$ and $s_j(t)$ are generally correlated, the equal time correlation disappears in the thermodynamic limit $N \to \infty$ for $J_{ij} \sim O(1/N)$,

$$\langle x_j(t-1)s_j(t-1)\rangle = \langle x_j(t-1)\rangle\langle s_j(t-1)\rangle. \tag{14}$$

We define the thermal averages of $s_i(t)$ and $x_i(t)$ as

$$m_i(t) \equiv \langle s_i(t)\rangle, \quad X_i(t) \equiv \langle x_i(t)\rangle. \tag{15}$$

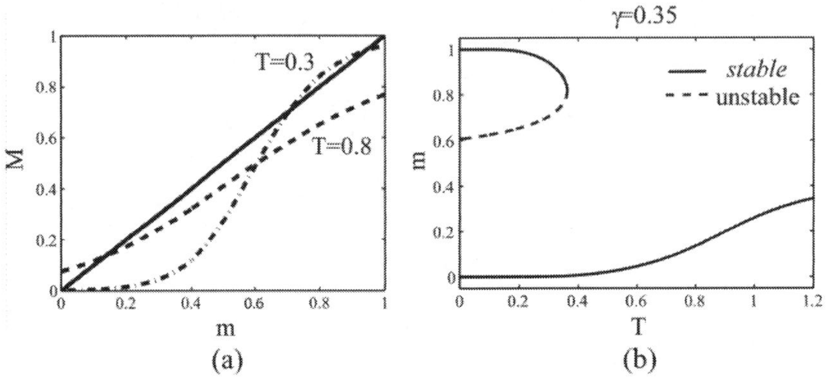

Figure 1. (a): Graphical solutions to equation (22). (b): *Stable* and unstable solutions to equation (21).

As a result, we have succeeded in obtaining the dynamical mean field equations of $m_i(t)$ and $X_i(t)$ as

$$m_i(t+1) = g_\beta(\sum_{j\neq i}^{N} J_{ij}(2X_j(t)m_j(t) - 1)) \tag{16}$$

$$X_j(t+1) = X_j(t) + \frac{1 - X_j(t)}{\tau} - UX_j(t)m_j(t). \tag{17}$$

These equations for the present stochastic neural network model coincide with those of an analog neural network model with the synaptic depression [9].

The steady state equation for the thermal average $X_i = X_i(\infty)$ is

$$\langle x_i \rangle = X_j = \frac{1}{1+\gamma m_j}, \quad \langle x_j s_j \rangle = X_j m_j = \frac{m_j}{1+\gamma m_j}, \quad \gamma = \tau U, \tag{18}$$

which is a finite temperature version of equation (8) obtained in the $T = 0$ case [6]. Finally, we obtain a microscopic steady state equation of $m_i = m_i(\infty)$ for the system with the synaptic depression, i.e.,

$$m_i = g_\beta(\sum_{j\neq i}^{N} J_{ij}(\frac{2m_j}{1+\gamma m_j} - 1)). \tag{19}$$

3. Ferromagnetic system

In this section, we treat a system with uniform synaptic weights,

$$J_{ij} = J_0/N, \tag{20}$$

which corresponds to the mean field model of the ferromagnetic system called the Husimi-Temperly (HT) model.

3.1. Macroscopic steady state equation

We derive a macroscopic steady state equation for the ferromagnetic system based on the mean field theory proposed in §2.2. According to the symmetry of synaptic weights in equation (20),

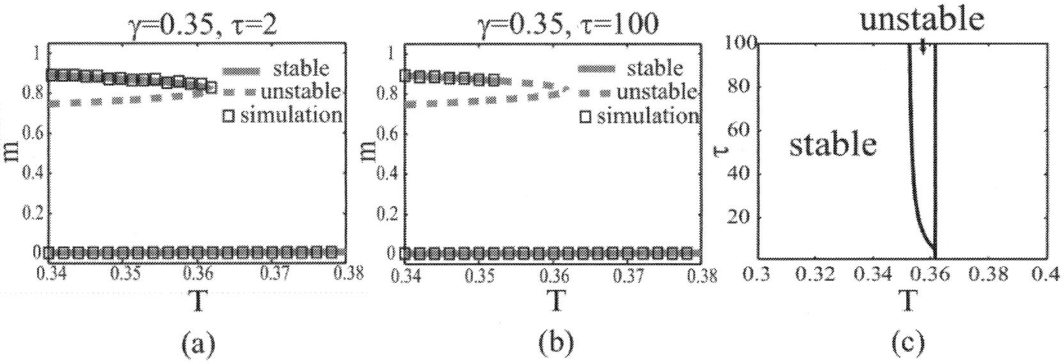

Figure 2. Squares in figures (a) and (b) represent thermal average m numerically obtained by computer simulation with $N = 10^3$. Solid lines denote stable solutions while dashed lines denote unstable solutions in (a) and (b). (c) represents phase diagram for stability of $m > 0.5$ state.

the thermal average of each neuron m_i can be set to $m_i = m$. Substituting this condition into equation (19), we obtain the macroscopic steady state equation,

$$m = g_\beta\big(J_0(\frac{2m}{1+\gamma m} - 1)\big) = \frac{1}{2}\big\{\tanh(\beta J_0(\frac{2m}{1+\gamma m} - 1) + 1\big\}. \tag{21}$$

Equation (21) reduces to the order parameter equation of the HT model, when the synaptic depression is absent $\gamma = 0$. Figure 1 (a) shows graphical solutions to,

$$M = m, \quad M = \frac{1}{2}\big\{\tanh\beta(J_0(\frac{2m}{1+\gamma m} - 1) + 1\big\}. \tag{22}$$

for $T = 0.3$ (dot-dashed line) and $T = 0.8$ (dashed line) when $\gamma = \tau U = 0.35$. A saddle node bifurcation occurs at $T_C = 0.36$. The solid lines in Figure 1 (b) denote *stable* solutions to equation (21), while the dashed line denotes an unstable solution in the sense of the equilibrium statistical mechanics.

3.2. Stability analysis of macroscopic steady state

We numerically checked the stability of *stable* solutions using computer simulations. The number of neurons is $N = 10^3$ and the initial condition of the neuron state and the dynamical amplitude factor is $s_i(0) = 1$, $x_i(0) = 1$, for all i. We numerically obtained the steady states and calculated the thermal averages m under various condistions. The simulation results are represented by the squares in Figure 2. The *stable* solution is stable for $\tau = 2$ plotted in Figure 2 (a), while it becomes unstable for $\tau = 100$ in Figure 2 (b), where τ is the time constant of the dynamical amplitude factor in equation (8). These two cases have the same value of $\gamma = 0.35$; thus, they have the same steady state solution to equation (21). These results imply that the stability of solution strongly depends on the time constant τ as shown in Figure 2.

To examine the stability of this homogenous macroscopic state m and $X = 1/(1+\gamma m)$, let us derive the dynamics of a homogenous mode, $m_i(t) = m(t)$, $X_i(t) = X(t)$, for all i. Substituting this mode and the $J_{ij} = J_0/N$ of equation (20) into the microscopic dynamical equations (16) and (17), we obtain macroscopic dynamical equations for the homogenous mode,

$$m(t+1) = F_m(X(t), m(t)) = \frac{1}{2}\big\{\tanh(\beta J(2X(t)m(t) - 1)) + 1\big\} \tag{23}$$

$$X(t+1) = F_X(X(t), m(t)) = X(t) + \frac{1 - X(t)}{\tau} - UX(t)m(t). \tag{24}$$

Next, we obtain dynamical equations for small deviations,

$$\delta m(t) \equiv m(t) - m, \quad \delta X(t) \equiv X(t) - X, \quad X \equiv \frac{1}{1+\gamma m}, \tag{25}$$

around the steady state of equation (21),

$$\begin{pmatrix} \delta m(t+1) \\ \delta X(t+1) \end{pmatrix} = H \begin{pmatrix} \delta m(t) \\ \delta X(t) \end{pmatrix}, \quad H \equiv \begin{pmatrix} \frac{\partial F_m}{\partial m} & \frac{\partial F_m}{\partial X} \\ \frac{\partial F_X}{\partial m} & \frac{\partial F_X}{\partial X} \end{pmatrix} = \begin{pmatrix} 4\beta JX(m(1-m)) & 4\beta J(m^2(1-m)) \\ -UX & 1-\frac{1}{\tau}-Um \end{pmatrix}. \tag{26}$$

The two eigen values of the Hessian matrix H are,

$$\lambda_\pm = \frac{1}{2}\Big[\pm(4\beta JX(m(1-m)) + (1-\frac{1}{\tau}-Um)) + \sqrt{\{4\beta JX(m(1-m)) - (1-\frac{1}{\tau}-Um)\}^2 - 16UX\beta J(m^2(1-m))}\Big]. \tag{27}$$

We found from equation (27) that all solutions with $m \leq 0.5$ are stable. For solutions with $m > 0.5$, we found that a supercritical Hopf bifurcation occurs at $m = 0.865$, $T = 0.353$. The solid lines in Figures 2 (a) and (b) denote stable solutions while the dashed lines denote unstable ones for the case $m > 0.5$. These figures indicate that the results from the present analysis of stability and computer simulations coincide with each other. We obtain a phase diagram for the stability of the $m > 0.5$ state shown in Figure 2 (c). The state tends to be unstable when the time constant τ of the dynamic amplitude factor increases. The steady state is stable in the $\tau \to 0$ limit. The unstable region increases until $\tau \simeq 50$, then settles in an approximately constant region.

4. Associative memory model

The associative memory model was independently proposed by Nakano [11], Anderson [12], and Kohonen [13]. Hopfield reformulated this model from the statistical-mechanical point of view [14]. It is one of the most important attractor neural network models. The p memory patterns $\boldsymbol{\xi}^\mu$ $(\mu = 1, 2, \cdots, p)$ are stored as attractors in the model through the rule,

$$J_{ij} = \frac{1}{N}\sum_{\mu=1}^{p} \xi_i^\mu \xi_j^\mu. \tag{28}$$

Each element ξ_i^μ of memory pattern $\boldsymbol{\xi}^\mu$ is independently generated by the probability,

$$\mathrm{Prob}[\xi_i^\mu = \pm 1] = \frac{1}{2}, \quad i = 1, 2, \cdots, N, \quad \mu = 1, 2, \cdots, p. \tag{29}$$

In this paper, we treat a finite loading case where the number of memory patterns p is on the order of $O(1)$ with respect to the number of neurons N.

4.1. Macroscopic steady state equation

Substituting equation (28) into microscopic steady state equation (19), we obtain the following microscopic equation,

$$m_i = g_\beta(h_i), \quad h_i = \frac{1}{N}\sum_{\mu=1}^{p} \xi_i^\mu \sum_{j\neq i}^{N} \xi_j^\mu (\frac{2m_j}{1+\gamma m_j} - 1). \tag{30}$$

We employ the notion of sublattice to obtain a macroscopic description of the model. We define p dimensional dual memory patterns $\boldsymbol{\xi}_i = (\xi_i^1, \xi_i^2, \cdots, \xi_i^p) \in \{-1,1\}^p$ with respect to the memory patterns $\boldsymbol{\xi}^\mu$. A set of neuron indices $\{1,2,\cdots,N\}$ is divided into 2^p groups according to the sublattice index $\boldsymbol{\eta} = (\eta^1, \eta^2, \cdots, \eta^p) \in \{-1,1\}^p$,

$$\mathcal{I}_{\boldsymbol{\eta}} = \{i | \boldsymbol{\xi}_i = \boldsymbol{\eta}\}, \quad \{1,2,\cdots,N\} = \bigcup_{\boldsymbol{\eta}} \mathcal{I}_{\boldsymbol{\eta}}, \tag{31}$$

where $\mathcal{I}_{\boldsymbol{\eta}}$ is called a sublattice. $|\mathcal{I}_{\boldsymbol{\eta}}|$ is defined as the number of neurons in the sublattice $\mathcal{I}_{\boldsymbol{\eta}}$. Each $|\mathcal{I}_{\boldsymbol{\eta}}|$ is $O(N)$ since 2^p is $O(1)$ with respect to the number of neurons N. If the i-th and i'-th neurons belong to the same sublattice $\mathcal{I}_{\boldsymbol{\eta}}$, we can introduce sublattice firing rate (magnetization) $m_{\boldsymbol{\eta}}$, input (local field) $h_{\boldsymbol{\eta}}$ and dynamic amplitude factor $X_{\boldsymbol{\eta}}$ as $m_{\boldsymbol{\eta}} \equiv m_i = m_{i'}$, $h_{\boldsymbol{\eta}} \equiv h_i = h_{i'}$ and $X_{\boldsymbol{\eta}} \equiv X_i = X_{i'}$ from the equations in (30). Their relation is,

$$m_{\boldsymbol{\eta}} = g_\beta(h_{\boldsymbol{\eta}}), \quad h_{\boldsymbol{\eta}} = \sum_{\boldsymbol{\eta}'} p_{\boldsymbol{\eta}} \boldsymbol{\eta} \cdot \boldsymbol{\eta}' \left(\frac{2m_{\boldsymbol{\eta}'}}{1+\gamma m_{\boldsymbol{\eta}'}} - 1\right), \quad p_{\boldsymbol{\eta}} = \frac{|\mathcal{I}_{\boldsymbol{\eta}}|}{N}, \quad X_{\boldsymbol{\eta}} = \frac{1}{1+\gamma m_{\boldsymbol{\eta}}}. \tag{32}$$

This equation is a macroscopic steady state equation for the associative memory model with the synaptic depression, which corresponds to the macroscopic equation (21) for the Husimi-Temperly model. $p_{\boldsymbol{\eta}}$ is 2^{-p} in the case of equation (29). The ovelap m^μ is defined as the direction cosine between the neuron state $\boldsymbol{m} = (m_1, m_2, \cdots, m_N)$ and the μ-th memory pattern $\boldsymbol{\xi}^\mu$, which can be calculated by using the sublattice firing rate $m_{\boldsymbol{\eta}}$,

$$m^\mu \equiv \frac{1}{N} \sum_{i=1}^{N} \xi_i^\mu (2m_i - 1) = \sum_{\boldsymbol{\eta}} p_{\boldsymbol{\eta}} \eta^\mu (2m_{\boldsymbol{\eta}} - 1). \tag{33}$$

The overlap m^μ represents the retrieval quality of neuron state $\boldsymbol{m}$ for the μ-th memory pattern $\boldsymbol{\xi}^\mu$. The self-consistent equation for the overlap can be obtained in the case without the synaptic depression, i.e.,

$$m^\mu = \langle\langle \tanh \beta (\sum_{\nu=1}^{p} \xi^\nu m^\nu) \rangle\rangle, \tag{34}$$

where $\langle\langle\cdot\rangle\rangle$ represents the average with respect to equation (29). This equation coincides with that of the conventional associative memory model [15, 16].

4.2. Stability analysis of macroscopic steady state

To evaluate the stability of this homogenous macroscopic state in equation (32), we derive the dynamics of a homogenous mode within the sublattice $\boldsymbol{\eta}$,

$$m_i(t) = m_{\boldsymbol{\eta}}(t), \quad X_i(t) = X_{\boldsymbol{\eta}}(t), \quad i \in \mathcal{I}_{\boldsymbol{\eta}}. \tag{35}$$

Substituting equations (28) and (35) into the dynamical mean field equations (16) and (17), we obtain macroscopic dynamical equations for the homogenous mode for the 2^p sublattices, i.e.,

$$m_{\boldsymbol{\eta}}(t+1) = G_{m_{\boldsymbol{\eta}}}(\{m_{\boldsymbol{\eta}}\}, \{X_{\boldsymbol{\eta}}\}) = \frac{1}{2}(1 + \tanh(\beta h_{\boldsymbol{\eta}}(t)), \tag{36}$$

$$h_{\boldsymbol{\eta}}(t) = \sum_{\boldsymbol{\eta}'} p_{\boldsymbol{\eta}} \boldsymbol{\eta} \cdot \boldsymbol{\eta}' (2X_{\boldsymbol{\eta}'}(t) m_{\boldsymbol{\eta}'} - 1), \tag{37}$$

$$X_{\boldsymbol{\eta}}(t+1) = G_{X_{\boldsymbol{\eta}}}(\{m_{\boldsymbol{\eta}}\}, \{X_{\boldsymbol{\eta}}\}) = X_{\boldsymbol{\eta}}(t) + \frac{1 - X_{\boldsymbol{\eta}}(t)}{\tau} - U X_{\boldsymbol{\eta}}(t) m_{\boldsymbol{\eta}}(t). \tag{38}$$

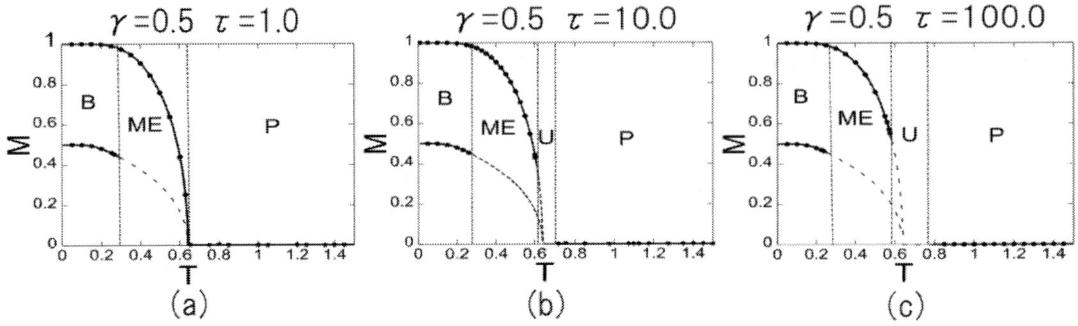

Figure 3. Value of $\gamma = \tau U$ is $\gamma = 0.5$ for all three cases. Vertical lines denote the temperature $T = 1/\beta$, while horizontal lines represent the overlap m for the memory state $(M, 0, 0)$ and for the mixed state (M, M, M). Solutions on solid lines are stable, while those on dashed lines are unstable.

The dynamical behavior around the steady state in equation (32) is governed by the $2^{p+1} \times 2^{p+1}$ Hessian matrix H of these equations, i.e.,

$$H = \begin{pmatrix} A & B \\ C & D \end{pmatrix}, \tag{39}$$

where A, B, C and D are $2^p \times 2^p$ block matrices. Elements of each $2^p \times 2^p$ block matrices are given by

$$A_{\eta\eta'} = \frac{\partial G_{m_\eta}}{\partial m_{\eta'}} = 4\beta m_\eta (1 - m_\eta) p_\eta \boldsymbol{\eta} \cdot \boldsymbol{\eta}' X_{\eta'}, \tag{40}$$

$$B_{\eta\eta'} = \frac{\partial G_{m_\eta}}{\partial X_{\eta'}} = 4\beta m_\eta (1 - m_\eta) p_\eta \boldsymbol{\eta} \cdot \boldsymbol{\eta}' m_{\eta'}, \tag{41}$$

$$C_{\eta\eta'} = \frac{\partial G_{X_\eta}}{\partial m_{\eta'}} = -U X_\eta \delta_{\eta\eta'}, \tag{42}$$

$$D_{\eta\eta'} = \frac{\partial G_{X_\eta}}{\partial X_{\eta'}} = (1 - \frac{1}{\tau} - U m_\eta) \delta_{\eta\eta'}. \tag{43}$$

We examined a case where there were $p = 3$ memory patterns. It is known that there are two kinds of attractors in the model [15, 17]. The first is near the memory pattern, while the second is near the mixed state. The i-th element of the mixed state is given by $\mathrm{sgn}(\xi_i^1 + \xi_i^2 + \xi_i^3)$ for $p = 3$. We call the first attractor a memory state and the second a mixed state. The macroscopic state equations (32) and (33) have two kinds of solutions corresponding to this finding. The first corresponding to the memory state has the form of solution $(m^1, m^2, m^3) = (M, 0, 0)$, while the second corresponding to the mixed state has a symmetric solution of $(m^1, m^2, m^3) = (M, M, M)$.

Figure 3 shows how the overlap M for each attractor depends on the synaptic depression. The value of $\gamma = \tau U$ has been fixed at $\gamma = 0.5$ from figures 3 (a) to (c). Since the value of γ has the fixed value, the steady state solutions to equations (32) and (33) have the same form for the three cases; however, their stabilities depend on the time constant τ of the synapse depression. The lines at the top in Figures 3 (a), (b) and (c) denote the solutions to M for the memory state, while those at the bottom represent the mixed state. The solutions on the solid lines are stable, while those on the dashed lines are unstable. Region "B" is a bi-stable region for the memory and mixed states. Only the memory state is stable in the "ME" region. All of the

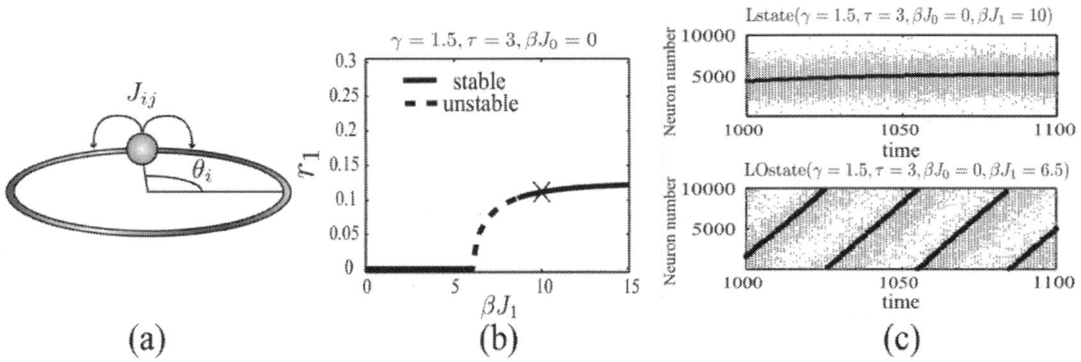

Figure 4. (a): Ring neural network model. i-th neuron is located at angle $\theta_i = (2\pi/N)i$. The synaptic weight J_{ij} has translational invariance with respect to neuron position θ as in equation (44). Solid and dashed lines in (b) represent stable and unstable solutions to the amplitude of localized activity r_1. (c) has raster plots of neuron activities for numerical simulation ($N = 10^4$). Solid lines denote temporal behaviors of bump position ϕ.

overlaps are 0 in the paramagnetic "P" region. There are no steady states in the region "U". The dotts represent steady state overlaps numerically obtained by computer simulation with $N = 9.6 \times 10^4$. The results from theory and computer simulation coincide with each other.

5. Ring network

We discuss a ring neural network model in this section, where the N neurons are located on a one-dimensional ring parametrized with $\theta \in (0, 2\pi]$. The i-th neuron is located at angle $\theta_i = \frac{2\pi}{N}i$, $i = 1, 2, \cdots, N$ as shown in Figure 4 (a). The synaptic weight,

$$J_{ij} = \frac{J_0}{N} + \frac{J_1}{N}\cos(\theta_i - \theta_j), \tag{44}$$

consists of the Fourier 0-th component of the J_0 term and the first component of the J_1 term [18, 19, 20]. The model without the J_1 term reduces to the HT model discussed in §3. Substituting equation (44) into the microscopic steady state equation (19), we obtain the following microscopic equation,

$$m_i = g_\beta(h_i), \quad r_i = \frac{m_i}{1+\gamma m_i}, \quad h_i = J_0 r_0 + J_1(r_1^{\mathrm{C}}\cos\theta_i + r_1^{\mathrm{S}}\sin\theta_i),$$
$$r_0 = \frac{1}{N}\sum_{i=1}^{N} 2r_i - 1, \quad r_1^{\mathrm{C}} = \frac{1}{N}\sum_{i=1}^{N}\cos\theta_i(2r_i - 1), \quad r_1^{\mathrm{S}} = \frac{1}{N}\sum_{i=1}^{N}\sin\theta_i(2r_i - 1), \tag{45}$$

where $r_i = m_i/(1+\gamma m_i)$ is considered to be a renormalized output of the i-th neuron due to the synaptic depression. We can obtain the following self-consistent equation for the macroscopic steady state in the thermodynamic limit $N \to \infty$,

$$r_0 = \frac{1}{2\pi}\int_0^{2\pi} \mathrm{d}\theta r(\theta) - 1, \quad r_1^{\mathrm{C}} = \frac{1}{2\pi}\int_0^{2\pi} \mathrm{d}\theta \cos\theta r(\theta), \quad r_1^{\mathrm{S}} = \frac{1}{2\pi}\int_0^{2\pi} \mathrm{d}\theta \sin\theta r(\theta) \tag{46}$$

$$r(\theta) = \frac{g_\beta(J_0 r_0 + J_1(r_1^{\mathrm{C}}\cos\theta + r_1^{\mathrm{S}}\sin\theta)}{1+\gamma g_\beta(J_0 r_0 + J_1(r_1^{\mathrm{C}}\cos\theta + r_1^{\mathrm{S}}\sin\theta)}. \tag{47}$$

Here, r_0 is the renormalized magnetization or global firing rate. r_1^{S} and r_1^{C} are the 1-st order Fourier components of the renormalized neuron output $r(\theta)$, and show how localized activity, i.e., the bump, is aligned around $\theta = 0$ and $\theta = \pi/2$, respectively. The amplitude of the bump is $r_1 \equiv \sqrt{(r_1^{\mathrm{C}})^2 + (r_1^{\mathrm{S}})^2}$, and its position ϕ is $\phi = \tan^{-1}(r_1^{\mathrm{S}}/r_1^{\mathrm{C}})$.

Figure 4 (b) shows how the amplitude of localized activity r_1 depends on βJ_1 for $\beta J_0 = 0$. The solid and dashed lines represent stable and unstable solutions to equation (47). The stability was numerically evaluated using the Hessian matrix around the steady state solution of equation (47). The top of Figure 4 (c) shows a raster plot of neuron activities for the numerical simulation ($N = 10^4$), and the solid line denotes temporal behavior of bump position ϕ for $\beta J_0 = 0$ and $\beta J_1 = 10$ indicated by $\times$ in Figure 4 (b), where the bump state is stable. We numerically found that bumps propagated stably, i.e., traveling wave [9] for $\beta J_0 = 0$ and $\beta J_1 = 6.5$, as seen at the bottom of Figure 4 (c).

6. Summary

We explored the macroscopic properties of various types of attractor neural networks with synaptic depression in this paper. To do this, we first proposed a microscopic dynamical mean field theory. Next, we derived their macroscopic steady state equations, and assessed how stable these steady states were. The results obtained from the present theory and computer simulation coincided with each other.

We only focused on non-frustrated systems in this paper. A further direction of this study will extend the microscopic dynamical mean field theory to frustrated systems, such as the Sherrington and Kirkpatrick (SK) model [21].

Acknowledgments

This work was partially supported by Grants-in-Aid for Scientific Research (18079003, 20240020, 20650019, and 08J08950).

References

[1] Thomson A and Deuchars J 1994 *Trends Neurosci.* **17** 119–26
[2] Abbott L, Varela J, Sen K and Nelson S 1997 *Science* **275** 220–4
[3] Tsodyks M and Markram H 1997 *Proc. Natl. Acad. Sci. USA* **94** 719–23
[4] Bibitchkov D, Herrmann J and Geisel T 2002 *Network* **13** 115
[5] Torres J, Pantic L and Kappen H 2002 *Phys. Rev.* E **66** 061910
[6] Matsumoto N, Ide D, Watanabe M and Okada M 2007 *J. Phys. Soc. Japan* **76** 084005
[7] Mejias J F and Torres J J 2009 *Neural Comput.* **21** 851–71
[8] Pantic L, Torres J, Kappen H and Gielen S 2002 *Neural Comput.* **14** 2903–23
[9] York L C and van Rossum M C W 2009 *J. Comput. Neurosci.* 0929–5313
[10] Shiino M and Fukai T 1992 *J. Phys. A: Math. Gen.* **25** L375–81
[11] Nakano K 1972 *IEEE Trans. Syst. Man Cybern.* **2** 381–8
[12] Anderson J A 1972 *Math. Biosci.* **14** 197–220
[13] Kohonen T 1972 *IEEE Trans. Comput.* **21** 353–9
[14] Hopfield J J 1982 *Proc. Natl. Acad. Sci. USA* **79** 2554–8
[15] Amit D J, Gutfreund H and Sompolinsky H 1985 *Phy. Rev.* A **32** 1007–18
[16] Okada M 1996 *Neural Networks* **9** 1429–58
[17] Okada M 2006 *New Gener. Comput.* **24** 185–201
[18] Ben-Yishai R, Bar-Or R and Sompolinsky H 1995 *Proc. Natl. Acad. Sci. USA* **92** 3844–8
[19] Hamaguchi K, Hatchett JPL and Okada M 2006 *Phy Rev* E **73** 051104
[20] Hamaguchi K, Urano H and Okada M 2008 *Phy Rev* E **78** 051124
[21] Sherrington D and Kirkpatrick S 1975 *Phys. Rev. Lett.* **35** 1792–6

A Frank mixture copula family for modeling higher-order correlations of neural spike counts

Arno Onken[1,2] **and Klaus Obermayer**[1,2]

[1] School of Electrical Engineering and Computer Science, Technische Universität Berlin, Berlin, Germany

[2] Bernstein Center for Computational Neuroscience Berlin, Berlin, Germany

E-mail: `aonken@cs.tu-berlin.de`

Abstract. In order to evaluate the importance of higher-order correlations in neural spike count codes, flexible statistical models of dependent multivariate spike counts are required. Copula families, parametric multivariate distributions that represent dependencies, can be applied to construct such models. We introduce the Frank mixture family as a new copula family that has separate parameters for all pairwise and higher-order correlations. In contrast to the Farlie-Gumbel-Morgenstern copula family that shares this property, the Frank mixture copula can model strong correlations. We apply spike count models based on the Frank mixture copula to data generated by a network of leaky integrate-and-fire neurons and compare the goodness of fit to distributions based on the Farlie-Gumbel-Morgenstern family. Finally, we evaluate the importance of using proper single neuron spike count distributions on the Shannon information. We find notable deviations in the entropy that increase with decreasing firing rates. Moreover, we find that the Frank mixture family increases the log likelihood of the fit significantly compared to the Farlie-Gumbel-Morgenstern family. This shows that the Frank mixture copula is a useful tool to assess the importance of higher-order correlations in spike count codes.

1. Introduction

How do populations of neurons in the brain communicate with each other? Single neurons transmit information via all-or-none action potentials, called spikes. Investigating spike counts of populations of neurons within short periods of time is one possibility to gain insight into the neural code of populations. Spike counts between different neurons are typically correlated [1, 2]. It is not clear, however, how important pairwise and higher-order correlations between spike counts of different neurons are in terms of the information that is conveyed. On the one hand, it is debated whether correlations matter at all [3, 4, 5, 6]. On the other hand, it is debated whether pairwise correlations versus higher-order correlations are sufficient for obtaining good estimates of the information [7, 8, 9]. New data acquisition techniques make it possible to record from many neurons simultaneously and to investigate the population coding questions by applying information theoretic methods. In order to apply measures from information theory statistical models of the neural responses are required. Recently, the Ising model was proposed as a model of parallel spike trains [8]. This model parametrizes only pairwise interactions between the responses of different neurons and assumes a very strict distribution for the single neuron spike counts.

Flexible parametric distributions with separate parameters for higher-order correlations are required in order to investigate the impact of higher-order correlations on information measures. The copula approach provides a promising method to construct such distributions [10, 11]. Copulas are multivariate distributions that can be used to generate spike count distributions with arbitrary marginal distributions and different dependence structures. One example of such a copula is the so-called Farlie-Gumbel-Morgenstern (FGM) copula family which has separate parameters for all pairwise and higher-order correlations [10]. However, there are tight constraints on the parameters and only weak correlations can be modeled using the FGM copula.

In this paper we introduce a new copula family and compare it to the FGM copula and to the Ising model. The family is constructed as a finite mixture of copula families of different orders. It has separate parameters for all pairwise and higher-order correlations and can model dependence strengths between independence and perfect deterministic dependence. We show that this model provides a much better fit to data generated by a network of leaky integrate-and-fire neurons. Moreover, we compare the copula-based models to the Ising model. The Ising model has an implicit probability distribution of spike counts, which has smaller variance than mean. We explore the relation to Poisson-like spike count distributions and overdispersed distributions that are more frequently observed experimentally [10, 12, 2]. We show that the deviation to these distributions is considerable in terms of the Shannon information.

In the next section we describe several models of neural responses, including the Ising model and copula-based models of dependent multivariate spike counts. Moreover, we introduce a new copula family, the Frank mixture family, that is capable of modeling strong higher-order correlations. We construct a network of leaky integrate-and-fire neurons that can be used to generate spike trains with detailed higher-order correlations. In section 3 we apply the statistical models to data generated by the leaky integrate-and-fire network and demonstrate the superior flexibility of the Frank mixture model compared to the Ising model and the FGM copula family. The paper concludes with a discussion of the advantages and limitations of the presented models.

2. Methods

Synthetic data are generated by a network of leaky integrate-and-fire neurons. We will first describe the network and then introduce several statistical models of correlated neural responses.

2.1. Network models

To generate data we applied the leaky integrate-and-fire neuron as a very simple yet successful neuron model that is based on membrane potentials. The membrane potential V is determined by the differential equation

$$\tau_m \frac{dV}{dt} = E_L - V + R_m I_s,$$

where τ_m denotes the time constant, E_L denotes the resting membrane potential, R_m is the total membrane resistance, and I_s is the synaptic input current. V is initialized by a value V_{init}. When V reaches a threshold V_{th}, the neuron emits a spike and V is reset to V_{reset} [13]. We parametrized the model with $\tau_m = 20\,\text{ms}$, $R_m = 20\,\text{M}\Omega$, $V_{th} = -50\,\text{mV}$, and $E_L = V_{reset} = V_{init} = -65\,\text{mV}$, which are typical values found in [13].

We applied the α-function to model synaptic input [13]: $I_s = I_{max} \frac{t}{\tau_s} \exp(1 - t/\tau_s)$, where the presynaptic release occurs at time $t = 0$. The function has its peak I_{max} at time $t = \tau_s$. There is a decay which can be varied by the time constant τ_s. We used $I_{max} = 1\,\text{nA}$ for excitatory synapses, $I_{max} = -1\,\text{nA}$ for inhibitory synapses, and $\tau_s = 5\,\text{ms}$.

We constructed a small network which allows us to vary pairwise correlations and third order correlations of the spike counts generated by the leaky integrate-and-fire neurons. The network is depicted in figure 1. Poisson spike generators project to the leaky integrate-and-fire neurons

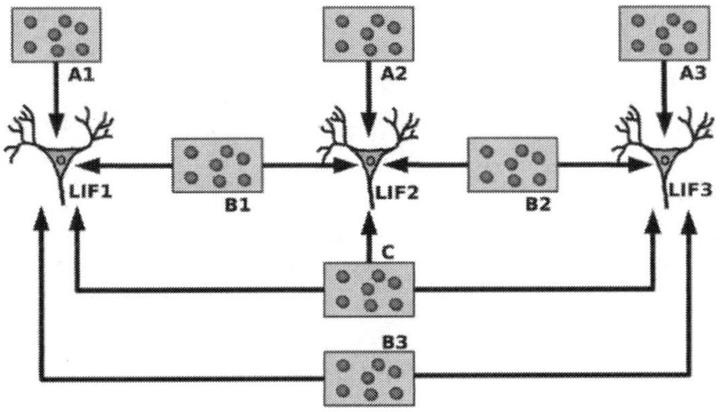

Figure 1. A network of leaky integrate-and-fire neurons to generate spike count data with higher-order correlations. Each rectangle with circles (A1-A3, B1-B3, C) represents two Poisson spike generators: one for the excitatory and one for the inhibitory input. The Poisson spike generators represent populations of neurons that project to three leaky integrate-and-fire neurons (LIF1-LIF3). Connections are illustrated as arrows.

LIF1-LIF3 via inhibitory and excitatory Poisson spike generators. The generators projecting to two neurons (B1-B3) vary the pairwise interactions, whereas the generator projecting to three neurons (C) varies the third order interaction. The rate of excitatory input is always four times the rate of the inhibitory input. We use the generators projecting to only one neuron (A1-A3) to keep the total input rates of the leaky integrate-and-fire neurons constant in spite of varying correlations.

2.2. Ising model

The Ising model is a very popular model in statistical mechanics. It is a maximum entropy model of binary variables called spins that have only pairwise interactions [14]. Recently, it was introduced to neuroscience as a model of spike responses of retinal ganglion cells [8, 9]. In this setting, spike trains are binned into short intervals. A value of either 1 or 0 is assigned to each bin depending on whether at least one spike falls into the bin or not. The bin size should be short enough such that no more than one spike falls into each bin. Successive bins are assumed to be independent, while correlations of parallel spike trains of other neurons are taken into account. The probability of these multivariate binary spike variables $\mathbf{r}$ can be modeled by the Ising distribution [8, 9]

$$P^{\text{Ising}}(\mathbf{r}) = \frac{1}{Z} \exp\left(\sum_i h_i r_i + \sum_{i<j} J_{i,j} r_i r_j \right),$$

where Z is the partition function $Z = \sum_{\mathbf{r}} P_I(\mathbf{r})$ and $r_i \in \{0, 1\}$.

Based on the Ising distribution we can calculate the implicit spike count distribution that arises because of the independence assumption within single neuron spike trains. For spike count bins of length $T \times$ (bin size of binary variables) we get:

$$\begin{aligned} P_T^{\text{Ising}}(\mathbf{r}) &= \sum_{\{\mathbf{Y} | \sum_t Y_{ti} = r_i\}} \prod_{t=1}^{T} P^{\text{Ising}}(\mathbf{Y_t}) \\ &= \frac{1}{Z^T} \exp\left(\sum_i h_i r_i \right) \sum_{\{\mathbf{Y} | \sum_t Y_{ti} = r_i\}} \exp\left(\sum_{i<j} J_{i,j} Y_{ti} Y_{tj} \right), \end{aligned}$$

where $r_i \in \{0, \ldots, T\}$. Due to the independence assumption of successive bins the marginals of P_T^{Ising} (i.e. the spike count distributions of single neurons) take the form of a binomial distribution $P^{\mathrm{Bin}}(x) = \binom{T}{x} p^x (1-p)^{T-x}$, where p is the probability of a spike in the bin.

2.3. Copulas

A d-copula C is defined as a cumulative distribution function (CDF) on the d-dimensional unit hypercube $[0,1]^d$ that has uniform marginals [15]. There is a theorem by Sklar which states that copulas can be used to construct multivariate distributions with arbitrary marginal distributions. If C is a copula and $F_1, \ldots, F_d$ are marginal CDF's then the function $F_{\mathbf{X}}(r_1, \ldots, r_d) \equiv C(F_1(r_1), \ldots, F_d(r_d))$ is again a cumulative distribution function and has the marginal distributions $F_1, \ldots, F_d$ [15].

Different dependence structures can be modeled by different copulas. Many parametric copula families can be found in the literature. One example is the Frank copula family [16]. Its CDF takes the form

$$C_\theta^{\mathrm{Frank}}(\mathbf{u}) = \begin{cases} -\frac{1}{\theta} \ln\left(1 + \left(\prod_{i=1}^d (e^{-\theta u_i} - 1)\right)(e^{-\theta} - 1)^{1-d}\right), & \text{if } \theta \neq 0, \\ \prod_{i=1}^d u_i, & \text{if } \theta = 0. \end{cases} \tag{1}$$

The bivariate Frank family is commutative and radial symmetric: its probability density c_θ abides by $\forall(u,v) \in [0,1]^2 : c_\theta(u,v) = c_\theta(1-u, 1-v)$ [16]. The scalar parameter θ controls the strength of dependence. As $\theta \to \pm\infty$ the copula approaches deterministic positive/negative dependence: knowledge of one variable implies knowledge of the other (so-called Fréchet-Hoeffding bounds [15]). The linear correlation coefficient is capable of measuring this dependence. θ must be non-negative for $d > 2$. Copulas of the Frank family are shown in figure 2 for different parameters.

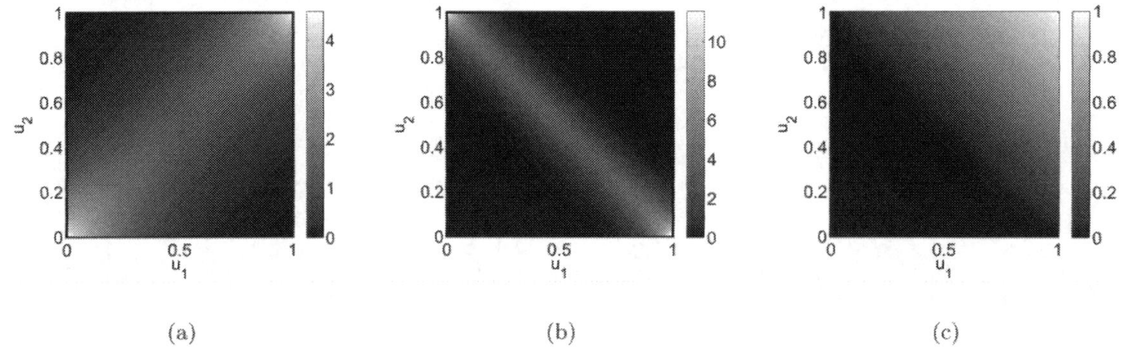

Figure 2. Distribution functions of selected bivariate Frank copulas. (a): Frank copula probability density function for $\theta = 5$. (b): Frank copula probability density function for $\theta = -15$. (c): Frank copula cumulative distribution function for $\theta = -15$.

Another copula family is the Farlie-Gumbel-Morgenstern (FGM) copula family [15]. This family has $2^d - d - 1$ parameters that vary the pairwise and higher-order interactions. Its CDF is given by

$$C_\alpha^{\mathrm{FGM}}(\mathbf{u}) = \left(1 + \sum_{k=2}^{d} \sum_{1 \le j_1 < \cdots < j_k \le d} \alpha_{j_1 j_2 \ldots j_k} \prod_{i=1}^{k} (1 - u_{j_i})\right) \prod_{i=1}^{d} u_i$$

subject to

$$\forall \varepsilon_1, \varepsilon_2, \ldots \varepsilon_d \in \{-1, 1\}: \quad 1 + \sum_{k=2}^{d} \sum_{1 \le j_1 < \cdots < j_k \le d} \alpha_{j_1 j_2 \ldots j_k} \prod_{i=1}^{k} \varepsilon_{j_i} \ge 0.$$

These constraints turn out to be very tight. Due to the constraints for α, the correlations that the FGM copula can model are small in terms of their absolute value.

For a given realization $\mathbf{r}$ which can represent the counts of two neurons, we can set $u_i = F_i(r_i)$ and $F_X(\mathbf{r}) = C_\theta(\mathbf{u})$, where F_i can be arbitrary univariate CDF's. Thereby, we can generate a multivariate distribution with specific marginals F_i and a dependence structure determined by C.

Copulas allow us to have different discrete marginal distributions [10, 11]. Typically, the Poisson distribution is a good approximation to spike count variations of single neurons [17]. For this distribution the CDF's of the marginals take the form

$$F_i^{\text{Pois}}(r; \lambda_i) = \sum_{k=0}^{\lfloor r \rfloor} \frac{\lambda_i^k}{k!} e^{-\lambda_i},$$

where λ_i is the mean spike count of neuron i for a given bin size. We will also use the negative binomial distribution as a generalization of the Poisson distribution:

$$F_i^{\text{NBin}}(r; \lambda_i, \upsilon_i) = \sum_{k=0}^{\lfloor r \rfloor} \frac{\lambda_i^k}{k!} \frac{1}{(1 + \frac{\lambda_i}{\upsilon_i})^{\upsilon_i}} \frac{\Gamma(\upsilon_i + k)}{\Gamma(\upsilon_i)(\upsilon_i + \lambda_i)^k},$$

where Γ is the gamma function. The additional parameter υ_i controls the degree of overdispersion: the smaller the value of υ_i, the greater the Fano factor. The variance is given by $\lambda_i + \frac{\lambda_i^2}{\upsilon_i}$. As υ_i approaches infinity, the negative binomial distribution converges to the Poisson distribution.

Now we have three parametric distributions for the spike count marginals at the disposal: the Poisson distribution with variance equal to the mean, the negative binomial distribution with variance greater than the mean, and the binomial distribution with variance less than the mean. Moreover, we can use the empirical marginals to construct a semi-parametric model with a parametric dependence function and non-parametric marginals. The empirical cumulative marginal of a sample $x_1, \ldots, x_n$ is given by $\hat{F}_i(r_i) = \frac{1}{n} \sum_{j=1}^{n} \mathrm{I}(x_j \le r_i)$, where I is the indicator function.

Likelihoods of discrete vectors can be computed by applying the inclusion-exclusion principle of Poincaré and Sylvester [10]. The probability of a realization $\mathbf{r}$ is given by

$$P_{\mathbf{X}}(\mathbf{r}) = \sum_{k=0}^{d} (-1)^k \sum_{\substack{\mathbf{m} \in \{0,1\}^d, \\ \sum m_i = k}} F_{\mathbf{X}}(r_1 - m_1, \ldots, r_d - m_d). \quad (2)$$

Thus, we can compute the probability mass of a realization $\mathbf{r}$ using only the CDF of $\mathbf{X}$.

For spike count data at hand, we can apply the line-search algorithm for constrained nonlinear optimization [18] to estimate the copula parameters using equation 2 as the objective function.

2.4. Frank mixture copula family

The correlation strengths that can be modeled by the FGM family are small due to the tight constraints for the parameters. We will now introduce a new copula family C_θ^{ho} that overcomes this limitation. We define

$$C_\theta^{\text{ho}}(\mathbf{u}) \equiv \sum_{k=2}^{d} \sum_{1\le j_1<\dots<j_k\le d} z_{j_1\dots j_k} C_{\theta_{j_1\dots j_k}}^{j_1\dots j_k}(\mathbf{u}), \tag{3}$$

where

$$C_{\theta_{j_1\dots j_k}}^{j_1\dots j_k}(\mathbf{u}) \equiv C_\theta^{\text{Frank}}(u_{j_1},\dots,u_{j_k}) \prod_{\substack{i\in\{1,\dots,d\}\\ \setminus\{j_1\dots j_k\}}} u_i. \tag{4}$$

The family has $2^d - d - 1$ mixture components, each consisting of a Frank copula of different order. These Frank copulas are augmented by independent elements such that each mixture component has dimension d. Hence, each of these mixture components has a single scalar parameter that controls the respective pairwise or higher-order correlation.

The model complexity can be reduced by omitting mixture components that are not necessary or that cannot be estimated reliably. E.g. a model can easily be constructed that has only pairwise and third order parameters.

The Frank mixture family is constructed as a mixture of copulas and thus is a mixture of distributions. A natural choice for an inference algorithm is therefore the expectation-maximization algorithm [19, 20]. We have latent variables z_i that need to be estimated in addition to the Frank copula parameters. The expectation-maximization algorithm has two steps that are repeated until convergence of the parameter values. In the expectation step, the weights z_i can be updated according to

$$z_i^{t+1} = \frac{1}{n}\sum_{s=1}^{n} \frac{z_i^t P_{C_{\theta_i}^i}(\mathbf{r_s} \mid \alpha_i^t)}{\sum_j z_j^t P_{C_{\theta_j}^j}(\mathbf{r_s} \mid \alpha_j^t)},$$

where n is the number of trials and $P_{C_{\theta_i}^i}$ is the probability mass function of the model based on the copula $C_{\theta_i}^i$ (see equation 2.) In the maximization step the copula parameters α_i^{t+1} are optimized for fixed values of z_i^{t+1}. Both steps are repeated until parameter values converge.

The Frank mixture copula can be used to investigate the importance of generalized correlation coefficients on information theoretic quantities. There is no generally accepted definition of higher-order correlation coefficients. There are several candidates, however, one of them being the higher-order cumulants, which generalizes the pairwise covariance [21]. Another candidate is based on a generalization of the pairwise normalized product expectation. We adopt the definition given in [7]:

$$\rho_{i_1,\dots,i_k} \equiv \frac{\langle (r_{i_1} - \langle r_{i_1}\rangle)\dots(r_{i_k} - \langle r_{i_k}\rangle)\rangle}{\langle r_{i_1}\rangle\dots\langle r_{i_k}\rangle},$$

where $\langle\cdot\rangle$ denotes the expectation operator. For the Frank mixture family the impact of a copula parameter $\theta_{i_1,\dots,i_k}$ on this higher-order correlation is given by

$$\left|\frac{\langle (r_{i_1} - \langle r_{i_1}\rangle)\dots(r_{i_k} - \langle r_{i_k}\rangle)\rangle - \langle (r_{i_1} - \langle r_{i_1}\rangle)\dots(r_{i_k} - \langle r_{i_k}\rangle)\rangle_{\theta_{i_1,\dots,i_k}=0}}{\langle r_{i_1}\rangle\dots\langle r_{i_k}\rangle}\right|$$
$$= \left|\frac{\sum_{x1,\dots,x_d}(x_{i_1} - \lambda_{i_1})\dots(x_{i_k} - \lambda_{i_k})\psi_{i_1,\dots,i_k}\prod_{\substack{n\in\{1,\dots,d\}\\ \setminus\{i_1\dots i_k\}}} F_n(x_n - m_n)}{\lambda_{i_1}\dots\lambda_{i_k}}\right|,$$

where

$$\psi_{i_1,\dots,i_k} = \sum_{s=0}^{d}(-1)^s \sum_{\substack{\mathbf{m}\in\{0,1\}^d,\\ \sum m_i = s}} \left(C_{\theta_{i_1,\dots,i_k}}^{\text{Frank}}(F_{i_1}(x_{i_1} - m_{i_1}),\dots,F_{i_k}(x_{i_k} - m_{i_k})) - \prod_{j=1}^{k} F_{i_j}(x_{i_j} - m_{i_j}) \right)$$

Here we assume that the weight vector $\mathbf{z}$ is the canonical unit vector with all components set to zero except the $i_1, \ldots, i_k$ component. For this weight vector the impact of the copula parameter is maximal. Note that the impact on this higher-order correlation depends on $\theta_{i_1 \ldots i_k}$ only.

2.5. Shannon entropy

The Shannon entropy [22] of dependent spike counts $\mathbf{X}$ is a measure of the information that a decoder is missing when it does not know the value $\mathbf{x}$ of $\mathbf{X}$. It is given by

$$H(\mathbf{X}) = \langle I(\mathbf{X}) \rangle = \sum_{\mathbf{x} \in \mathbb{N}^d} P_{\mathbf{X}}(\mathbf{x}) I(\mathbf{x}),$$

where $I(\mathbf{x}) = -\log_2(P_{\mathbf{X}}(\mathbf{x}))$ is the self-information of the realization $\mathbf{x}$.

We can estimate the expectation by drawing samples $\mathbf{x}_i$ with probability $P_{\mathbf{X}}(\mathbf{x}_i)$. The term $\frac{1}{k}\sum_{i=1}^{k} I(\mathbf{x}_i)$ will then converge to the entropy almost surely, as k approaches infinity [23]. For k sufficiently large the confidence interval of this term is approximately normal distributed.

3. Results

In the following, we will explore the effects of violations of the Ising model assumptions on the Shannon information and the goodness of fit and compare it to the performance of the copula-based models.

3.1. The cost of ignoring short-term non-stationarity

The Ising distribution yields a binomial distribution for single neuron spike counts between 0 and T due to the independence assumption between successive bins of the binary variables. Thus, statistics of successive bins are assumed to be stationary. For the binomial distribution with parameters T and p the variance $Tp(1-p)$ is always less than the mean Tp. This is at odds with many experimental findings of spike count statistics in the cortex. More frequently, spike count variances are observed that are equal to or greater than the mean [10, 12, 2].

The negative binomial distribution was shown to provide a good parametric model of counts with overdispersion [10]. If an Ising model and therefore a binomial distribution is assumed instead of Poisson or negative binomial distributions, then the entropy estimations based on the inappropriate distribution will deviate from the true entropy. The deviation should increase for increasing variance, since the distributions become more distinct.

We quantify the normalized deviation of the entropy between two distributions P and Q by:

$$D^{P,Q} \equiv \frac{H_P - H_Q}{H_P} = 1 - \frac{H_Q}{H_P},$$

where H_X is the entropy of distribution X. For the binomial and the Poisson distribution this yields

$$D^{\text{Bin,Pois}} = 1 + \frac{e^{-\lambda}\sum_{x=0}^{\infty}\frac{\lambda^x}{x!}(x\log_2(\lambda) - \lambda - \log_2(x!))}{\phi},$$

where

$$\phi = \sum_{x=0}^{T}\binom{T}{x}\left(\frac{\lambda}{T}\right)^x\left(1-\frac{\lambda}{T}\right)^{T-x}\left(x\log_2\left(\frac{\lambda}{T}\right) + (T-x)\log_2\left(1-\frac{\lambda}{T}\right) + \log_2\binom{T}{x}\right).$$

For the binomial and negative binomial distribution we get

$$D^{\mathrm{Bin,NBin}} = 1 + \left(1+\frac{\lambda}{\upsilon}\right)^{-\upsilon} \sum_{x=0}^{\infty} \frac{\lambda^x}{x!} \frac{(\upsilon+x-1)!}{(\upsilon-1)!(\upsilon+\lambda)^x} \cdot$$
$$\frac{\left(-\log_2(x!) + x\log_2\lambda + \log_2((\upsilon+x-1)!) - \log_2((\upsilon-1)!) - x\log_2(\upsilon+\lambda) - \upsilon\log_2\left(1+\frac{\lambda}{\upsilon}\right)\right)}{\phi}.$$

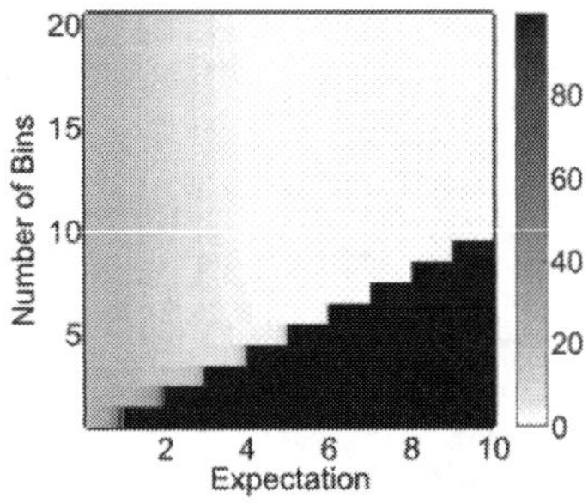

(a)

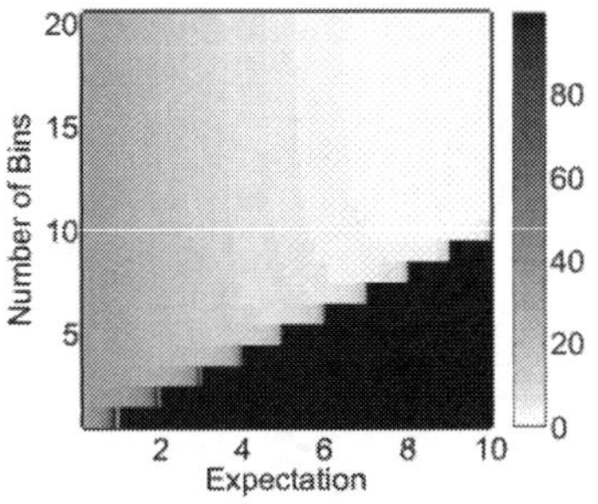

(b)

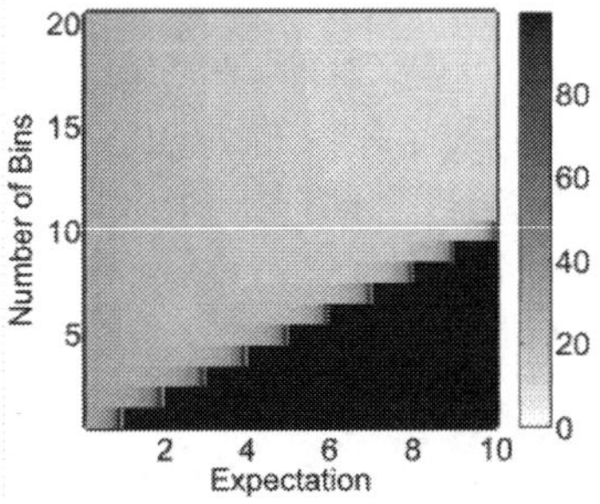

(c)

Figure 3. Normalized deviation of entropy in percent between the binomial distribution and different single neuron spike count distributions. The mean and the number of successive independent bins T of the Ising distribution were varied on the axes. The binomial distribution is not defined for parameter combinations in the lower right corners, because for the binomial distribution the mean cannot exceed the number of bins T. (a): Entropy deviations to the Poisson distribution. (b): Entropy deviations to the negative binomial distribution with overdispersion parameter $\upsilon = 5$. (c): Entropy deviations to the negative binomial distribution with $\upsilon = 1$.

We investigated these deviations of the entropy for different spike rates and bin sizes. The results are shown in figure 3. The variance of the distribution that was used for the comparison increased from (a) to (c). In concert with our expectation the entropy deviation also increased from (a) to (c). Moreover, the deviation increased for decreasing mean. This can be explained in light of the asymptotic distributions: for increasing mean all the distributions approach the normal distribution. The entropy deviation therefore vanishes for increasing means.

3.2. Modeling higher-order correlations

The FGM family can be used to construct multivariate spike count distributions with separate parameters for all pairwise and higher-order correlations. However, the strengths of the correlations that can be modeled using this family are small. To overcome this limitation and to improve the fit further, we introduced the Frank mixture family in this paper (see section 2.4).

We compared the Frank mixture family (see section 2.4) to the FGM family (see section 2.3) on synthetic data generated by a network of leaky integrate-and-fire neurons (see section 2.1). The results are cross-validated, i.e. evaluated on a test set separate of the data that were used to fit the parameters.

The comparison on the test set is shown in figure 4. Since we are interested in the impact of the dependence structure we subtracted the log likelihood that is due to the marginals, i.e. the log likelihood of the product distribution with negative binomial marginals. The remaining residues are due to the copula family. Figure 4 (a) shows the log likelihoods of the residues

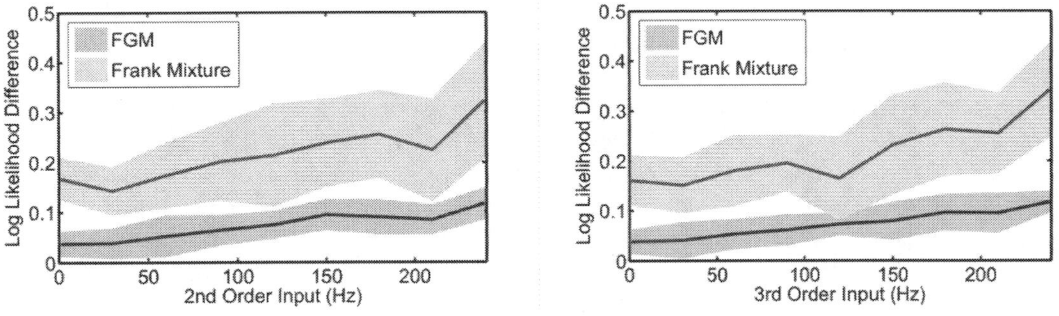

(a) (b)

Figure 4. Copula model fits to the data generated by the leaky integrate-and-fire network presented in figure 1 and evaluated on a separate test set. The log likelihood increase is shown for the FGM copula and the Frank mixture copula with negative binomial marginals compared to the independence model with negative binomial marginals. The total excitatory input rate of each neuron was fixed at 840 Hz while the total inhibitory input rate was fixed at 210 Hz. The correlations were increased by increasing the rates of populations B1-B3 and C (see figure 1) while keeping the total input rates constant. The network was simulated for 10 s for the training set and for the test set in each parameter combination. Spikes were counted in 100 ms bins. (a): Log likelihood increase for different pairwise interaction strengths. Each point is averaged over all third order interactions. The shaded area is the standard deviation over third order interactions. (b): Log likelihood increase for different third order interaction strengths. Each point is averaged over all pairwise interactions. The shaded area is the standard deviation over pairwise interactions.

as a function of pairwise interaction strengths in the network and averaged over different third order interaction strengths. The likelihood of the Frank mixture copula family is clearly greater than the likelihood of the FGM family. The likelihood of the fit is increased by the new family in every single point. Figure 4 (b) depicts the log likelihoods of the residues as a function of the third order interaction strength in the network averaged over different pairwise interaction strengths. Again, the likelihood of the proposed Frank mixture family is superior.

On average, the log likelihood is increased by 214 % in terms of the residues. Here it becomes apparent that the Frank mixture family can capture the structure of pairwise and higher-order correlations much better than the FGM family.

4. Discussion

We introduced the Frank mixture family as a new copula family for modeling strong higher-order correlations. As a copula it can be applied to construct multivariate spike count distributions with biologically appropriate marginal distributions such as the negative binomial distribution. We compared the model to the Ising model, a distribution of binary spike trains which has recently gained a lot of attention in neuroscience. Moreover, we compared the model to the FGM copula family, another family that is capable of modeling higher-order correlations. Based on synthetic data generated by a network of leaky integrate-and-fire neurons we found that the fit and the entropy estimation with regard to the marginals and the dependence structure can be improved considerably by applying the Frank mixture family together with appropriate marginal distributions.

The Frank mixture distribution does not overcome the principle problem of models with detailed higher-order correlations, namely the curse of dimensionality. The problem of estimating all higher-order correlations is infeasible even for moderate numbers of neurons. First, the amount of data that is necessary for this estimation is too large and cannot be collected. Second, it is infeasible to compute the likelihood function for larger numbers of neurons. Nevertheless, the complexity of the Frank mixture family can easily be reduced. Since it has separate mixture components for the different orders of correlations, components that can be estimated reliably can be selected. For instance, a model can be constructed that takes correlations up to order three into account but does not have separate parameters for correlations of order greater than three. Directions for future research include the validation of population coding principles that were obtained on the assumption of the Ising model.

Acknowledgments
This work was supported by BMBF grant 01GQ0410.

References
[1] Bair W, Zohary E and Newsome W T 2001 *J. Neurosci.* **21** 1676–97
[2] Kohn A and Smith M A 2005 *J. Neurosci.* **25** 3661–73
[3] Averbeck B B and Lee D 2003 *J. Neurosci.* **23** 7630–41
[4] Panzeri S, Pola G, Petroni F, Young M P and Petersen R S 2002 *Biosystems* **67** 177–85
[5] Nirenberg S, Carcieri S M, Jacobs A L and Latham P E 2001 *Nature* **411** 698–701
[6] Sompolinsky H, Yoon H, Kang K and Shamir M 2001 *Phys. Rev.* E **64** 051904
[7] Roudi Y, Nirenberg S and Latham P E 2009 *PLoS Comput. Biol.* **5** e1000380+
[8] Schneidman E, II M J B, Segev R and Bialek W 2006 *Nature* **440** 1007–12
[9] Shlens J, Field G D, Gauthier J L, Grivich M I, Petrusca D, Sher E, Litke A M and Chichilnisky E J 2006 *J. Neurosci.* **26** 8254–66
[10] Onken A, Grünewälder S, Munk M and Obermayer K 2009 *Advances in Neural Information Processing Systems* vol 21 ed Koller D, Schuurmans D, Bengio Y and Bottou L (Cambridge: MIT press) pp 1233–40
[11] Berkes P, Wood F and Pillow J 2009 *Advances in Neural Information Processing Systems* vol 21 ed Koller D, Schuurmans D, Bengio Y and Bottou L (Cambridge: MIT press) pp 129–36
[12] Osborne L C, Bialek W and Lisberger S G 2004 *J. Neurosci.* **24** 3210–22
[13] Dayan P and Abbott L F 2001 *Theoretical Neuroscience* (Cambridge: MIT Press)
[14] Ising E 1925 *Z. Phys.* 253–8
[15] Nelsen R B 2006 *An Introduction to Copulas* 2nd ed (New York: Springer)
[16] Frank M J 1979 *Aequationes Math.* **19** 194–226
[17] Tolhurst D J, Movshon J A and Dean A F 1982 *Vision Res.* **23** 775–85
[18] Waltz R A, Morales J L, Nocedal J and Orban D 2006 *Math. Program.* **107** 391–408
[19] Dempster A P, Laird N M and Rubin D B 1977 *J. R. Stat. Soc.* **39** 1–38
[20] Hu L 2006 *Appl. Finan. Econ.* **16** 717–29
[21] Stratonovich R L 1967 *Probl. Inf. Transm.* **3**(2) 1–17
[22] Shannon C E 1948 *AT&T Tech. J.* **27** 379–423
[23] Robert C P and Casella G 2004 *Monte Carlo Statistical Methods* 2nd ed (New York: Springer)

How fast can we learn maximum entropy models of neural populations?

Elad Ganmor[1], Ronen Segev[2] and Elad Schneidman[1]

[1] Department of Neuroscience, Weizmann Institute of Science, Rehovot 76100, Israel
[2] Department of Life Sciences & Zlotowski Center for Neuroscience, Ben-Gurion University of the Negev, Beer-Sheva 84105, Israel

Email: {elad.ganmor, elad.schneidman}@weizmann.ac.il

Abstract. Most of our knowledge about how the brain encodes information comes from recordings of single neurons. However, computations in the brain are carried out by large groups of neurons. Modelling the joint activity of many interacting elements is computationally hard because of the large number of possible activity patterns and limited experimental data. Recently it was shown in several different neural systems that maximum entropy pairwise models, which rely only on firing rates and pairwise correlations of neurons, are excellent models for the distribution of activity patterns of neural populations, and in particular, their responses to natural stimuli. Using simultaneous recordings of large groups of neurons in the vertebrate retina responding to naturalistic stimuli, we show here that the relevant statistics required for finding the pairwise model can be accurately estimated within seconds. Furthermore, while higher order statistics may, in theory, improve model accuracy, they are, in practice, harmful for times of up to 20 minutes due to sampling noise. Finally, we demonstrate that trading accuracy for entropy may actually improve model performance when data is limited, and suggest an optimization method that automatically adjusts model constraints in order to achieve good performance.

1. Introduction

Sensory information is encoded in the brain by the joint activity of groups of neurons that emit stereotyped electric pulses, termed spikes, as their output [1, 2]. Importantly, when sensory systems are presented repeatedly with identical stimuli, responses vary over presentations [3, 4]. This led many researchers to use a probabilistic approach to describe and study neural coding [5, 6]. Even if we simplify neural responses into a sequence of spikes (1's) and silence (-1's), we are still faced with a potentially exponential number of activity patterns.

Clearly it is not feasible to directly sample the joint distribution of large groups of neurons. Accordingly, many studies of neural populations have either focused on small sets of neurons, or tried to base models on statistics that can be accurately estimated from limited data. Recently it was shown, in several different neural systems, that for groups of ~10 neurons the maximum entropy model which takes into account only firing rates and pairwise correlation, out of the exponential number of possible correlation functions among neurons, provides an extremely accurate approximation of neural activity pattern distribution [7-9].

Correlation based models have a natural hierarchy, according to the order of correlation they rely on. This allows us to quantify the contribution of different orders of correlation to the total correlation in the network [10, 11]. The pairwise model is one member of this correlation based hierarchy, but other more complex models which take into account higher order correlations may also be considered. As we move up in the hierarchy, adding higher orders of correlations, the resulting models are guaranteed to give more accurate results. However, building such complex models becomes more challenging – from computational and statistical standpoints. Here we examine the sampling properties of such models and the empirical statistics they rely on. Our results indicate that the pairwise model can be accurately learned within seconds, and furthermore, taking into account higher order correlations proves harmful due to sampling noise.

2. Maximum entropy based hierarchy of models of correlated neural population activity

To investigate the responses of groups of neurons we recorded the simultaneous activity of dozens of retinal ganglion cells in the isolated vertebrate retina presented with natural visual stimuli [12]. We discretize time into 20 ms bins and represent the state of the network at any moment by an n-bit binary word $\{\sigma_1, \sigma_2, \dots, \sigma_n\}$ (n being the number of neurons), where each bit corresponds to the firing of a single neuron.

Since pairs of neurons are typically weakly correlated [7, 13], it is tempting to assume that larger neural populations can be well described by an independent model, denoted $P^{(1)}$, in which the probability of a state $\{\sigma_i\}$ is given by $P^{(1)}(\{\sigma_i\}) = \prod_{i=1}^{n} P(\sigma_i)$. $P^{(1)}$ is the minimal model that takes into account only the firing rates of individual neurons. It allows for efficient learning and inference, while avoiding the exponential complexity of the full joint distribution. However, small networks of ~10 neurons already display significantly correlated activity [7], and so $P^{(1)}$ makes large errors in describing the activity of even small networks. As neural correlations have been shown to carry information about stimuli, which may be behaviorally relevant [14-18] , a model that will take correlations into account is needed.

A natural extension of the independent model is to take into account correlations between pairs of neurons, in addition to the firing rates of individual neurons. The unique distribution which takes into account only pairwise correlations and firing rates, but makes no other implicit assumptions is the maximum entropy pairwise model ($P^{(2)}$) . The maximum entropy pairwise model can be found by methods of constrained optimization, specifically by finding the maximum of the following function:

$$\Lambda\left(P^{(2)}, \lambda\right) = H\left(P^{(2)}\right) - \sum_{i=1}^{n} h_i(<\sigma_i>_{P^{(2)}} - <\sigma_i>_{data}) - \sum_{i<j\le n} J_{ij}(<\sigma_i\sigma_j>_{P^{(2)}} - <\sigma_i\sigma_j>_{data}) - \theta\left(\sum_{\{\sigma\}} P^{(2)}(\{\sigma\}) - 1\right) \tag{1.1}$$

where H denotes the entropy function $H(P) = -\sum_{\{\sigma\}} P(\{\sigma\}) \log_2 P(\{\sigma\})$ [19]. This is the unique model whose firing rates $<\sigma_i>$ and coincident firing rates $<\sigma_i\sigma_j>$ fit those in the data ($<\cdot>$ denotes averages over the empirical distribution P_{data} unless mentioned otherwise), and has maximal entropy. The h_i's and J_{ij}'s here are Lagrange multipliers, while θ multiplies the normalization constraint (giving rise to the partition function Z in the following formula). The solution is known to take the form [20, 21]:

$$P^{(2)} = \frac{1}{Z}\exp\left(\sum_{i=1}^{n} h_i\,\sigma_i + \sum_{i<j\le n} J_{ij}\,\sigma_i\sigma_j\right) \tag{1.2}$$

Although a joint distribution over 10 neurons may be governed by higher orders of interactions [10, 22-25], it was found that pairwise models give a surprisingly accurate approximation of the joint distribution of neural responses in different neural systems [7-9]. The contribution of network correlation, or deviation from independence, can be quantified by $I_n = H\left(P^{(1)}\right) - H(P)$, and can be decomposed into a sum of contributions by different orders of correlations $I_n = \sum_{k=2}^{n} I^{(k)}$, where $I^{(k)}$ corresponds to the contribution of correlations of order k (see [11] for details). The contribution of pairwise correlations to the total correlation structure is then quantified by $I^{(2)}/I_n = \left(H(P^{(1)}) - H(P^{(2)})\right) / \left(H(P^{(1)}) - H(P)\right)$ (where P is the true distribution). Studies in several different neural systems have shown that pairwise correlations account for approximately 90% of the total correlation in networks of ~10 neurons [7-9].

3. Sampling accuracy and robustness of maximum entropy pairwise models of neural responses

As was previously shown [7-9], we find that although pairwise correlations are typically weak, the independence assumption fails. The distribution of synchronous spiking events (the number of spiking neurons in a single time bin, figure 1(b)) as measured in our experimental data, differs considerably from the distribution predicted by the independent model $P^{(1)}$. Again, in agreement with previous reports, the pairwise model gives a very good prediction to the empirically measured quantities (figure 1(b), (c)). We show that this result is robust to specific bin size selection. For bin sizes ranging from 5 to 40 ms, the contribution of pairwise correlations to the total network correlation (measured by $I^{(2)}/I_n$) is approximately 90%, and varies very little (figure 1(d)).

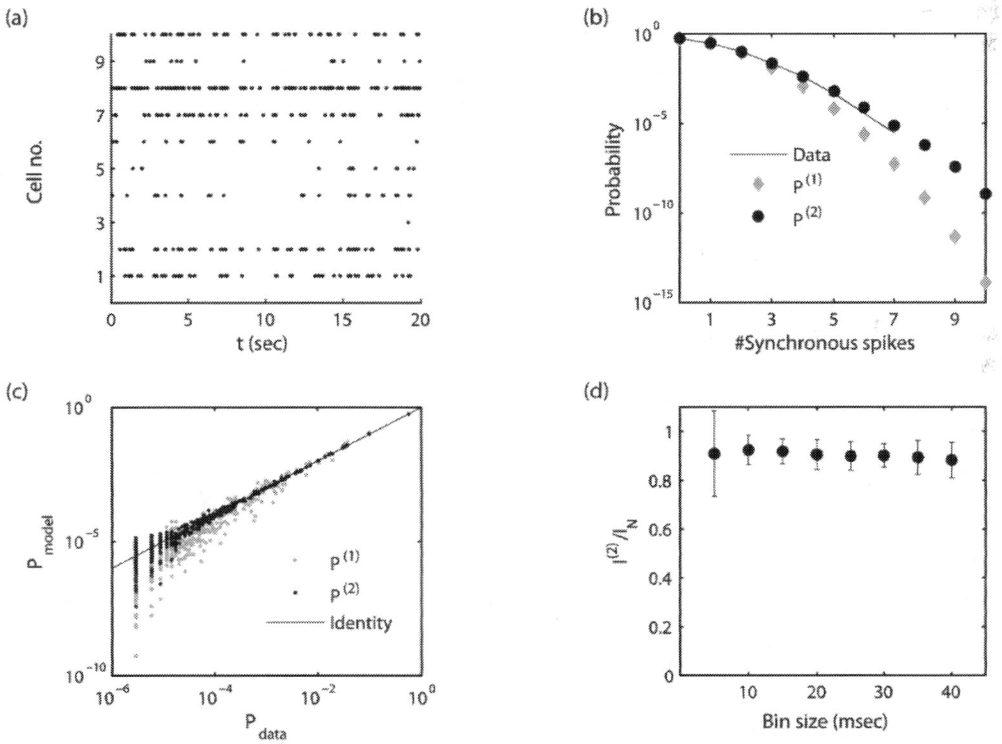

Figure 1. Comparison of independent and pairwise maximum entropy models of the joint spiking of a group of 10 neurons and the empirical distribution. **(a)** Raster plot of the response of 10 neurons to 20 sec of natural movie presentation. Each dot is a spike, the x-axis represents time, and the y-axis represents different neurons. **(b)** Distribution of synchronous spiking events. The probability of an event is plotted against the number of participating neurons. Black line connects the actual data points, black dots are the predictions of $P^{(2)}$, and gray diamonds are the predictions of $P^{(1)}$. **(c)** Predictions of $P^{(2)}$ (black dots) and $P^{(1)}$ (gray dots) for the observation frequency of each pattern in the data (ordinate), plotted against the empirically measured frequency (abscissa). Each dot represents a single pattern. Black line corresponds to identity. **(d)** The contribution of pairwise correlations to the total network correlation as measured by $I^{(2)}/I_n$ is plotted as a function of bin size used to collect spikes (error bars represent STD). Bin size has little effect on this quantity.

The only quantities we need to measure in order to uniquely determine $P^{(2)}$ are the firing rates $< \sigma_i >$ and coincident firing rates $< \sigma_i \sigma_j >$. These quantities can be accurately estimated using far fewer samples than are required in order to directly estimate the full joint distribution. Interestingly, due to the low firing rates of neurons, a characteristic common to many neural systems [26], the relative estimation error, measured by the Fano factor (variance over mean), is similar for both firing rates and coincident firing rates (figure 2(a)). This result does not hold for higher firing rates (figure 2(b), (c)).

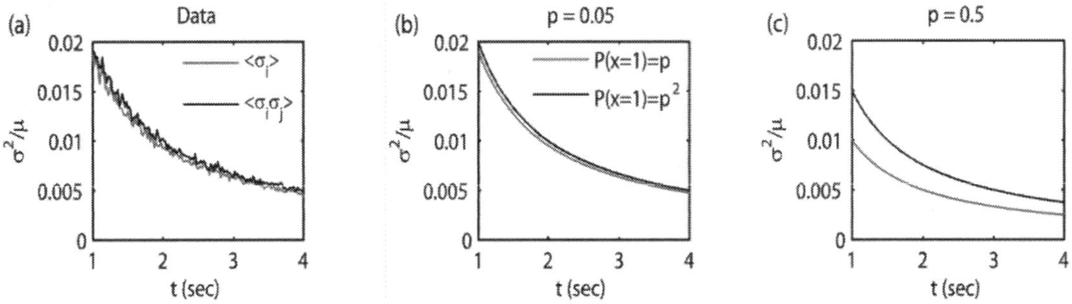

Figure 2. Sampling error for different firing rates. **(a)** Fano factor for the empirical estimates of firing rates (gray line) and coincident firing rates (black line). **(b)** The Fano factor as a function of sampling time (assuming 50 samples per second as in the experiment), for a random variable $Y = N^{-1}\sum_{i=1}^{N} X_i$, where $X \sim$ Bernoulli(p=0.05) (black line), and $X \sim$ Bernoulli(p^2) (gray line). The parameter p (which is analogous to firing rate) is set to be similar to the actual firing rates of neurons. **(c)** Same as **(b)** but for a higher probability of success (analogous to higher firing rate), $p = 0.5$.

Studies so far have used very large data sets in order to construct maximum entropy pairwise models. Clearly a large data set is necessary in order to construct the empirical distribution P_{data}, which enables us to measure such quantities as $I^{(2)}/I_n$. But how much data do we really need in order to construct an accurate pairwise model?

To answer this question, we constructed independent $P^{(1)}$, pairwise maximum entropy $P^{(2)}$ and triplewise maximum entropy $P^{(3)}$ models, for different amounts of data, and compared their performance. Figure 3(a) shows that the pairwise model converges to its final parameters after ~100 sec, as measured by the Kullback-Leibler divergence (D_{KL}, [21]) between $P^{(2)}$ and the empirical distribution constructed from the entire data set. A 50 minute natural movie was presented in a loop a little over two times, resulting in approximately 2 hours of continuous stimulus presentation, thus ~350,000 samples were collected from a distribution over 1024 states. We find that $P^{(2)}$ becomes a more accurate model of the population activity patterns than $P^{(1)}$ after only 35 sec, on average. Moreover, $P^{(2)}$ proves superior to the next model in the correlation based hierarchy, $P^{(3)}$, which is defined analogously to $P^{(2)}$, but with the addition of triplewise correlations $< \sigma_i \sigma_j \sigma_k >$, for times of up to 20 minutes. Thus, for a wide range of behaviorally relevant timescales, the most accurate model in the hierarchy one can construct for neural responses, and the stimuli they represent, is the pairwise model, as considering higher order correlations mainly introduces noise.

Can a pairwise maximum entropy model be learned in behaviorally relevant tasks on timescales of seconds? To test this, we presented a 50 sec long natural movie to the retina repetitively for 101 times. Offline we segmented the movie into 5 clips, each 10 sec long. We constructed a pairwise model for each clip presentation, i.e. 505 different models (101 models for each of the 5 clips). We then

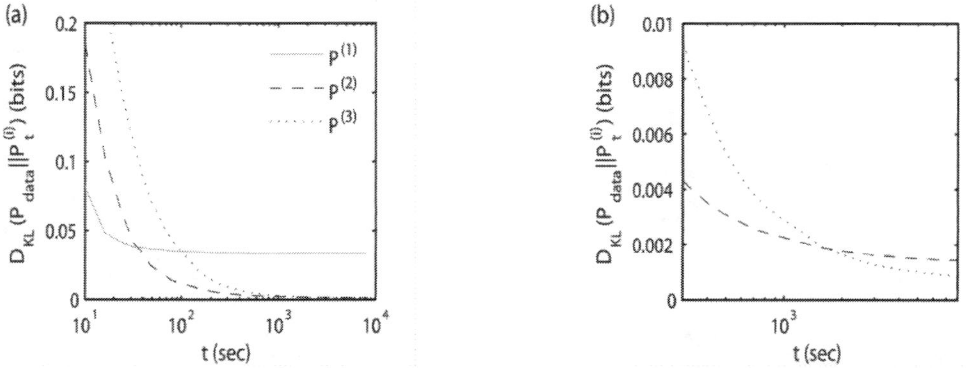

Figure 3. Sampling properties of maximum entropy models of different orders. **(a)** The Kullback-Leibler divergence of the different models in the correlation based hierarchy (see text) from the empirical distribution (estimated using approximately 2 hours of data, or 350,000 samples), is plotted as a function of the time used to estimated the relevant expected values (50 samples/sec). The pairwise model proves to be the best model in a wide range of sampling times. **(b)** Zoom in on longer time scales. $P^{(3)}$ passes $P^{(2)}$ only after over 1000 sec. Legend as in (a).

calculated the Jensen-Shannon divergence (D_{JS}[3], [21, 27]) between each pair of models. When grouping models corresponding to the same visual stimulus together in the distance matrix, we clearly see that models constructed from the same stimulus are much more similar to each other than to models of other stimuli (figure 4(a)). This result demonstrates that pairwise models constructed using a very limited amount of data (only 10 sec) can be utilized in discrimination tasks. Nevertheless, as mentioned earlier, for such short periods of time the independent model is more accurate than the pairwise model. This can be seen by the higher ratio of the average distance between models corresponding to the same stimuli and the average distance between models of differing stimuli (figure 4(b)). Based on the results of figure 3, we expect that for clips which are a little longer than 35 seconds, $P^{(2)}$ will give a more accurate description of the response distribution than $P^{(1)}$. We emphasize that this discrimination does not rely on averaging across repeated presentations of the same stimulus, as is commonly done, but rather stimuli can be discriminated based on a single presentation.

4. Relaxing constraints according to sampling certainty

Acknowledging that our empirical measurements of any observables are noisy may allow us to improve our model by trading model accuracy for increased entropy.

As evident from eq. (1.2), finding $P^{(2)}$ is reduced to finding the h_i's and J_{ij}'s. Since Λ is concave this can be done by gradient ascent based on the relevant derivatives:

$$\frac{\partial \Lambda}{\partial h_i} = <\sigma_i>_{data} - <\sigma_i>_{P^{(2)}}$$
$$\frac{\partial \Lambda}{\partial J_{ij}} = <\sigma_i \sigma_j>_{data} - <\sigma_i \sigma_j>_{P^{(2)}} \tag{4.1}$$

[3] The Jensen Shannon divergence is defined as $D_{JS}(P||Q) = \frac{1}{2}D_{KL}(P||Q) + \frac{1}{2}D_{KL}(Q||P)$. Unlike the Kullback-Leibler divergence, it provides a symmetric measure of similarity between probability distributions, which is bounded between 0 (identical distributions), and 1 (non-overlapping distributions).

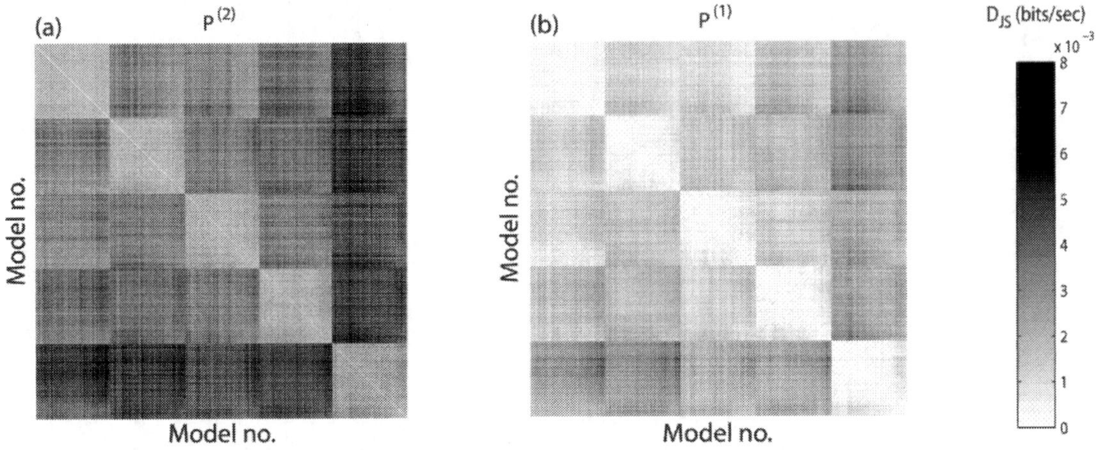

$1.9\cdot10^{-3}$	$3.2\cdot10^{-3}$	$3.1\cdot10^{-3}$	$3.6\cdot10^{-3}$	$4.6\cdot10^{-3}$
	$2.0\cdot10^{-3}$	$3.4\cdot10^{-3}$	$3.7\cdot10^{-3}$	$4.7\cdot10^{-3}$
		$2.0\cdot10^{-3}$	$3.3\cdot10^{-3}$	$4.2\cdot10^{-3}$
			$2.1\cdot10^{-3}$	$4.5\cdot10^{-3}$
				$2.5\cdot10^{-3}$

$3\cdot10^{-4}$	$1.3\cdot10^{-3}$	$1.1\cdot10^{-3}$	$1.7\cdot10^{-3}$	$2.5\cdot10^{-3}$
	$5\cdot10^{-4}$	$1.6\cdot10^{-3}$	$1.8\cdot10^{-3}$	$2.8\cdot10^{-3}$
		$4\cdot10^{-4}$	$1.2\cdot10^{-3}$	$2.0\cdot10^{-3}$
			$5\cdot10^{-4}$	$2.4\cdot10^{-3}$
				$5\cdot10^{-4}$

Figure 4. Similarity map of maximum entropy models for different visual stimuli, learned using single stimulus presentation. **(a)** Top: The distance matrix of different models to different clips of a repeating natural movie. Distance was measured by the Jensen-Shannon divergence between the pairwise models generated from the expected values estimated during each 10 sec segment. Models for the same visual stimuli are grouped together. Colorbar on the right. Bottom: Average distance between models of movie clips (each clip corresponds to the same visual stimulus). Models are grouped as in above matrix. **(b)** Same as **(a)**, but using the independent model. Clearly models of the same visual stimulus are more similar than models of differing visual stimuli.

Denoting the vector of parameters by $\bar{\lambda}$, i.e. $\bar{\lambda} = (h_1 \dots h_n, J_{12} \cdots J_{(n-1)n})$, we can iteratively update the model parameters according to he following formula - $\bar{\lambda}_{t+1} = \bar{\lambda}_t + \eta\nabla\Lambda$ (where $\nabla\Lambda$ denotes the gradient of Λ, $(\nabla\Lambda)_i = \frac{\partial\Lambda}{\partial\bar{\lambda}_i}$. For sufficiently small learning rate (η) this process is guaranteed to converge, thus we can find a solution that is arbitrarily close to the true maximum entropy distribution. In practice, to terminate learning in finite time we must set some convergence threshold, a predetermined condition which indicates that we are close enough to the desired solution. Such a condition is commonly set according to the magnitude of the gradient, i.e., the optimization halts when the overall gradient is smaller than some threshold, $\|\nabla\Lambda\| < \delta$. Generally, we would like δ to be as small as possible in order to achieve maximal accuracy, while the only reason to increase δ is to allow for faster termination of the optimization procedure. However, in our case it proves beneficial to increase δ when sampling noise is high. We see that when only a few samples are available, corresponding to short learning periods, a model with higher δ is in fact closer to the empirical distribution than a more stringent model with smaller δ (figure 5(a)). This is because higher δ results in higher model entropy (figure 5(b)), thus less of an effort is made to concentrate the probability mass according to the (noisy) observations, and more of the probability mass is distributed over all possible patterns. Higher δ values result in higher entropy because of the initial conditions used in our optimization procedure. Since we initialize our parameters so they correspond to a uniform distribution, higher δ values tend to lead to higher entropy distributions upon termination of the optimization.

As model accuracy improves if we adjust our model tolerances according to the noise in our estimations, we would like to automatically adjust the convergence threshold according to the confidence in our data. Every expected value estimated from our data can be approximated by a mean of a sum of Bernoulli random variables with a parameter p equal to the expected value itself. Therefore, if we require that our model's expected values are within one standard deviation from the empirical estimates, we can set the convergence threshold as follows $(\nabla\Lambda)_i \leq \sqrt{<\sigma_i>_{data} \cdot (1-<\sigma_i>_{data})/N}$, where N is the number of data samples. The above is correct for h_i's, to adjust for J_{ij}'s the firing rates need to be substituted for the appropriate coincident firing rates. Hence, we have an individual convergence threshold for each parameter, which takes into account the confidence in the estimate of the relevant statistic. We denote this automatic adjustment of constraint flexibility *AutoFlex*, and show that the *AutoFlex* model is more accurate than the fixed threshold models we tested throughout our sampling range (figure 5(a)). We note that a more general form of using soft constraints to improve the model was presented recently by [28], where they directly maximize the entropy given the flexibility of the constraints, while here the increase in entropy is a result of the initial conditions.

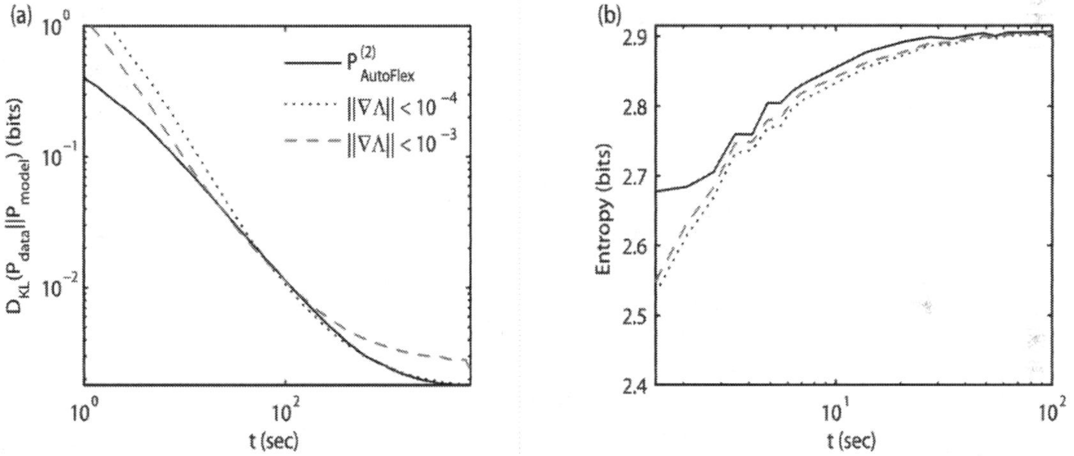

Figure 5. Flexible constraints result in a more accurate model with higher entropy. **(a)** Comparison of the accuracy of different pairwise models calculated using either fixed convergence thresholds (dashed, dotted lines) or using sample size adjusted flexible constraints (solid line), as a function of the time used to estimate firing rates and coincident firing rates (50 samples/sec). Accuracy was measured using the Kullback-Leibler divergence from the empirical distribution estimated using the entire data set (estimated using approximately 2 hours of data, or 350,000 samples). **(b)** Entropy of the corresponding models from **(a)**. Evidently, for noisy data, higher entropy results in better accuracy. The *AutoFlex* approach adjusts the convergence thresholds to get the best results throughout the sampling range.

5. Summary

Modelling the joint distribution of many interacting elements is known to be a hard computational problem encountered in many fields, such as genetics [29], statistical physics [30], machine learning [31], and image processing [32] among others [33]. As experimental technology advances, simultaneous recordings of many neurons become more readily available [34, 35]. Understanding and modelling the joint activity of groups of neurons is therefore becoming a central question in neuroscience.

Here we asked how much data is actually needed in order to construct maximum entropy models of neural population activity patterns, depending on the order of correlations that they rely on. We showed that accurate pairwise models of the responses of groups of 10 neurons can be constructed within several seconds, and that higher order correlations, which can, in theory, improve model accuracy by about 10%, are not useful for times of up to 20 minutes, due to sampling noise. Furthermore we demonstrate that by trading the accuracy of expected value reconstruction for model entropy, we can actually improve model accuracy. Finally we suggested how to adjust optimization parameters in order to rapidly construct accurate models for neural responses.

In this study we considered groups of 10 neurons, which may display up to 1024 different activity patterns. Our experimental data which consisted of 350,000 samples from a single experiment was sufficient to estimate the full probability distribution of activity patterns of this population. However, functional networks in the brain are comprised of much larger groups of neurons. It is possible that for larger networks the contribution of higher order correlations will become much more significant and pairwise models will no longer suffice, which would require different, more subtle ways to learn the nature of population activity patterns [24, 25, 36-39].

Acknowledgements
This work was supported by grants from the Israel Science Foundation (1525/08 to ES, and 502/07 and 1619/07 to RS), fellowships from the Center for Complexity Sciences (to ES and to RS), the Clore center for Biological Physics and a Peter and Patricia Award (ES), and the Zlotowski center for Neuroscience (RS).

References

[1] Adrian E 1928 *The Basis of Sensation* (London: Christophers)
[2] Georgopoulos A P, Schwartz A B and Kettner R E 1986 *Science* **233** 1416-9
[3] Mainen Z F and Sejnowski T J 1995 *Science* **268** 1503-6
[4] de Ruyter van Steveninck R R, Lewen G D, Strong S P, Koberle R and Bialek W 1997 *Science* **275** 1805-8
[5] Rieke F, Warland D, de Ruyter van Steveninck R and Bialek W 1999 *Spikes: Exploring the Neural Code* (Cambridge, MA: The MIT Press)
[6] Yang T and Shadlen M N 2007 *Nature* **447** 1075-80
[7] Schneidman E, Berry II M J, Segev R and Bialek W 2006 *Nature* **440** 1007-12
[8] Shlens J, Field G D, Gauthier J L, Grivich M I, Petrusca D, Sher A, Litke A M and Chichilnisky E J 2006 *J. Neurosci.* **26** 8254
[9] Tang A, et al. 2008 *J. Neurosci.* **28** 505-18
[10] Amari S 2001 *ITIT* **47** 1701-11
[11] Schneidman E, Still S, Berry II M J and Bialek W 2003 *Phys. Rev. Lett.* **91** 238701
[12] Segev R, Goodhouse J, Puchalla J and Berry II M J 2004 *Nat. Neurosci.* **7** 1154-61
[13] Schneidman E, Bialek W and Berry II M J 2003 *J. Neurosci.* **23** 11539
[14] Dan Y, Alonso J M, Usrey W M and Reid R C 1998 *Nat. Neurosci.* **1** 501-7
[15] Maynard E M, Hatsopoulos N G, Ojakangas C L, Acuna B D, Sanes J N, Normann R A and Donoghue J P 1999 *J. Neurosci.* **19** 8083-93
[16] Romo R, Hernandez A, Zainos A and Salinas E 2003 *Neuron* **38** 649-57
[17] Stark E, Globerson A, Asher I and Abeles M 2008 *J. Neurosci.* **28** 10618-30
[18] Fujisawa S, Amarasingham A, Harrison M T and Buzsaki G 2008 *Nat. Neurosci.* **11** 823-33
[19] Shannon C E 1948 *Bell System Tech. J.* **27** 379-423
[20] Jaynes E T 1957 *Phys. Rev.* **106** 620-30
[21] Cover T M and Thomas J A 1991 *Elements of Information Theory* (New York: Wiley-Interscience)
[22] Bohte S M, Spekreijse H and Roelfsema P R 2000 *Neural Comput.* **12** 153-79
[23] Martignon L, Deco G, Laskey K, Diamond M, Freiwald W and Vaadia E 2000 *Neural Comput.* **12** 2621-53
[24] Amari S, Nakahara H, Wu S and Sakai Y 2003 *Neural Comput.* **15** 127-42

[25] Montani F, Ince R A A, Senatore R, Arabzadeh E, Diamond M E and Panzeri S 2009 *Phil. Trans. Royal Soc.* A **367** 3297-310
[26] Olshausen B A and Field D J 2004 *Curr. Opin. Neurobiol.* **14** 481-7
[27] Lin J 1991 *ITIT* **37** 145-51
[28] Dudík M, Phillips S J and Schapire R E 2004 *Proc. 17th Ann. Conf. Comput. Learning Theory* (New York: ACM Press) pp 655-62
[29] Shen-Orr S S, Milo R, Mangan S and Alon U 2002 *Nat. Genet.* **31** 64-8
[30] Binder K 1986 *Monte Carlo Methods in Statistical Physics* (New York: Springer-Verlag)
[31] Ackley D H, Hinton G E and Sejnowski T J 1985 *Cogn. Sci.* **9** 147-69
[32] Li S Z 2001 *Markov Random Field Modeling in Image Analysis* (Tokyo: Springer-Verlag)
[33] Milo R, Shen-Orr S, Itzkovitz S, Kashtan N, Chklovskii D and Alon U 2002 *Science* **298** 824-7
[34] Meister M, Pine J and Baylor D A 1994 *J. Neurosci. Methods* **51** 95-106
[35] Nicolelis M A L 1999 *Methods for Neural Ensemble Recordings* (Boca Raton, FL: CRC press)
[36] Tkacik G, Schneidman E, Berry II M J and Bialek W 2006 *q-bio/0611072*
[37] Bethge M and Berens P 2008 *Advances in Neural Information Processing Systems* vol 20 ed Platt J C, Koller D, Singer Y and Roweis S T (Cambridge: MIT press) pp 97-104
[38] Shlens J, Field G D, Gauthier J L, Greschner M, Sher A, Litke A M and Chichilnisky E J 2009 *J. Neurosci.* **29** 5022-31
[39] Roudi Y, Nirenberg S and Latham P E 2009 *PLoS Comput. Biol.* **5** e1000380

Journal of Physics: Conference Series **197** (2009) 012021 doi:10.1088/1742-6596/197/1/012021

Visual image reconstruction from human brain activity: A modular decoding approach

Yoichi Miyawaki[1,2], Hajime Uchida[3,2], Okito Yamashita[2], Masa-aki Sato[2], Yusuke Morito[4,5], Hiroki C Tanabe[5,4], Norihiro Sadato[5,4] and Yukiyasu Kamitani[2,3]

[1] National Institute of Information and Communications Technology, 2-2-2, Hikaridai, Seika, Soraku, Kyoto 619-0288, Japan.
[2] ATR Computational Neuroscience Laboratories, 2-2-2, Hikaridai, Seika, Soraku, Kyoto 619-0288, Kyoto, Japan.
[3] Nara Institute of Science and Technology, 8916-5, Takayamacho, Ikoma, Nara 630-0192, Japan.
[4] The Graduate University for Advanced Studies, Hayamacho, Miura, Kanagawa 240-0193, Japan.
[5] National Institute for Physiological Sciences, 38, Nishigonaka, Myodaiji, Okazaki, Aichi 444-8585, Japan

E-mail: `kmtn@atr.jp`

Abstract. Brain activity represents our perceptual experience. But the potential for reading out perceptual contents from human brain activity has not been fully explored. In this study, we demonstrate constraint-free reconstruction of visual images perceived by a subject, from the brain activity pattern. We reconstructed visual images by combining local image bases with multiple scales, whose contrasts were independently decoded from fMRI activity by automatically selecting relevant voxels and exploiting their correlated patterns. Binary-contrast, 10 x 10-patch images (2^{100} possible states), were accurately reconstructed without any image prior by measuring brain activity only for several hundred random images. The results suggest that our approach provides an effective means to read out complex perceptual states from brain activity while discovering information representation in multi-voxel patterns.

1. Introduction

Prediction of perceptual contents from brain activity represents a major challenge in human neuroimaging. Previous fMRI studies have shown that visual features, such as orientation and motion direction [1,2], and visual object categories [3,4] perceived by a subject can be predicted from fMRI activity patterns by a statistical 'decoder,' which learns the relationship between brain activity patterns and pre-specified stimulus categories. However, such a simple classification approach is insufficient to capture the complexity of perceptual experience, since our perception consists of numerous possible states, and it is impractical to measure brain activity for all possible states. A recent study [5] has demonstrated that a presented image can be identified among a large number of candidate images using a receptive field model that predicted fMRI activity for visual images (see also [6] for a related approach). But the image identification was still constrained by the candidate image set. Even more challenging is visual image reconstruction (see the earlier work [7] using cat LGN spikes for

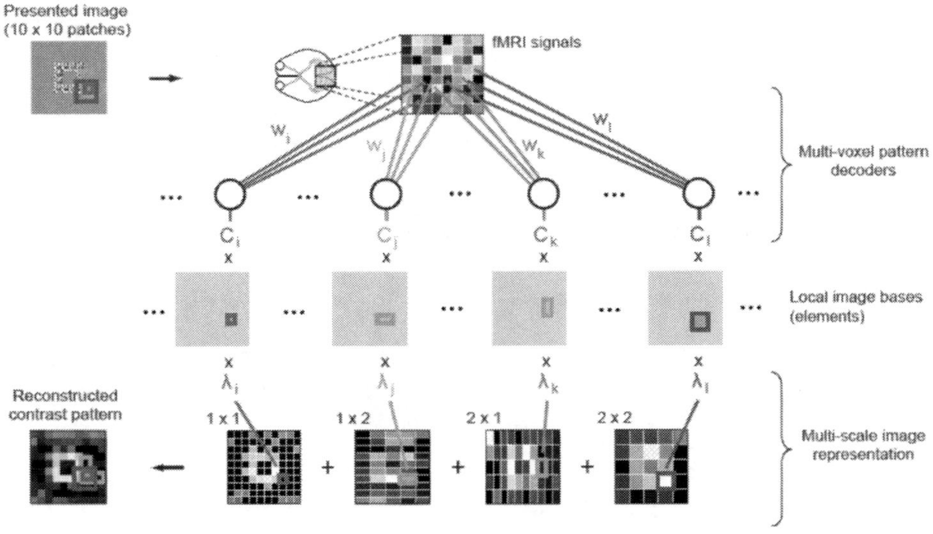

Figure 1. Reconstruction procedure. fMRI activity is measured while a contrast-defined 10 x 10 patch image is presented. 'Local decoders' use linearly-weighted multi-voxel fMRI signals (voxel weights, w_i, w_j, ...) to predict the contrasts (contrast values, C_i, C_j, ...) of 'local image bases (or elements)' of multiple scales (1 x 1, 1 x 2, 2 x 1, and 2 x 2 patch areas, defined by colored rectangles). Local image bases are multiplied by the predicted contrasts, and linearly combined using 'combination coefficients' (λ_i, λ_j, ...) to reconstruct the image. Contrast patterns of the reconstructed images are depicted by a gray scale. Image bases of the same scale (except the 1 x 1 scale) partially overlapped with each other, though the figure displays only non-overlapping bases for the purpose of illustration. Modified from [8].

reconstruction), which translates brain activity into an image, free from the pre-specified stimulus categories.

A possible approach is to utilize the retinotopy in the early visual cortex. Retinotopy associates a specific visual field location to an active cortical location, or voxel, providing a mapping from the visual field to the cortical voxels. Thus, one may predict local contrast information by monitoring the fMRI signals corresponding to the retinotopy map of the target visual field location. The retinotopy can be further elaborated using a voxel receptive-field model. By inverting the receptive-field model, a presented image can be inferred given the brain activity consistent with the retinotopic map [9].

However, it may not be optimal to use retinotopy or the inverse of the receptive field model to predict local contrast in an image. These methods are based on the model of individual voxel responses given a visual stimulus, and multi-voxel patterns are not taken into account for the prediction of the visual stimulus. Recent studies have demonstrated the importance of the activity pattern, in particular the correlation among neurons or cortical locations in the decoding of a stimulus [10, 11]. Since even a localized small visual stimulus elicits spatially-spread activity over multiple cortical voxels, multi-voxel patterns may contain information useful for predicting the presented stimulus.

In addition, a visual image is thought to be represented at multiple spatial scales in the visual cortex, which may serve to retain the visual sensitivity to fine-to-coarse patterns at a single visual field location [12, 13]. The conventional retinotopy, by contrast, does not imply such multi-scale representation, as it simply posits a location-to-location mapping. It may be possible to extract multi-scale information from fMRI signals and use it to achieve better reconstruction.

In this paper, we summarize our recent work on visual image reconstruction (see [8] for full details). We present an approach to visual image reconstruction using multi-voxel patterns of fMRI signals and multi-scale visual representation (figure 1). We assume that an image is represented by a linear combination of local image elements of multiple scales, whose contrasts are independently predicted from multi-voxel patterns. As each local element has fewer possible states than the entire image, the training of local decoders requires only a small number of training samples. Hence, each local decoder serves as a 'module' for a simple image component, and the combination of the modular decoders allows us to represent numerous variations of complex images. The decoder is not explicitly informed about the retinotopy mapping, but automatically selects relevant voxels from the visual areas. We show that arbitrary visual images defined by 10 x 10 binary contrast (2^{100} variations), including geometric and alphabet letter shapes, can be reconstructed accurately, using only several hundreds of images for training the model. Quantitative analyses reveal that the multi-voxel pattern decoder, which exploits voxel correlations especially in V1, and the multi-scale reconstruction model both significantly contribute to the high quality of reconstruction.

2. Methods

2.1. Subjects

We first screened four subjects for head motion in preliminary scans, and two of them (male adults with normal or corrected-to-normal visual acuity) who showed the least head motion underwent the full experimental procedure. The subjects gave written informed consent. The study was approved by the Ethics Committee of ATR and National Institute for Physiological Sciences.

2.2. Visual stimulus and experimental design

We had three types of experimental sessions to measure the fMRI responses of the visual cortex: 1) the random image session, 2) the figure image session, and 3) the conventional retinotopy mapping session.

In the random image session, subjects observed images consisting of 12 x 12 small square patches (1.15 x 1.15 deg each) presented on a gray background. Each patch was either a flickering checkerboard (spatial frequency, 1.74 cycles/deg; temporal frequency, 6 Hz) or a homogeneous gray area, with equal probability. Each image presentation lasted for 6 s, followed by a 6-s rest period. To avoid the effects of the stimulus frame, the central 10 x 10 area was used for analysis. A total of 440 different random patterns were presented to each subject.

In the figure image session, stimulus images consisted of flickering checkerboard patches as in the random image session, but formed geometric shapes ('square,' 'small frame,' 'large frame,' 'plus,' and 'X') or alphabet letters ('N,' 'E,' 'U,' 'R,' and 'O'). Each image was presented for 12 s followed by a 12-s rest period. A total of 6 -- 8 trials per stimulus were performed.

The retinotopy mapping session followed the conventional procedure [14] using a rotating wedge and an expanding ring of flickering checkerboard. The data were used to delineate the borders between visual cortical areas and to identify the retinotopy map on the flattened cortical surfaces. We used fMRI signals from areas V1 and V2 for analyses (unless otherwise stated).

Throughout these experiments, subjects were requested to view the stimulus sequence while maintaining fixation on a spot presented at the center of the screen.

2.3. MRI acquisition and data preprocessing

Preliminary experiments were performed using 3.0-Tesla Siemens MAGNETOM Allegra located at National Institute for Physiological Sciences. MRI data for the presented results were all obtained using a 3.0-Tesla Siemens MAGNETOM Trio A Tim scanner located at the ATR Brain Activity Imaging Center. We measured 30 slices of functional images around the visual cortex (TR, 2 s; voxel size, 3 cubic mm).

Journal of Physics: Conference Series **197** (2009) 012021 doi:10.1088/1742-6596/197/1/012021

The acquired fMRI data were preprocessed and coregistered to the high-resolution anatomical image with standard procedures using SPM2. After voxels of extremely low signal amplitudes were removed, the fMRI data underwent linear trend removal, amplitude normalization, baseline correction, hemodynamic delay compensation (4-s data shift), and within-stimulus-block averaging. See [8] for details.

2.4. Labelling of fMRI data

Each fMRI data sample was labeled by the mean contrast values of local image elements in the corresponding stimulus image. Local image elements were 1 x 1, 1 x 2, 2 x 1, and 2 x 2 patch areas covering the entire 10 x 10 patch area with overlaps (a total of 361 elements; 1 x 1, 100; 1 x 2, 90; 2 x 1, 90; 2 x 2, 81). The mean contrast value of each local image element was defined as the number of flickering patches divided by the total number of patches (1 x 1, [0 or 1]; 1 x 2 and 2 x 1, [0, 0.5, or 1]; 2 x 2, [0, 0.25, 0.5, 0.75, or 1]).

2.5. Training of local decoders

Local decoders were defined to predict the mean contrast of each local image element. They were individually trained with fMRI data and the corresponding class labels representing the mean contrast values. Each local decoder consisted of a multi-class classifier, which classified fMRI data samples into the classes defined by the mean contrast values.

Our classification model is based on multinomial logistic regression, in which each contrast class has a linear discriminant function that calculates the weighted sum of the inputs (voxel values). The discriminant function for contrast class k in a local decoder is expressed as,

$$y_{\mathbf{w}_k}(\mathbf{r}) = \sum_{d}^{D} w_k^d r^d + w_k^0,$$

where w_k^d is a weight parameter for voxel d and contrast class k, r^d is the fMRI signal of voxel d, w_k^0 is the bias, and D is the number of voxels. The probability that an fMRI signal pattern $\mathbf{r} = [r^1, r^2, \cdots, r^D]$ belongs to the contrast class k is defined using the softmax function,

$$p_{\mathbf{w}}(k \mid \mathbf{r}) = \frac{\exp[y_{\mathbf{w}_k}(\mathbf{r})]}{\sum_{j}^{K} \exp[y_{\mathbf{w}_j}(\mathbf{r})]},$$

where K is the number of the contrast classes. The predicted contrast class for m-th local image element, $C_m(\mathbf{r})$, is chosen as the contrast class with the highest probability.

In conventional multinomial logistic regression, the weight parameters are determined by finding the values that maximize the likelihood function of the weight parameters given a training data set,

$$p_{\mathbf{w}}(\mathbf{S} \mid \mathbf{w}_1, \cdots, \mathbf{w}_K) = \prod_{n}^{N} \prod_{k}^{K} p_{\mathbf{w}}(k \mid \mathbf{r}_n)^{s_{nk}},$$

where $\mathbf{S}$ represents a class label matrix whose element s_{nk} is 1 if the trial n corresponds to the contrast class k otherwise 0, $\mathbf{w}_k$ is the weight vector for contrast class k including the bias term ($(D+1)\times 1$ vector), and N is the number of trials.

In this study, we adopted a full-Bayesian approach to the estimation of weight parameters ('sparse logistic regression' [15]). The above likelihood function was combined with a prior distribution for each weight to obtain the posterior distribution. Weight parameters were estimated by taking the expectation of the posterior distribution for each weight.

The prior distribution of a weight parameter is described by a zero-mean normal distribution with a variance, whose inverse is treated as a hyperparameter,

$$p(w_k^d \mid \alpha_k^d) = \mathrm{N}\left(0, \frac{1}{\alpha_k^d}\right),$$

where N represents a normal distribution, and α_k^d is the hyperparameter denoting the inverse of the variance, or precision, of the weight value for voxel d and contrast class k. The hyperparameter α_k^d is also treated as a random variable, whose distribution is defined by,

$$p(\alpha_k^d) = \frac{1}{\alpha_k^d}.$$

These prior distributions are known to lead to 'sparse estimation' in which only a small number of parameters have non-zero values and the remaining parameters are estimated to be zero [16]. Thus, the prior distributions implement the assumption that only a small number of voxels are relevant for the decoding of each local image element. This sparseness assumption may be validated by the fact that a spatially localized visual stimulus gives rise to neural activity only in small regions of the early visual cortex. The sparse parameter estimation avoids overfitting to noisy training data by pruning irrelevant voxels, thereby helping to achieve high generalization performance [15].

Since the direct evaluation of the posterior distribution is analytically intractable, we used a variational Bayesian method to approximate the distribution. The algorithm for the parameter estimation is presented in [8, 15].

2.6. Combination of local decoders

The outputs of the local decoders were combined by a linear model of the corresponding local image elements,

$$\hat{I}(x \mid \mathbf{r}) = \sum_m^M \lambda_m C_m(\mathbf{r}) \phi_m(x),$$

where $\phi_m(x)$ represents a local image element, or a basis, ($\phi_m(x) = 1$ if location x is contained in the area of the local image element, otherwise $\phi_m(x) = 0$), $C_m(\mathbf{r})$ is the predicted contrast, and λ_m is the combination coefficient. Combination coefficients, λ_m, were determined so as to minimize square errors of reconstructed images for the training data set.

2.7. Evaluation of performance

The trained reconstruction model was tested with independent samples. We performed two types of reconstruction tests. First, to obtain a quantitative and unbiased evaluation, we conducted cross-validation analysis using the samples in the random image session. Second, to illustrate the quality of reconstructed images, the model obtained from the random image session was used to reconstruct the images presented in the figure image session.

3. Results

3.1. Reconstructed visual images

Reconstructed images from all trials of the figure image session are illustrated in figure 2. Even though the geometric and alphabet shapes were not used for the training of the reconstruction model, the reconstructed images reveal essential features of the original shapes.

We also found that reconstruction was possible even from 2-s single-volume data without block averaging. The results show the temporal evolution of volume-by-volume reconstruction including the rest periods. All reconstruction sequences are shown as a movie data. The movie data can be found on [17].

In the following sections, we examine how multi-voxel patterns and multi-scale image representation contributed to the high reconstruction performance.

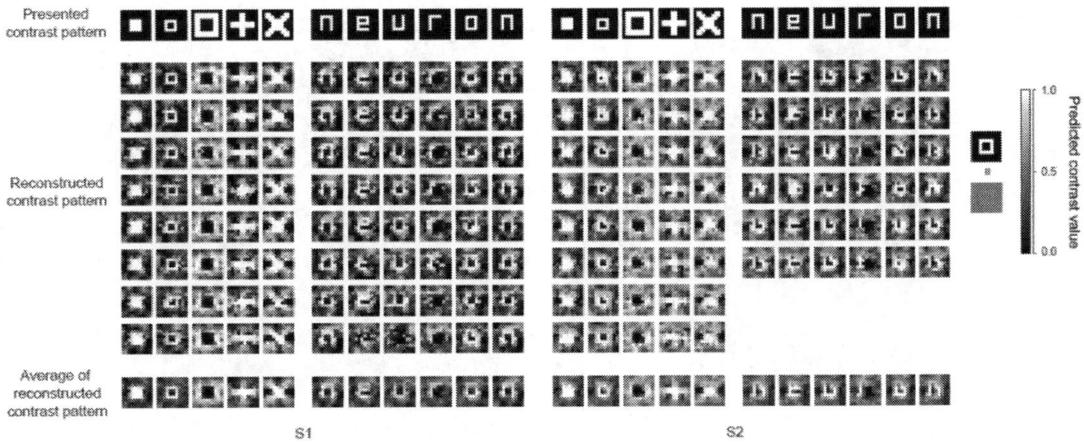

Figure 2. Reconstructed visual images. The reconstruction results of all trials for two subjects (S1 and S2) are shown with the presented images from the figure image session. The reconstructed images are sorted in ascending order of the mean square error. For the purpose of illustration, each patch is depicted by a homogeneous square, whose intensity represents the contrast of the checkerboard pattern. The mean images of the reconstructed images are presented in the bottom row. The same images of the alphabet letter 'n' are displayed in the rightmost and leftmost columns for each subject. Modified from [8].

3.2. Weight distribution on the cortical surface

We first examined the distributions of voxel weights of local decoders in comparison with the conventional retinotopy. Cortical surface maps show the distributions of weight magnitudes for a foveal and a peripheral patch (figure 3). The largest weight values are found around the cortical locations consistent with the retinotopic representation of the patch locations, showing that local decoders mainly used voxels corresponding to the retinotopic locations for their target patches. The weight distribution tended to be blurred for peripheral patches, suggesting that peripheral decoders failed to select retinotopic voxels.

3.3. Advantage of multi-voxel pattern decoders

To examine whether our local decoders exploit multi-voxel patterns for the prediction of target contrast, we devised other types of local decoders that only used retinotopic voxels ('retinotopic decoders'). By applying the standard general linear model to the data from the random image session, we identified a single voxel with the highest t-value, or a group of significantly responsive voxels ($p < 0.05$, false discovery rate (FDR) corrected for multiple comparisons) for each patch. We used 1) the most responsive voxel, and 2) the average of the significantly responsive voxels. The decoders consisted of the standard univariate logistic regression model. The performance of these decoders was compared with that of the multi-voxel pattern decoder.

Cross-validation analysis using the random image trials revealed that the multi-voxel pattern decoder achieved significantly higher correct rates than either of the two retinotopic decoders (two-way ANOVA, Bonferroni-corrected $p < 0.05$ for multiple comparisons), while the difference gradually diminished at the periphery approaching the chance level (figure 4a). Although the figure illustrates the performance only for the 1 x 1 scale, the decoders of other scales showed similar results. The number of the significantly responsive voxels was larger than the number of the voxels selected by the multi-voxel pattern decoder for the foveal to middle eccentricity. Since in this range of eccentricity the multi-voxel pattern decoder largely outperformed the retinotopic decoders, the higher performance of

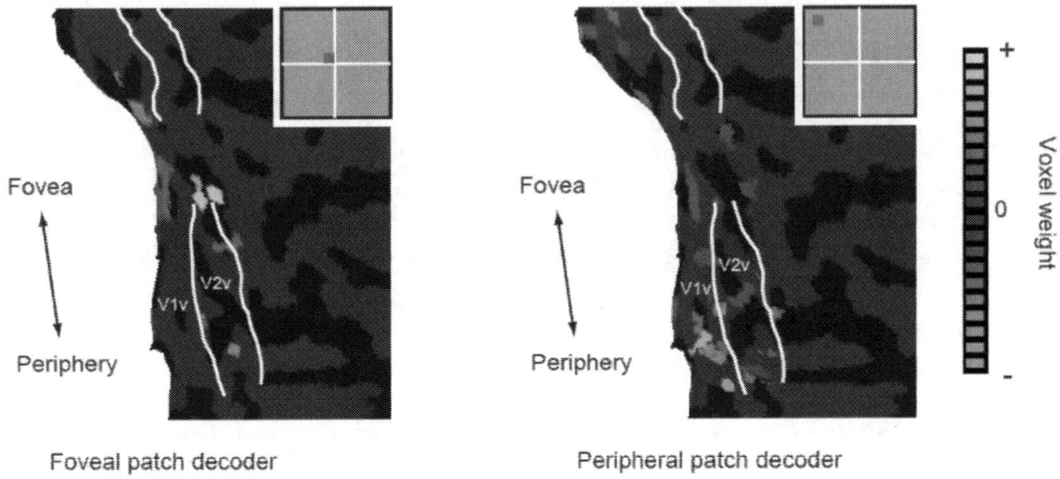

Figure 3. Distributions of voxel weights on the flattened cortex for a foveal and a peripheral decoder. Voxel weights are shown on the right visual cortical surface of subject S1. The location of each patch (1 x 1) is indicated in the inset of the top-right corner. The white lines are the V1 and V2 borders. Modified from [8].

the multi-voxel pattern decoder is not merely due to noise reduction by pooling multi-voxel signals. These results indicate that our local decoders did not simply depend on the mapping between a cortical location and a stimulus location, but that they effectively exploited multi-voxel patterns.

One of the key features of multi-voxel patterns is the correlation between voxels. To examine how voxel correlations contribute to decoding accuracy, we trained the decoder with fMRI data in which voxel correlations were removed, and compared the performance with that of the original decoder. The data were created by shuffling the order of the trials with the same stimulus label in each voxel [10]. This shuffling procedure removes voxel correlations that are independent of the stimulus label. Note that since the stimuli were random images, the voxel correlations observed in the original training data do not reflect the correlations between stimulus patches. The trained decoder was tested with independent non-shuffled data.

The performance with shuffled data was significantly lower than that with the original data (two-way ANOVA, $p < 0.05$), particularly at the middle range of eccentricity (figure 4b). The results suggest that the multi-voxel pattern decoder makes effective use of voxel correlation to achieve high decoding performance.

3.4. Reconstruction using individual visual areas

We next compared the reconstruction between individual visual areas by using the voxels in each of V1, V2, and V3 as the input. As illustrated in figure 5a, reconstruction quality progressively deteriorated along the visual cortical hierarchy. Quantitative comparison was performed by calculating the reconstruction errors for the images from the random image session (squared difference between the presented and the reconstructed contrast in each patch averaged over each entire image). Higher visual areas showed significantly larger errors than V1 (figure 5b; ANOVA, Bonferroni-corrected $p < 0.05$ for multiple comparisons), indicating that V1 contains most reliable information for reconstructing visual images.

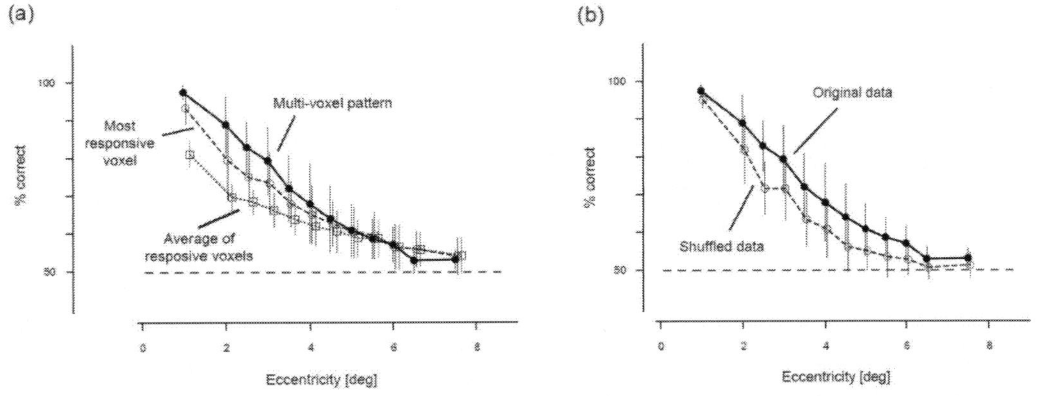

Figure 4. Advantage of multi-voxel pattern decoders. (a) Performance of the multi-voxel pattern decoder and retinotopic decoders. The binary classification performance for 1 x 1 patches is plotted as a function of eccentricity. Classification was performed using 1) a multi-voxel pattern, 2) the most responsive voxel for each patch (with the highest t-value), and 3) the mean of significantly responsive voxels for each patch ($p<0.05$, FDR corrected for multiple comparisons). The performance was evaluated by cross-validation using data from the random image session. The average performance was calculated in each 0.5-deg eccentricity bin (two subjects pooled; error bars, s.d., dashed line, chance level). (b) Effect of voxel correlation in training data. Performance is compared between the multi-voxel pattern decoders trained with the original data and the same decoders trained with 'shuffled' data, in which voxel correlations were removed. The results for the multi-voxel pattern decoder are the same as those displayed in (a). Modified from [8].

The lower reconstruction accuracy at higher visual areas could be accounted for largely by the less ordered retinotopic organization. However, we also found differences in the contribution of voxel correlations across visual areas. The shuffling analyses showed that the removal of voxel correlations impaired reconstruction performance most severely in V1. (figure 5c; ANOVA, Bonferroni-corrected $p < 0.05$ for multiple comparisons), indicating the critical role of voxel correlations in V1. These findings suggest that the reliable information available in V1 is represented not only in the retinotopic organization, but also in the correlated voxel patterns.

3.5. Advantage of a multi-scale reconstruction model

We then tested the significance of the multi-scale representation by comparing the multi-scale model with single-scale models that consisted of optimally combined, single-scale image bases (1 x 1, 1 x 2, 2 x 1, or 2 x 2; V1 and V2 voxels used as the input). Representative examples of the reconstructed images obtained from the figure image session are presented in figure 6a. The reconstructed image from the 1 x 1 scale model shows fine edges but exhibited patchy noise. By contrast, the 2 x 2 scale model produced a spatially blurred image. The images from the 1 x 2 and 2 x 1 scale models contained horizontally and vertically elongated components. The reconstructed image from the multi-scale model appears to have balanced features of these individual scales. The reconstruction error of the multi-scale model, calculated with the images from the random image session, was significantly smaller than those of the single scale models (figure 6b; ANOVA, Bonferroni-corrected $p < 0.05$ for multiple comparisons).

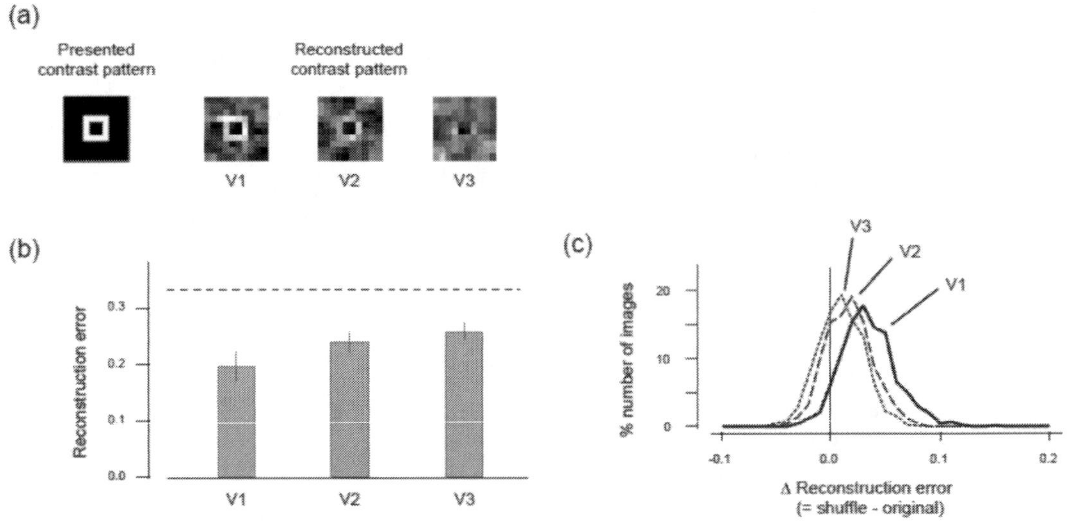

Figure 5. Reconstruction using individual visual areas. (a) Reconstructed images. Examples from the figure image session (S1, 'small frame') are shown. (b) Reconstruction performance with entire images. The bar graph shows reconstruction errors, averaged across all test images in the random image session (two subjects pooled; error bars, s.d.). The dashed line indicates the chance level (1/3), which is achieved when a contrast value for each patch is randomly picked from the uniform distribution of 0 to 1. (c) Effect of correlations on reconstructed images. The distribution of difference in reconstruction error (<error with the shuffled> - <error with the original>, for each image) is plotted for each visual area (two subjects pooled). Modified from [8].

We also calculated reconstruction errors at each eccentricity (figure 6c). For all scales, the reconstruction error increased with eccentricity, but the profiles were different. The error sharply increased with eccentricity for the 1 x 1 model, while the profile was rather flat for the 2 x 2 model. As a result, the errors for these models were reversed at the fovea and the periphery. The 1 x 2 and 2 x 1 models showed intermediate profiles. Statistical analysis revealed a significant interaction between scale and eccentricity ($p < 0.05$ for interaction between eccentricity and scale, two-way ANOVA). The multi-scale model exhibited an error profile matching the minimum envelope of those for the single scale models. Thus, the multi-scale model appears to optimally find reliable scales at each eccentricity.

Consistent with the above observation, analyses of the combination coefficients of the multi-scale model showed that the reconstruction of the foveal and peripheral regions relied on the fine- and coarse-scale decoders, respectively [8]. These results indicate that the optimization of combination coefficients indeed found reliable local decoders at each visual field location to achieve high reconstruction performance.

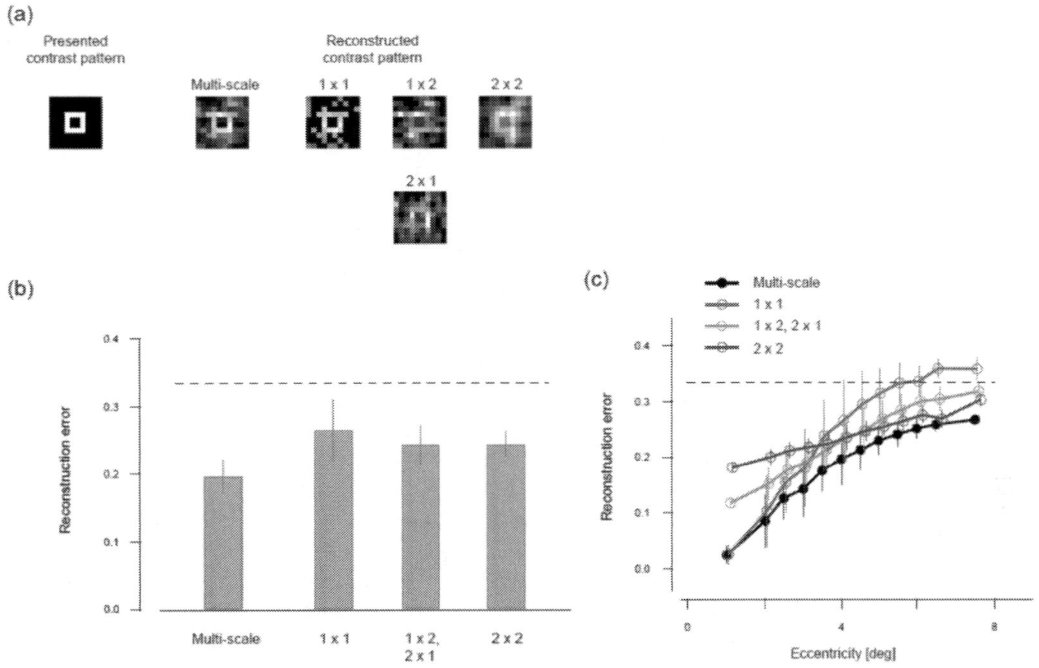

Figure 6. Advantage of multi-scale reconstruction model. (a) Reconstructed images. Examples from the figure image session (S1, 'small frame') are shown for the multi- and single-scale reconstruction models. (b) Reconstruction performance with entire images for the multi- and single-scale models. The results for the 1 x 2 and 2 x 1 scales are combined. (c) Reconstruction performance as a function of eccentricity. Patch-wise errors were averaged across all test images at each eccentricity (two subjects pooled; error bars, s.d.; dashed line, chance level). Modified from [8].

4. Discussion

We have shown that contrast-defined arbitrary visual images can be reconstructed from fMRI signals of the human visual cortex. By combining the outputs of local decoders that predicted local contrasts of multiple scales, we were able to reconstruct a large variety of images (out of 2^{100} possible images) using only several hundred random images to train the reconstruction model. Analyses revealed that both the multi-voxel and the multi-scale aspects of our method were essential to achieve the high accuracy. Our automatic method for identifying relevant neural signals uncovered information represented in correlated activity patterns, going beyond mere exploitation of known functional anatomy.

4.1. Decoding from multi-voxel patterns

A major difference of our approach from the previous studies is that we directly computed the decoding model, instead of elaborating or inverting an encoding model. In our decoding approach, the model is optimized so as to best predict individual stimulus parameters given a multi-voxel pattern while taking into account voxel correlations. In contrast, the encoding models in the previous studies were optimized so as to predict individual voxel responses given a stimulus without considering voxel correlations when estimating the model parameters [5, 9].

Recent imaging studies suggest that there is a better combination of population responses to decode a given visual stimulus than using a signal from the most responsive cortical location or an averaged signal over the responsive cortical locations [1, 11]. In particular, if signals from multiple locations are correlated, a successful decoder should optimally assign various weights, including negative ones, to each location depending on the correlation structure [10, 11].

Consistent with this observation, our decoder using a multi-voxel pattern outperformed a decoder using only a single responsive voxel or an average of responsive voxels (figure 4a). The shuffling of training data, which removed voxel correlations, impaired the decoding performance, indicating the critical role of voxel correlation for constructing an optimal decoder (figure 4b). Careful inspection of the weight distributions in figure 3 indicates that a decoder trained with the original data uses both positive and negative weights, which are found at nearby locations, particularly at the middle to peripheral range of eccentricity. Further analyses showed that the magnitudes of negative weights decreased after shuffling the training data, suggesting that negative weights served to exploit voxel correlation [8].

Although the previous animal study [11] suggested that neural activity in V1 contains significant spatial correlations that can be useful for decoding a visual stimulus, it has been unclear whether such informative correlations are present in other areas of the early visual cortex. Our analysis (figure 5) showed that much of the information available in V1 was represented in voxel correlations, while other areas were less dependent on them.

There are many possible sources of voxel correlation. As the neural populations in nearby voxels are likely to be synaptically coupled, correlated fMRI signals could be spontaneously induced. Nearby voxels might also show correlations through their vascular coupling. Physiological status (e.g., cardiac and respiratory noise) and fMRI scanner conditions (e.g., gradient coil heating) might also cause slow fluctuations correlated among voxels. However, they are unlikely to be major sources of the voxel correlations because the decoder's performance was not affected by filtering out slow components from the data [8]. In addition, head motions of a subject and spatial reinterpolation during preprocessing are also unlikely to be the source, since they cannot account for the area-specific effects of the voxel correlations (figure 5c).

4.2. *Multiple scales of visual representation*

Our multi-scale reconstruction model achieved higher reconstruction accuracy than single-scale models by combining reliable scales at each location. The reliable scales largely depended on eccentricity, which can be related to the receptive field size and the cortical magnification factor. The receptive field size of visual cortical neurons is known to increase with eccentricity [18] and, in parallel, the cortical magnification factor decreases with eccentricity [19, 20]. The receptive-field size for the human visual cortex was estimated at about 1 – 2 deg at 7 deg eccentricity, which is near the most peripheral patch in our stimulus image, while the cortical magnification factor at 7 deg is about 2 – 3 mm/deg. These estimates suggest that single voxels (3 cubic mm) for the peripheral representation carry retinotopic information about more than a single peripheral patch, and thus are not suitable for the decoding of fine-scale (1 x 1) patches, consistent with our reconstruction results (figure 6c). Such eccentricity-dependent changes in the scale of visual representation may partly account for the superior reconstruction by the multi-scale model.

However, it should also be noted that the reconstruction model did not exclusively select a single scale at each eccentricity. At every location except the most foveal region, all scales were effectively combined to improve the reconstruction accuracy (figure 6c) [8]. Previous studies have shown variability in receptive field size among neurons whose receptive fields overlap [13, 21]. Even though each fMRI voxel should contain numerous neurons with receptive fields of various sizes, it may be possible to extract scale-specific information by combining many voxels with a weak scale bias in each, analogous to the extraction of orientation information from coarse voxel sampling of cortical columns [1].

4.3. Modular decoding and its applications

Our approach provides a general procedure to deal with complex perceptual experience consisting of numerous possible states by using multiple decoders as modules. If a perceptual state can be expressed by a combination of elemental features, a modular decoder can be trained for each feature with a small number of data, but their combination could predict numerous states including those that have never been experienced.

Although we focused here on the reconstruction of contrast patterns, our approach could be extended to reconstruct visual images defined by other features, such as color, motion, texture, and disparity. Besides, the motor function may also be dealt with our approach. A large variety of motor actions could be described by a combination of putative modules [22]. Thus, the modular decoding approach may greatly improve the flexibility of prediction, which could also expand the capacity of neural prosthetics or brain-machine interfaces.

More interesting are attempts to reconstruct subjective states that are elicited without sensory stimulation, such as visual imagery, illusions, and dreams. Several studies have suggested that these subjective percepts occur in the early visual cortex [23], consistent with the retinotopy map [9]. One could address this issue by attempting to reconstruct a subjective state using a reconstruction model trained with physical stimuli. Reconstruction performance can also be compared among cortical areas and reconstruction models. Thus, our approach could provide valuable insights into the complexity of perceptual experience and its neural substrates.

Acknowledgements

The authors thank A. Harner and S. Murata for technical assistance, and T. Beck and M. Takemiya for manuscript editing. This research was supported in part by SRPBS, MEXT, the NICT-KARC, the Nissan Science Foundation, and the SCOPE, SOUMU.

References

[1] Kamitani Y and Tong F 2005 *Nat. Neurosci.* **8** 679-85
[2] Kamitani Y and Tong F 2006 *Curr. Biol.* **16** 1096-102
[3] Cox D D and Savoy R L 2003 *Neuroimage* **19** 261-70
[4] Haxby J V, Gobbini M I, Furey M L, Ishai A, Schouten J L and Pietrini P 2001 *Science* **293** 2425-30
[5] Kay K N, Naselaris T, Prenger R J and Gallant J L 2008 *Nature* **452** 352-5
[6] Mitchell T M, Shinkareva S V, Carlson A, Chang K M, Malave V L, Mason R A and Just M A 2008 *Science* **320** 1191-5
[7] Stanley G B, Li F F and Dan Y 1999 *J. Neurosci.* **19** 8036-42
[8] Miyawaki Y, Uchida H, Yamashita O, Sato M A, Morito Y, Tanabe H C, Sadato N and Kamitani Y 2008 *Neuron* **60** 915-29
[9] Thirion B, Duchesnay E, Hubbard E, Dubois J, Poline J B, Lebihan D and Dehaene S 2006 *Neuroimage* **33** 1104-16
[10] Averbeck B B, Latham P E and Pouget A 2006 *Nat. Rev. Neurosci.* **7** 358-66
[11] Chen Y, Geisler W S and Seidemann E 2006 *Nat. Neurosci.* **9** 1412-20
[12] Campbell F W and Robson J G 1968 *J. Physiol.* **197** 551-66
[13] De Valois R L, Albrecht D G and Thorell L G 1982 *Vision Res.* **22** 545-59
[14] Sereno M I, Dale A M, Reppas J B, Kwong K K, Belliveau J W, Brady T J, Rosen B R and Tootell R B 1995 *Science* **268** 889-93
[15] Yamashita O, Sato M A, Yoshioka T, Tong F and Kamitani Y 2008 *Neuroimage* **42** 1414-29
[16] Tipping M 2001 *J. Mach. Learn. Res.* **1** 211-44
[17] http://download.cell.com/neuron/mmcs/journals/0896-6273/PIIS0896627308009586.mmc2.mpg
[18] Dumoulin S O and Wandell B A 2008 *Neuroimage* **39** 647-60
[19] Dougherty R F, Koch V M, Brewer A A, Fischer B, Modersitzki J and Wandell B A 2003 *J. Vis.*

3 586-98
[20] Duncan R O and Boynton G M 2003 *Neuron* **38** 659-71
[21] Hubel D H and Wiesel T N 1968 *J. Physiol.* **195** 215-43
[22] Poggio T and Bizzi E 2004 *Nature* **431** 768-74
[23] Kosslyn S M, Thompson W L, Kim I J and Alpert N M 1995 *Nature* **378** 496-8

AUTHOR INDEX

Institute of Physics
Dirac House, Temple Back
Bristol BS1 6BE UK

CURRAN ASSOCIATES INC
proceedings.com